T0389658

Abyssal Channels in the Atlantic Ocean

Eugene G. Morozov • Alexander N. Demidov
Roman Y. Tarakanov • Walter Zenk

Abyssal Channels in the Atlantic Ocean

Water Structure and Flows

With a contribution by Anatoly Schreider and Vadim Sivkov
Editor: Georges Weatherly

Dr. Eugene G. Morozov
Physical Department
Shirshov Institute of Oceanology
Nakhimovsky 36
117997 Moscow
Russia
egmorozov@gmail.com

Dr. Roman Y. Tarakanov
Physical Department
Shirshov Institute of Oceanology
Nakhimovsky 36
117997 Moscow
Russia
rtarakanov@gmail.com

Dr. Alexander N. Demidov
Geographical Faculty
Moscow State University
Vorobievy Gory
119899 Moscow
Russia
tuda@mail.ru

Dr. Walter Zenk
Leibniz Institute of Marine Sciences
IFM-GEOMAR
Kiel University
Düsternbrooker Weg 20
24105 Kiel
Germany
wzenk@ifm-geomar.de

Additional material to this book can be downloaded from http://extra.springer.com

ISBN 978-90-481-9357-8 e-ISBN 978-90-481-9358-5
DOI 10.1007/978-90-481-9358-5
Springer Dordrecht Heidelberg London New York

Library of Congress Control Number: 2010934296

Cover illustration: Cover drawing by B. Filyushkin

Cover design: Cover photo by Morozov

Printed on acid-free paper

Springer is part of Springer Science+Business Media (www.springer.com)

Foreword

This book is dedicated to the study of structure and transport of deep and bottom waters above and through underwater channels of the Atlantic Ocean. The study is based on recent observations, analysis of historical data, and literature reviews. This approach allows us to understand how water transport and water mass properties have changed over the last years and decades. The focus of our study is on the propagation of bottom waters in the Atlantic Ocean based on new field data at key points.

At the end of the 1920s, the first integral study of water masses and bottom topography of the Central and South Atlantic was carried out from the German research vessel *Meteor*. This German Atlantic Expedition was one of the first cruises equipped with the newly developed echo sounder (fathometer): an obligatory prerequisite for the investigation of bottom morphology in the deep sea on an operational base. The results of the expedition were published by Wüst, Defant, and colleagues in the multivolume METEOR publication series starting with the cruise report by the ship's commander (Spiess 1928, 1932). Historically, this series of publications, intermittently interrupted by World War II, was the basis for many years of research into the development of modern concepts about Atlantic water masses and their circulation schemes.

Since then, many national and international programs were initiated to study the properties of bottom water in the Atlantic Ocean and their propagation through narrow channels in submarine ridges. The present book summarizes the results of many field experiments and describes our contribution to this research.

Recently, deep-water fractures, troughs, and channels between large oceanic basins became interesting objects of research. These investigations have allowed us to characterize specific features of water exchange in deep-water parts of the ocean through narrow channels. Such features cannot be obtained even at high resolution in modern global circulation models. The peculiar abyssal water exchange through narrow channels is extremely difficult to integrate in numerical global circulation models with a sufficiently high resolution. Long-term observations indicate that strong velocities reaching a few tenths of meters per second are not unusual at selected choke points in the deepest layers of the ocean. It is interesting that such high velocities are otherwise found primarily beneath the wind-driven surface layer of the ocean.

Intense research into the deep ocean was carried out during the field works that were a part of the World Ocean Circulation Experiment (WOCE) in the 1990s. The deepest fracture zones and channels of different origins in the ocean provide water exchange between deep basins. Since the beginning of the World Ocean Circulation Experiment in early 1991, sustained observations of Antarctic Bottom Water characteristics and transport were conducted at selected sites around the Brazil Basin. One of the WOCE core projects, the Deep Basin Experiment, was focused on the key passages that allow an equatorward interbasin exchange of bottom water. After the WOCE decade, field measurements in the Atlantic abyssal channels continued to broaden our understanding of these phenomena.

E. Morozov
A. Demidov
R. Tarakanov
W. Zenk

Preface

Our book reviews a number of studies carried out in different abyssal channels, combined with our recent oceanographic research and analysis of observations in the main channels of the Atlantic Ocean. We also analyze properties of water masses involved in flows through the channels and their variability on the decadal time scale.

The following abyssal channels are the most important in bottom water exchange in the Atlantic Ocean: Falkland Gap (Arhan et al. 1999), Vema Channel and Hunter Channel, Vema Fracture Zone, Romanche and Chain fracture zones, nameless Equatorial Channel, Kane Gap, and Charlie Gibbs Fracture Zone. Of course, numerous channels in the Scotia Sea and Drake Passage, as well as passages connecting the Weddell Sea with the Georgia Basin, which provide the initial outflow of Antarctic water from the Weddell Sea to the Atlantic abyssal zone, also play a significant role. Actually, water in the Weddell Sea is the source of the bottom waters in the Atlantic and Southern oceans, while the Scotia Sea is one of the basins responsible for the first appreciable transformation of bottom water. Locations of the main abyssal channels are shown in Fig. 1. Detailed charts of bottom topography in the regions of major channels are shown in individual topographic maps in Figs. 2–8.

Most of the Atlantic abyssal channels provide a unidirectional flow of bottom water. Progressive vector diagrams based on moored measurements of currents in the channels give a clear illustration of the prevailing flow directions, current strength, and steadiness. Figure 9 shows eleven progressive vector diagrams in the major channels around the Argentine and Brazil basins of the South Atlantic.

Images on the left side are inferred from observations across the Rio Grande Rise. Diagram 1 is related to the Deep Western Boundary Current on the continental rise off Santos. The strongest near-bottom flows through the Vema Channel were recorded on the west bank of the Sill (diagrams 2, 3) and on its east bank (diagrams 4, 5). Diagram 6 represents the main vein in the Hunter Channel. Diagram 7 shows examples of inflow for bottom water from the Shag Rocks Passage (Fig. 2) at the northern rim of the Scotia Sea. Diagram 8 shows the outflow of bottom water to the north, the Vema Extension (northern part of the Vema Channel). Diagrams 9 and 10 show outflow across the Mid Atlantic Ridge at the Romanche and Chain fracture

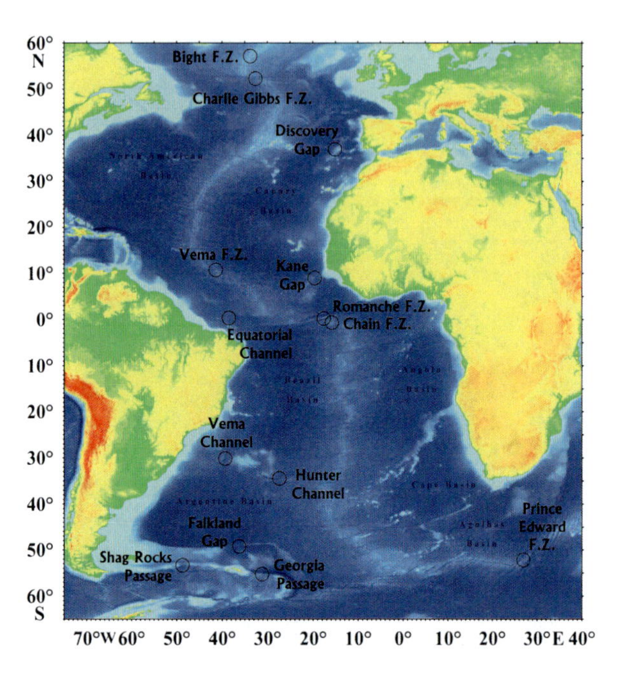

Fig. 1 Chart of the Atlantic Ocean showing the location of the major abyssal channels mostly associated with fracture zones (FZ)

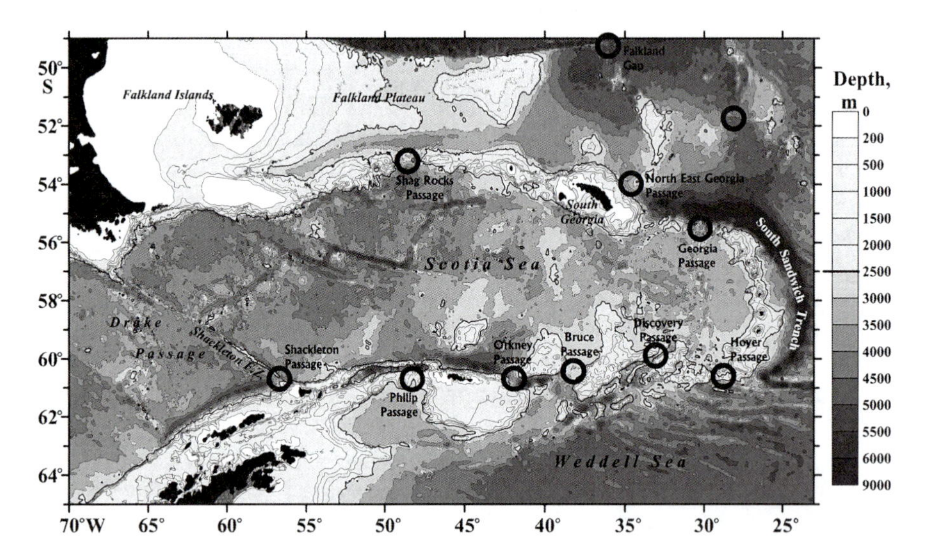

Fig. 2 Chart of the Scotia Sea and Drake Passage showing numerous channels that allow bottom water spreading

zones. All curves are barely 1.5 years long (527 days) except diagram 7, which is about half this duration time. Ticks are 30 days apart. No filters were applied to the data sets. Sampling interval is 2 h. The topographic steering effect is obvious at all locations.

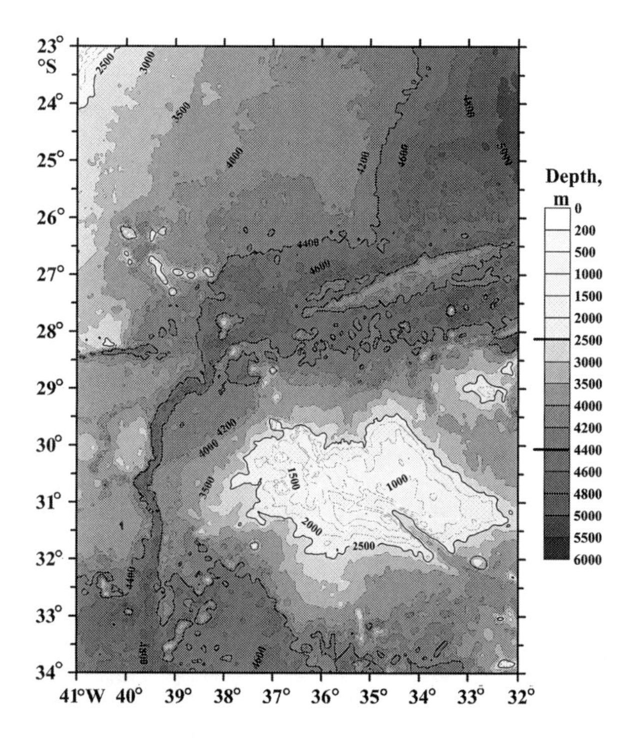

Fig. 3 Bathymetric chart of the Vema Channel

Moreover, specific similarity can be found between flows in different abyssal channels (Zenk et al. 1999). The authors constructed daily vertical profiles based on measurements in several channels. The vertical axis was normalized by the level, where the velocities decreased to 14% of their maximum (unit depth). The zero

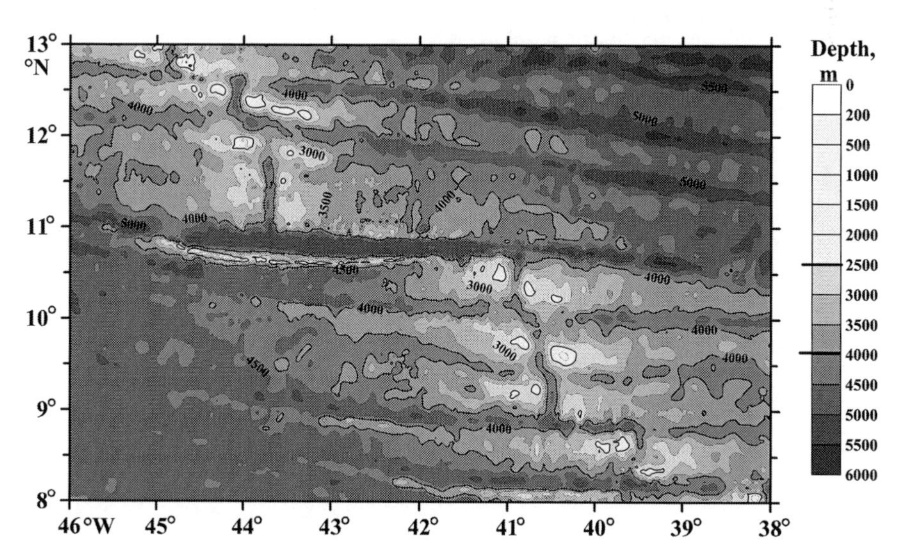

Fig. 4 Bathymetric chart of the Vema Fracture Zone

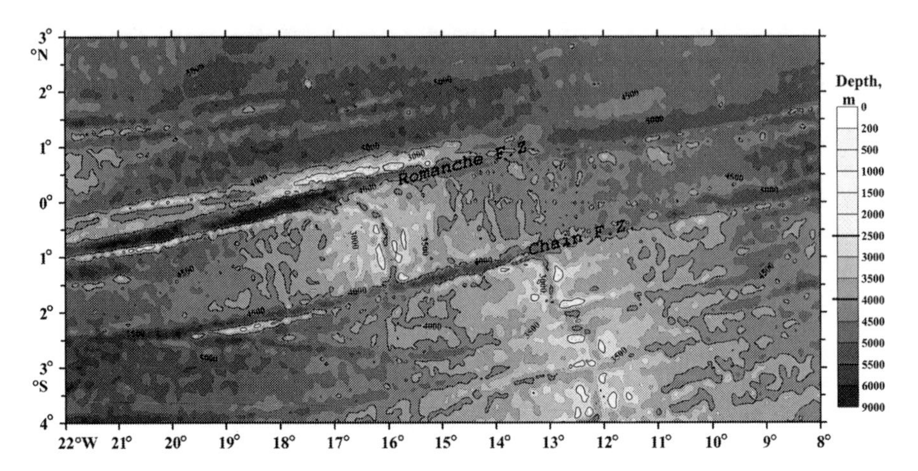

Fig. 5 Bathymetric chart of the Romanche and Chain fracture zones

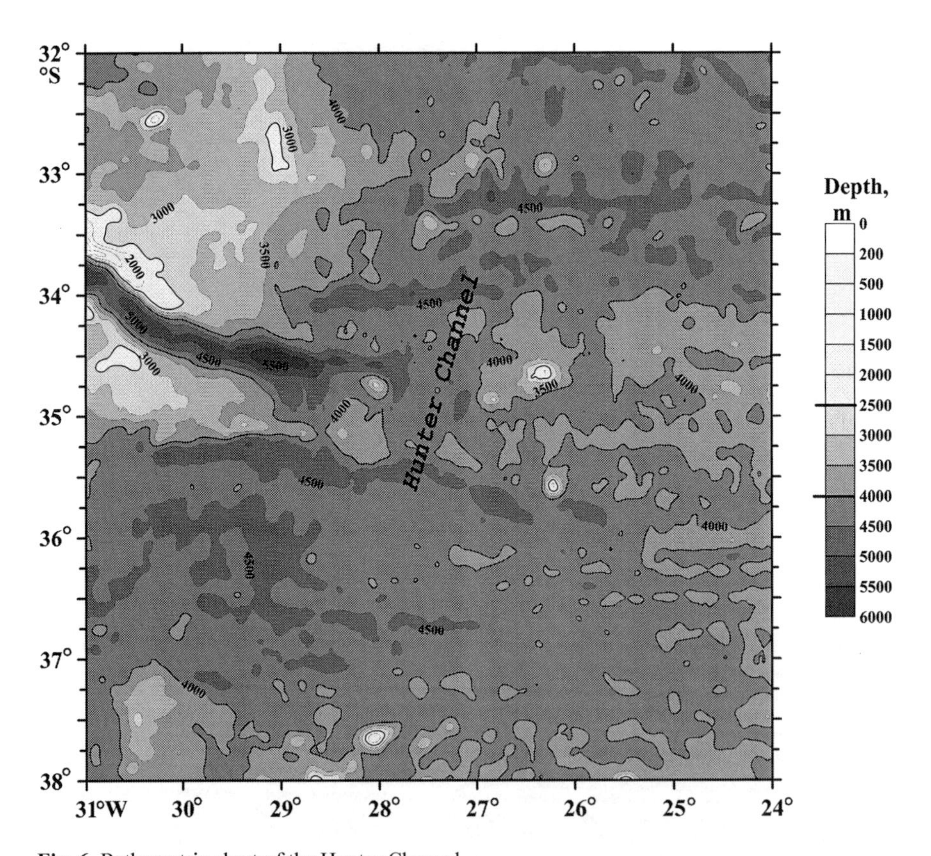

Fig. 6 Bathymetric chart of the Hunter Channel

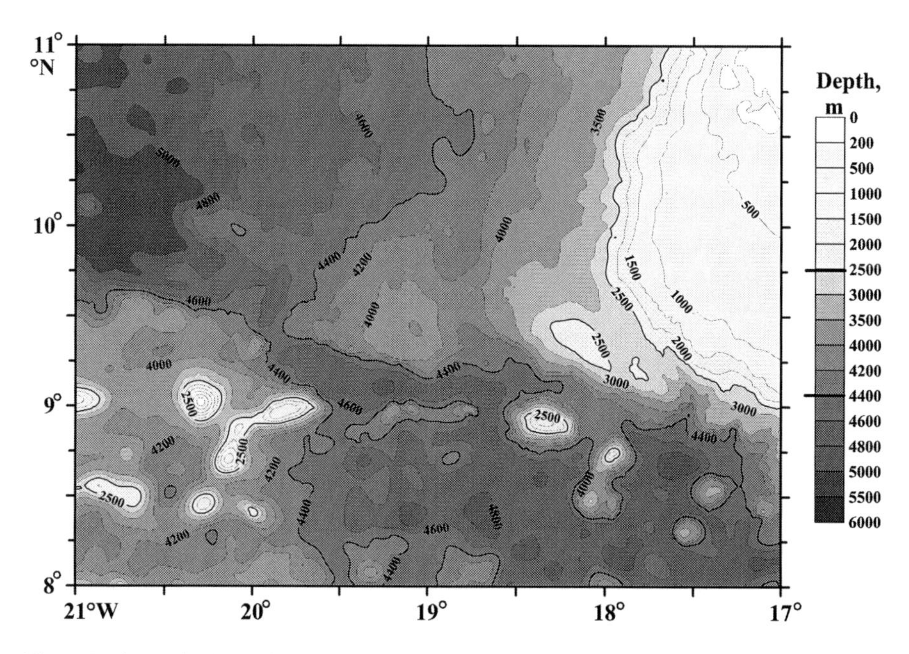

Fig. 7 Bathymetric chart of the Kane Gap

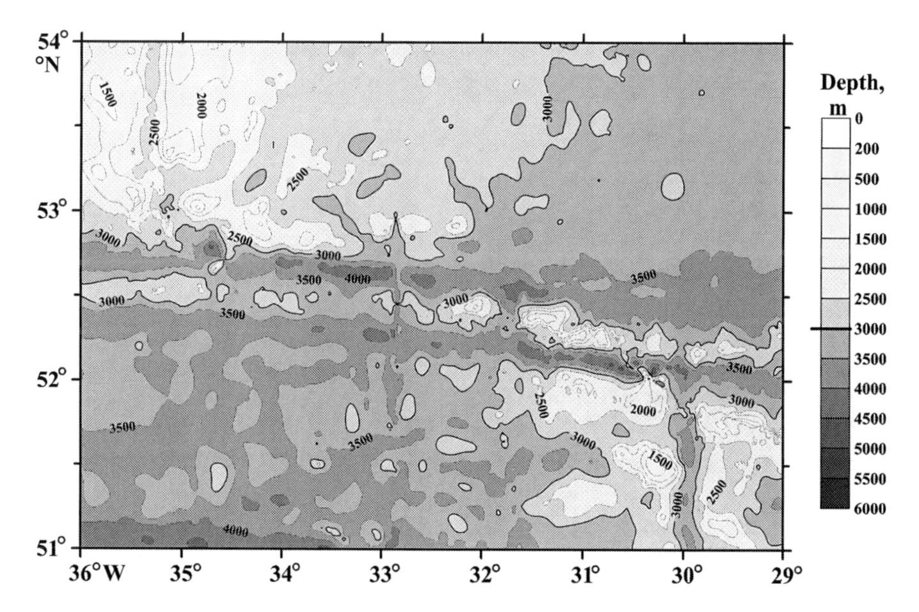

Fig. 8 Bathymetric chart of the Charlie Gibbs Fracture Zone

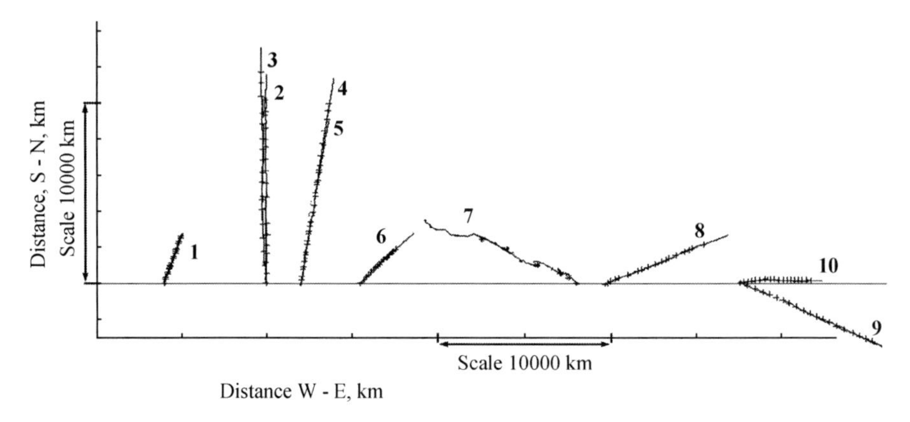

Fig. 9 Progressive diagrams based on 527-days long measurements in different abyssal passages around the Argentine and Brazil basins: (*1*) Santos Plateau; (*2–3*) west bank of the Vema Channel; (*4–5*) east bank of the Vema Channel; (*6*) Hunter Channel; (*7*) Shag Rocks Passage (1 year only); (*8*) Vema extension; (*9*) Romanche Fracture Zone; (*10*) Chain Fracture Zone. *Tick* marks are 30 days apart. See Fig. 1 for the location of passages. The figures are based on archive data of IFM-GEOMAR and WODB 2005

depth was defined as the level of the deepest current meter (15–50 m above the bottom). Mean velocity maxima served as scaling factors on the abscissa axis. Figure 10 shows time-averaged profiles from different abyssal channels: Vema Channel, Hunter Channel, Romanche Fracture Zone (Mercier and Speer 1998), and Vema Fracture Zone at 11° N (Vangriesheim 1980).

The general common property of abyssal flow in deep channels is location of the maximum velocity at a distance above the bottom. Velocities decrease in both upward and downward directions. This property of the flow was later confirmed by measurements with lowered acoustic current profilers (see Sect. 4.2.10). They demonstrated that the core of the flows is located at a distance of approximately 100 m above the bottom in different channels.

The major part of the book is dedicated to the propagation of Antarctic Bottom Water (AABW) and different modifications of this water in the Atlantic abyssal channels. Antarctic Bottom Water occupies a bottom position over the major part of the Atlantic Ocean. The bottom topography plays a crucial role in the propagation and properties of Antarctic Bottom Water. Therefore, only the upper part of this water can overflow each topographic obstacle. The characteristics and specific structure of abyssal waters in these regions may differ strongly due to entrainment and mixing with the overlying waters. Everywhere in the Atlantic Ocean, Antarctic Bottom Water is characterized by low potential temperature, low salinity, and high content of nutrients compared to the overlying deep waters. It is formed in several regions above the continental slope of Antarctica as the result of complex mixing of Antarctic Shelf Water and Circumpolar Deep Water. Formation of Antarctic Bottom Water spreading to the north in the Atlantic Ocean occurs mainly in the Weddell Sea.

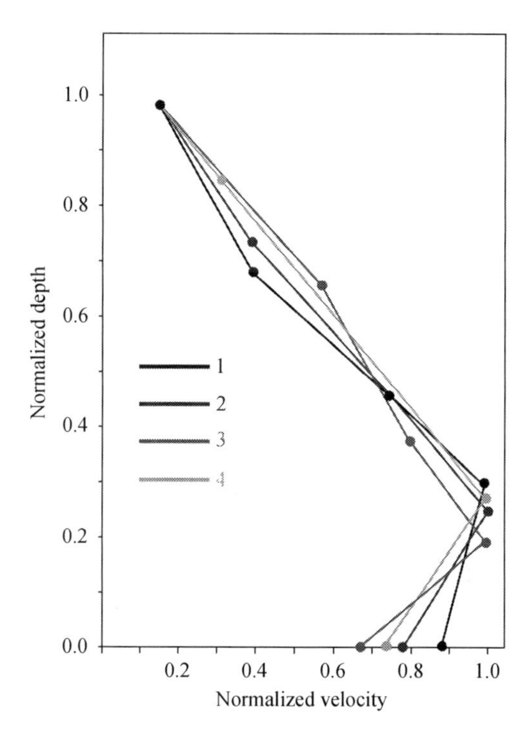

Fig. 10 Similarity of velocity vs. depth profiles in different channels. Averaged vertical current profiles of the recorded meridional current components in different regions (Modified on the basis of Zenk et al. (1999)): (*1*) Vema Channel, (*2*) Hunter Channel, (*3*) Romanche Fracture Zone, and (*4*) Vema Fracture Zone. The diagram shows velocities between 2°C potential temperature isotherm and the bottom

Abyssal channels in submarine ridges are key points for bottom water advection in the ocean. Locations of abyssal channels are geographically distant from each other. Therefore, the specific structure of waters is different in these regions. In addition, it is determined by complex processes of interaction between the waters of the North Atlantic and Antarctic origin. However, abyssal channels play an exclusive role in the formation of deep circulation, because they are the main pathways for Antarctic Bottom Water spreading to the north.

Although we are aware of a geographically unequal coverage of individual channels in the ocean, we believe that the Vema Channel is a test laboratory and the best example for analysis of the flows in underwater channels. This is the only channel that could be investigated deeply because many other abyssal channels are located very far from the leading oceanographic centers of the World. Our efforts were mainly concentrated on measurements in the Vema Channel, which is probably the best example of strong flow in a channel. Experiments in the Vema Channel constitute the most important part of our research. Our measurements in the Vema Channel provide insights into physical processes that occur in deep narrow channels with high velocities and intense mass transport.

Acknowledgements

The expeditions onboard research vessels *Akademik Ioffe* and *Akademik Sergey Vavilov* were carried out as part of projects "Southern Ocean", "Meridian" and "Meridian plus". The study was supported by the Russian Foundation for Basic Research, departments of Earth Sciences and Engineering Sciences (projects nos. 08-05-00943, 07-05-00657, 06-05-64634, 08-05-00120, 09-05-00788, 09-05-00802), and NWO-RFBR, project 047.017.2006.003; Federal Purposeful Program of Russia "World Ocean"; Program 17 of the Presidium of the Russian Academy of Sciences "Fundamental Problems of Oceanology: Physics, Geology, Biology, and Ecology"; Grant of the President of the Russian Federation for young scientists No MK-38.2009.5, and the Russian Foundation for the National Science Support. Some of the new observational results were provided by funds of the German Research Foundation (Bonn) and the German Federal Ministry of Education and Research, Berlin (CLIVAR Marin-2).

The authors are grateful to numerous colleagues who participated in the field work and helped to perform observations in the ocean: T. A. Demidova, E. Fahrbach, I. A. Gangnus, A. M. Gritsenko, V. A. Gritsenko, V. E. Morozov, T. J. Müller, S. V. Pisarev, I. S. Podymov, A. V. Remeslo, V. P. Shevchenko, S. G. Skolotnev, M. Visbeck, N. G. Yakovlev, V. S. Zapotylko, and Yu. A. Zyulyaeva. We are also grateful to the captains of research vessels for their contributions to the success of our expeditions: U. Pahl, G. A. Poskonny, L. V. Sazonov, and V. B. Lysak. We thank our colleagues who helped in the organization of expeditions: S. S. Lappo, V. B. Lapshin, R. I. Nigmatulin, S. M. Shapovalov, A. V. Sokov, L. I. Tolpygin, A. I. Diveev, and K. A. Pupkov. Comments by professors M. N. Koshlyakov and V. Yu. Liapidevsky were extremely helpful. Thanks go to colleagues who participated in data processing, analytical works, and preparation of the manuscript: D. R. Sakya, L. P. Filatova, and A. N. Boiko.

We also thank the crews of research vessels *Meteor, Polarstern, Akademik Ioffe, and Akademik Sergey Vavilov* for their contribution to this research.

Contents

Abbreviations

AABW	Antarctic Bottom Water (Note: AABW = WSBW + WSDW in classification of Orsi et al. (1999); AABW = WSDW + LCPW in classification of Wüst (1936) and Reid et al. (1977))
AAIW	Antarctic Intermediate Water
AASW	abbreviation used for both Antarctic Surface Water and Antarctic Shelf Water (AASurW and AAShW, respectively, in this book)
ACC	Antarctic Circumpolar Current
BF	Brazil Front
BFZ	Bight Fracture Zone
CBW	Circumpolar Bottom Water (term used in this book). Original term introduced in (Orsi et al. 1999) is ACCbw (ACC bottom water)
CFC	chlorofluorocarbon
CGFZ	Charlie Gibbs Fracture zone
DF	Deep Front (in the northern part of the Argentine Basin near the Vema Channel)
DSOW	Denmark Strait Overflow Water
FZ	Fracture zone
HFU	heat flux unit; 1 HFU = 42 m W/m^2
IfM	Institut für Meereskunde
ISOW	Iceland Scotland Overflow Water
LADCP	Lowered Acoustic Doppler Current Profiler
LCDW	Lower Circumpolar Deep Water (in the ACC zone)
LCPW	Lower Circumpolar Water (in the Argentine Basin (Reid et al. 1977)) (Note: LCPW is not the same as LCDW)
LOIW	Lower Intermediate Water
LSW	Labrador Sea Water
LWSDW	Lower Weddell Sea Deep Water
Ma	million years (age)
mbsf	meters below sea floor
NADW	North Atlantic Deep Water
PDW	Pacific Deep Water (sometimes SPDSW designates South Pacific Deep Slope Water)

psu	practical salinity unit
PVD	progressive vector (PVD) diagrams
SACCB	South Boundary of ACC
SACCF	South ACC Front
SAF	Subantarctic Front
SAMW	Subantarctic Mode Water
SB	South Boundary of ACC in some figures
SEPDW	South East Pacific Deep Water. Its regional modification CBW in Drake Passage, introduced by Sievers and Nowlin (1984)
SF	South ACC Front in some figures
SPF	South Polar Front; occasionally also Southern Polar Front
STF	Subtropic Front
Sv	Sverdrup (1 Sv = 10^6 m^3/s)
UCDW	Upper Circumpolar Deep Water (in the Antarctic Circumpolar Current zone)
UCPW	Upper Circumpolar Water (in the Argentine Basin (Reid et al. 1977))
UPIW	Upper Intermediate Water
UWSDW	Upper Weddell Sea Deep Water
WDW	Warm Deep Water (in the Weddell Gyre) (Note: WDW = LCDW + CBW if we use the approach in (Orsi et al. 1999))
WOCE	World Ocean Circulation Experiment in the 1990s (Siedler et al. 2001)
WSBW	Weddell Sea Bottom Water
WSDW	Weddell Sea Deep Water
WF	Weddell Front

Chapter 1
Geological and Geophysical Characteristics of the Transform Fault Zones[1]

1.1 General Description

From the point of view of the new concept of global tectonics, oceanic fracture zones consist of one or more transform faults. Their geomorphological descriptions were given for the first time in (Wilson 1965; Menard 1966; Menard and Chase 1970). The authors defined the transform faults as long narrow zones of strongly rugged bottom topography characterized by the existence of linear forms that usually separate topographic provinces with different regional depths. In the transform zones, fault-line ridges are extended parallel to the troughs. It is noteworthy that remains of shallow-water sediments are found on some of the fault-line ridges and bottom terraces. A fracture zone located between contiguous spreading axes is called an active zone. The fracture walls between neighboring spreading axes are characterized by opposite directions of motion. Passive parts of transform faults are located beyond the active zone, but the direction of their motion is the same. Transform faults are distinguished well not only in the ocean bottom topography, but also in anomalous geophysical fields. High fault-line ridges near the walls of faults (mainly between the neighboring parts of spreading axes), deep troughs, faults, and fissures are characteristic of the fracture zones, which represent an assemblage of deep and bottom structures. Anomalies of the magnetic and gravity fields, undulations of the heat flux, and other geophysical data testify to a complex dynamic regime of lithosphere in the fracture zones. Active parts of transform faults are also characterized by the most intense seismicity with clear manifestation of the shear component in earthquake sources.

Transform faults (Charlie Gibbs, Vema, Romanche, and Chain fracture zones) and the Vema Channel analyzed in this work are extremely different in terms of morphological and geological-geophysical characteristics, history of origin, and evolution. All transform faults in the Atlantic Ocean are characterized by their confinement to the slowly spreading Mid-Atlantic Ridge.

The Charlie Gibbs Fracture Zone is the southern boundary of the Reykjanes Ridge, a segment of the Mid-Atlantic Ridge, which extends in an anomalous oblique

[1] Author of this chapter is Anatoly Schreider.

E. G. Morozov et al., *Abyssal Channels in the Atlantic Ocean*,
DOI 10.1007/978-90-481-9358-5_1, © Springer Science+Business Media B.V. 2010

direction (at an angle of 30° relative to the spreading direction). The fracture is the northernmost major one, along which the ridge axis is displaced over many hundreds of kilometers. The Vema, Romanche, and Chain fracture zones are related to the equatorial segment of the Atlantic Ocean. This segment occupies a key position in the structure of the Atlantic, because only after its opening in the mid-Cretaceous did the Central and South Atlantic regions unite into a single oceanic basin. Before the final destruction of the continental bridge, which connected South America and Africa, the regions mentioned above were developing separately.

Concentration of large transform faults is a peculiarity of this segment. There are as many as 12 fractures over a distance of about 1,700 km along the spreading axis. The 15°20′ Fracture Zone (sometimes named the Cabo Verde Fracture Zone) is the northernmost one, while the Romanche Fracture Zone is the southernmost one. Owing to such concentration of transform faults, the spreading axis of the Mid-Atlantic Ridge is divided here into a large number of short, approximately 75-km-wide, segments. Total displacement of the spreading axis within this segment is more than 1,500 km to the east relative to its location north of the 15°20′ Fracture Zone (950 km of the total displacement is associated with the Romanche Fracture Zone).

The width of the Mid-Atlantic Ridge in the southern segment of the Charlie Gibbs Fracture Zone is only 500–650 km. The rift zone is represented by an echelon of troughs up to 4,000 m deep. Appearance of multiple (although small-scale) jumps and progradation of the spreading axis is a characteristic feature of this structure.

The transform faults have a quite complex structure mainly owing to variation in the location of rotation poles of the plates during the fracture development. The largest fracture zones (Romanche and Vema) are accompanied by troughs up to 4,500–5,200 m deep. The record depth of the trough (7,856 m) for the Mid-Atlantic Ridge is located in the Romanche Fracture Zone (Vema Deep). It is not surprising that such vertical motions led to exhumation of not only the entire section of the oceanic crust, but also the upper mantle layers. In addition, protrusions of serpentinized peridotites in the topography of these zones play an essential role in the fracture zones.

We shall use the US Geological Service (USGS) data [www.neic.usgs.gov/epic] (as of February 2009) to illustrate the seismological characteristic of the regions. The earthquake magnitudes are given according to the Gutenberg scale (body wave magnitude, mb). Morphology of the bottom topography in this research is based on the International database [www.topex.ucsd.edu/html/mar_topo.html] (as of February 1, 2009). All illustrations are made using the "Global Mapper" programming environment.

1.2 Charlie Gibbs Fracture Zone

The Charlie Gibbs transform fault is one of the well-studied fracture zones of the Atlantic. The rift valley of the Mid-Atlantic Ridge (Fig. 8) is displaced along the left-lateral fault plane over 340 km near 52°30′ N. Passive traces of the fracture zone are extended up to 49° N and 16°30′ W (Cherkis et al. 1973).

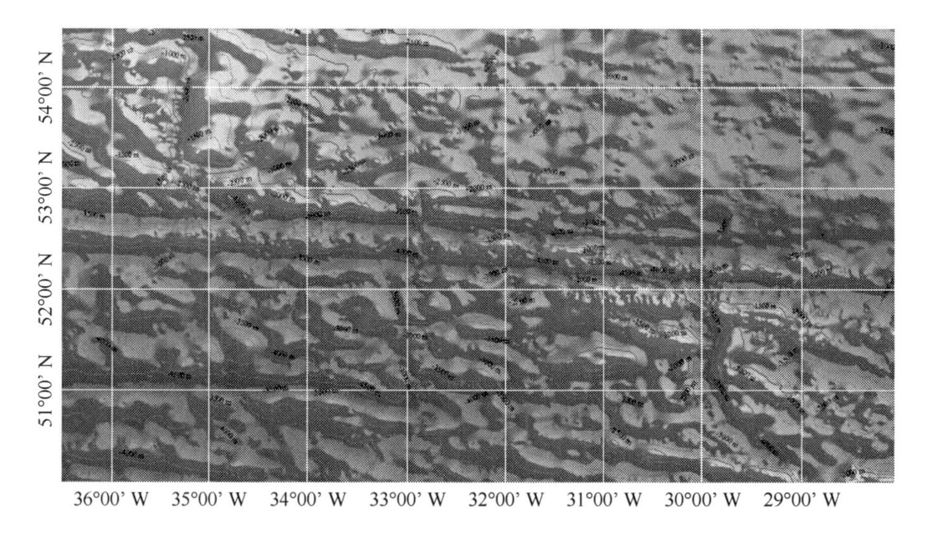

Fig. 1.1 Active part of the Charlie Gibbs Fracture Zone

The results of geological and geophysical investigations (Cherkis et al. 1973; Searle 1980; Dubinin 1987) allow us to state that a narrow 550-km-long ridge elongated along 52°30′ N, 28°30′–37° W is the main element of the bottom topography here (Fig. 1.1). The width of the ridge at its basement does not exceed 20 km. Individual peaks of the ridge are elevated above the ocean bed by 1,200–2,900 m. The ridge is limited from the north and south by troughs parallel to the ridge (Fig. 1.2). The ocean depth in the troughs is 4,000–4,800 m. The width of the steeper and deeper northern trough (Fig. 1.3) is 10–20 km (2–5 km at the bottom). The width of the southern trough is 20–30 km. The total width of the entire fracture zone reaches 80 km. A supposition was made on the basis of detailed studies in (Searle 1980; Lilwall and Kirk 1985) that the northern and southern trenches are connected in the active part of the fracture zone by a short segment of the spreading axis extending from north to south. This 31°48′ W segment, which is a narrow trench a few tens

Fig. 1.2 A perspective view of the active part of the Charlie Gibbs transform fault (view from the east)

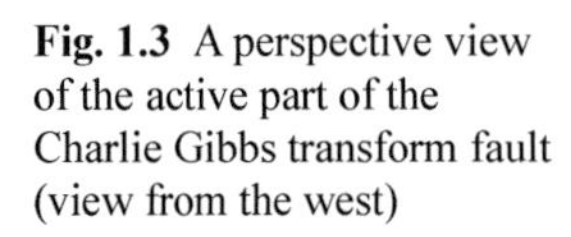

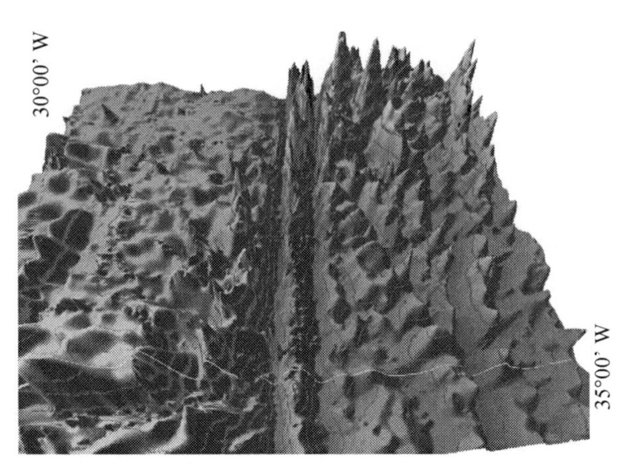

Fig. 1.3 A perspective view of the active part of the Charlie Gibbs transform fault (view from the west)

of kilometers wide with depths up to 4,500 m or more, can also serve as a channel for deep water exchange. Drilling and dredging in the Charlie Gibbs Fracture Zone revealed that the northern wall is composed of serpentinized peridotites, amphibolites, and basalts. Summits of the central ridge are basalt peaks and bare intrusions of gabbro and serpentinites (Oliver et al. 1974; Dubinin 1987).

It is likely that large ultrabasic mantle bodies intruded through the oceanic crust into the fracture zone. In the passive zones of the fracture zone, the main ridge and troughs become less prominent with increasing distance from the active zone. A sedimentary rock layer was found in the entire fracture zone except its southern valley. The thickness of dominating turbiditite sediments in the northern trench is 500–700 m, while the sedimentation rate is 2.5 mm/year (Blazhchishin and Lukashina 1977). Thickness of this bed increases with increasing distance from the active part of the fracture and reaches 1,000 m near 17° W. Morphological structures of the fracture zone are well expressed near the rift, but they are strongly smoothed and buried at a significant distance from the fracture zone (Smoot and Sharman 1985). Anomalies of the gravity force in the free-air reduction are minimal along the fracture zone. The greatest minimum Δg (−30 mGal) corresponds to deep depressions of the bottom topography (Oliver et al. 1974).

Anomalously high values of the heat flux were measured in the Charlie Gibbs Fracture Zone (Popova et al. 1984). Based on five measurements of the heat flux in the northern trough in the active zone of the fracture zone, the mean value is 3.8 ± 1.7 heat flux units (HFU), while the maximum value 6.5 HFU is confined to the middle part of the active zone of the transform fault.

The heat flux slightly decreases at the continuation of the northern trough to the inactive part of the transform fault west of the Reykjanes Ridge. The mean value of three measurements is 2.2 HFU. The heat flux in inactive parts of the southern trough of the Charlie Gibbs Fracture Zone is (1.3 HFU in the eastern part and 1.7 HFU in the western part). Two high values of heat flux were measured on the ridge dividing two depressions within the active part of the fracture: 1.3 and 1.7 HFU (Popova et al. 1984; Dubinin 1987).

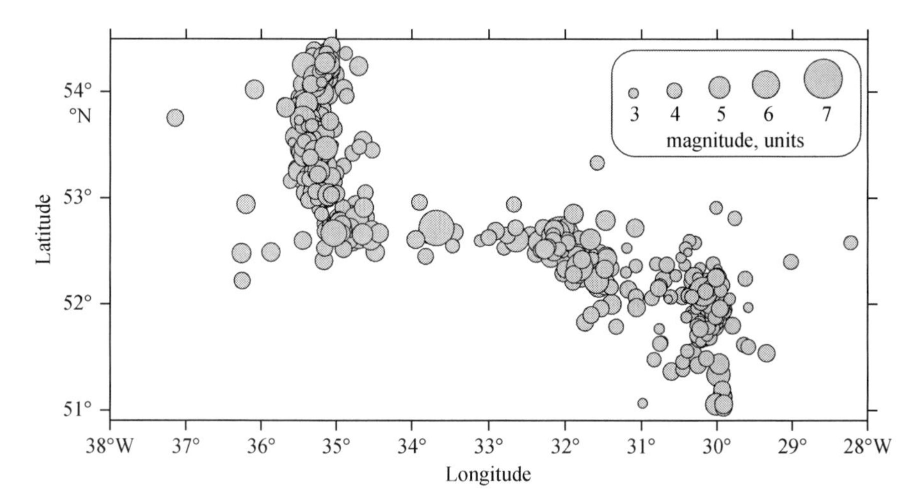

Fig. 1.4 Sites of recorded seismic activity in the Charlie Gibbs Fracture Zone

Intense seismicity was recorded between 30° W and 35° W along a 340-km-long segment of the transform fault (Fig. 1.4) (Kanamori and Steward 1976; Lilwall and Kirk 1985; Searle 1980; Dubinin 1987). The main earthquake events have a magnitude of 3–6. Their hypocenters are located at depths exceeding 35 km.

Thus, the active part of the Charlie Gibbs transform fault is characterized by anomalously high values of heat flux, while its values decrease significantly in inactive parts of the fracture zone. We suppose that the Charlie Gibbs Fracture Zone emerged on the continent in the Paleozoic and developed further during the opening of the Atlantic Ocean. The ocean bottom morphology and anomalously high values of the heat flux also suggest that widening occurs in the active part of the fracture.

1.3 Vema Fracture Zone

The active part of the Vema transform fault displaces the segments of the Mid-Atlantic Ridge by 320 km in the nearly latitudinal direction along 10°30′–11° N (Fig. 4). The fault has a prominent valley (up to 5,200 m deep) bounded by steep walls (20°–30°). The distance between walls varies from 10 to 20 km (Fig. 4). The southern wall of the valley is the slope of a narrow (approximately 30 km wide) monolithic ridge located parallel to the transform valley (Fig. 1.5). Some peaks of the ridge are as high as 500 m below surface, while the mean depth of the ridge is 2,000–2,500 m. Peaks of the Mid-Atlantic Ridge adjacent from north and south to the Vema Fracture Zone do not exceed 2,500 m. Steep walls and the general shape of the ridge suggest that this structure is an elevated block of the oceanic crust rather than a product of eruptions of several underwater volcanoes. This explanation is also valid for the northern slope. The ridge on the northern wall of the fracture is

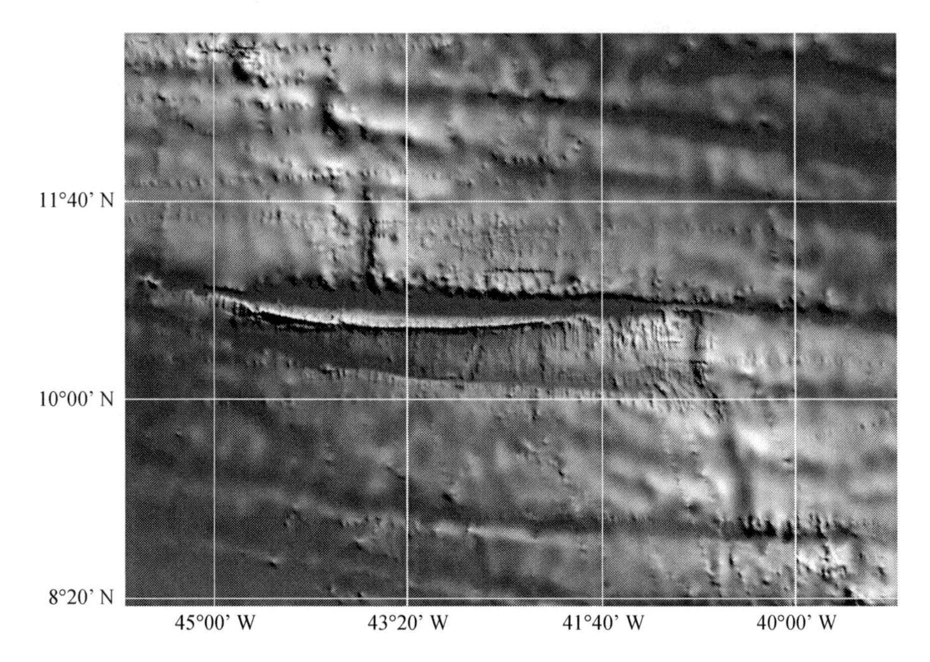

Fig. 1.5 Active part of the Vema Fracture Zone

relatively less prominent in its topography (ocean bottom depth 3,000–3,500 m). It smoothly grades into the abyssal plain, the depth of which is consistent with the general age dependence of the usual oceanic lithosphere with increasing distance from the mid-oceanic ridge axis. Thus, one can see a topographic asymmetry across the Vema Fracture Zone (Fig. 1.6). The fracture valley with adjacent ridges is also

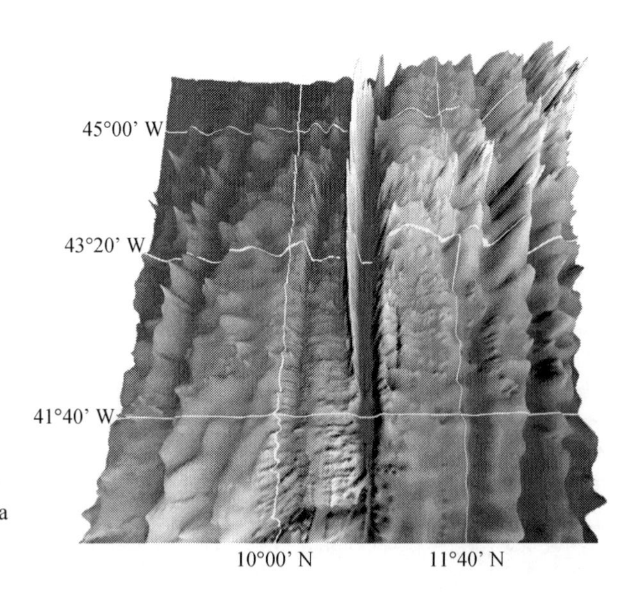

Fig. 1.6 A perspective view of the active part of the Vema Fracture Zone (view from the east)

prominent (although not so clearly) in the passive zones. The flat bottom of the valley extends west up to 44°30′ W. Then, the valley becomes narrower and its extension turns slightly to the north (van Andel et al. 1971). The transform valley is shallower near the southern segment of the Mid-Atlantic Ridge, while the eastern part of its passive zone extends far to the east into the Gambia Abyssal Plain as a solitary trough (Fig. 1.7).

In the 38° W zone, the trough is 600–900 m deep, while the heights of the southern and northern slopes are 1,900 and 1,000 m, respectively (Syrsky and Greku 1975).

Profiles based on the method of reflected waves (MRW) show that the Vema fracture valley is filled with a thick layer of sediments (1,000–1,200 m) lying on a highly reflective basement surface located at a depth of 6,200–6,500 mbsf (meters below sea floor) (van Andel et al. 1971; Kastens et al. 1998). The troughs of inactive valleys located south of the Vema Fracture Zone are covered with similar sediments up to 900 m thick. Drilling from R/V *Glomar Challenger* (cruise 4, Hole 26; 10°54′ N, 44°03′ W) showed that the sediments are represented mainly by Pleistocene turbidite sediments transported from the Demerara Abyssal Plane under the influence of the Amazon alluvial fan. The mean sedimentation rate is approximately 120 cm/ka (Bader et al. 1970). Sediments of the fracture valley are deposited uniformly as a flat layer. Insufficient overlapping and flat undulation can be related to erosion activity, different degrees of compaction, continuing rise of the fault-line ridges, and relative motion of contacting lithospheric blocks. The character of de-

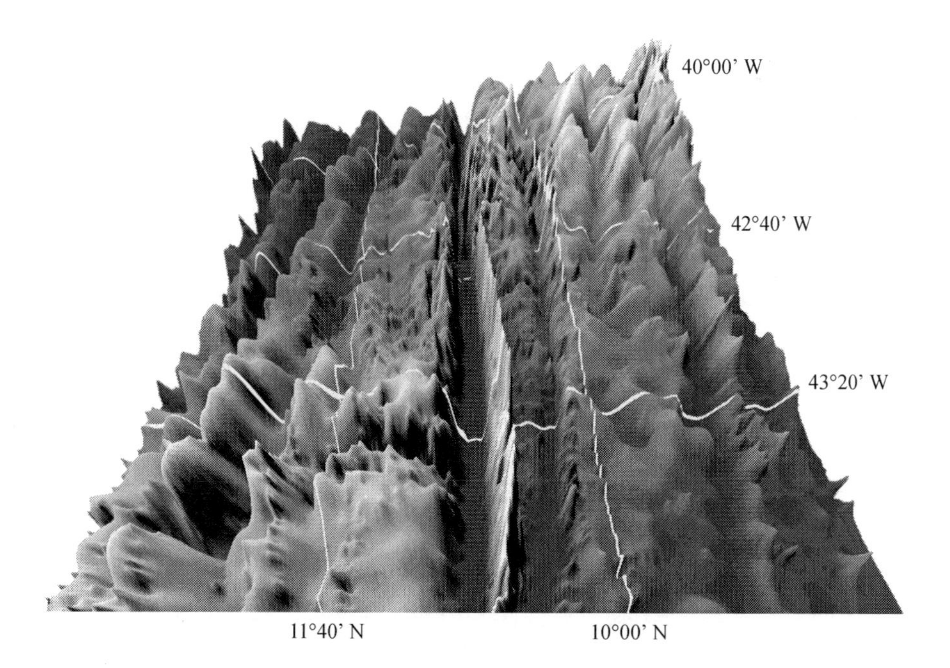

Fig. 1.7 A perspective view of the active part of the Vema Fracture Zone (view from the west)

formation suggests the existence of local (subordinate) contraction and extension zones, which can be indirectly related to the relative motion of contacting parts of plates.

Zones of sediment piling and rises make up a small median ridge. They can be related to local mantle protrusions under the sedimentary cover (Eittreim and Ewing 1978). Such distortions of the top of the sedimentary cover could have occurred during the last 500 ka. Based on the method of reflected wave (MRW) data, the valley bed under the sedimentary cover represents a series of rises and depressions (Eittreim and Ewing 1978; van Andel et al. 1971; Kastens et al. 1998; Bonatti et al. 2005). The deep structure of the oceanic lithosphere in the fracture zone based on the results of seismic investigations at 42°–43° W shows that seismic layer 3 of the oceanic crust (P = 5.9–6.3 km/s) under the fracture valley is significantly thinner (about 2.5 km) (Ludwig and Rabinowitz 1980).

This layer is underlain by rocks of the upper mantle (P = 8.12 km/s). Its roof is located above the expected Moho boundary. All these facts suggest the ascent of high-temperature mantle matter along a fracture in the study area of the fracture zone. It is likely that high-temperature mantle matter reached the hydrothermal circulation zone.

The active part of the Vema Fracture Zone is characterized by seismicity (Fig. 1.8). Earthquake epicenters are generally located along the southern footwall of the fault. Solutions of mechanisms in the sources suggest the predominance of shear dislocations along the transform fault (Sykes 1970).

Usually, the earthquake magnitude does not exceed 5, and the hypocenters are located at a depth of less than 35 km. Passive zones of the fracture are aseismic.

Rocks dredged along the walls of the fracture zone included basalts, metabasalts, gabbro, and metagabbro of the amphibolite stage of metamorphism, as well as peridotites subjected to different degree of serpentinization (Bonatti and Honnorez 1976; Bonatti et al. 2005). Compositions of rocks between the southern and northern walls are asymmetric. Although basalts dominate at all levels of the section, their chemical compositions are slightly different at the northern and southern

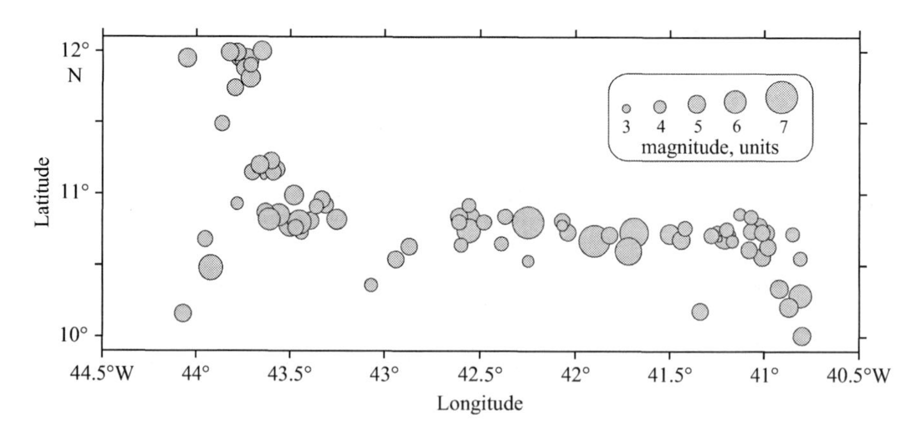

Fig. 1.8 Sites of recorded seismic activity in the Vema Fracture Zone

walls. The composition of basalts at the northern wall is close to that of basalts at the Mid-Atlantic Ridge axis (Bonatti 1976). The northern ridge is dominated by gabbro and serpentinites, while the southern slope is composed of highly metamorphosed amphibolites and mylonites. Serpentinites are found almost at all levels on walls of the fracture valley along both vertical and horizontal directions. Morphological, seismic, and petrological data on the Vema Fracture Zone indicate that the structure of the northern block is similar to the normal oceanic crust formed in the axial Mid-Atlantic Ridge (Bonatti 1976; Dubinin 1987).

The structure of the southern wall was strongly affected by processes in the fracture zone: diapir rise of serpentinized ultrabasic rocks responsible for high fault-line ridges (for example, the southern ridge of the Vema Fracture Zone). During their ascent, serpentinized rocks can capture fragments of surrounding rocks (harzburgites, peridotites, metagabbro, and gabbro) and transport them to higher layers of the Earth's crust.

The active part of the Vema Fracture Zone is characterized by increased heat flux with a mean value greater than 1 HFU. The maximum values of heat flux are confined to the fracture valley (3.42 ± 1.00 HFU), namely to its middle part located equidistantly from the Mid-Atlantic Ridge axis segments. Heat flux in some areas is as high as 6.2 HFU, although thick sediments decrease the heat flux in the trough through the basement by 20–30% (Langseth and Hobart 1976; Dubinin 1987).

The maximum free-air Δg anomalies (100–120 mGal) are confined to the southern ridge of the Vema Fracture Zone. Negative anomalies confined to the central valley reach −80 mGal (Robb and Kane 1975). Robb and Kane interpreted the anomalous gravity field as a two-layer lithospheric model with the following parameters: water density 1.03 g/cm^3, crust density 2.60 g/cm^3, and mantle density 3.15 g/cm^3. Density of the thick sedimentary cover in the local depressions was assumed to be 1.90 g/cm^3 (Bader et al. 1970). The lithosphere was bounded at a depth of 40 km by a flat isobaric surface. Anomalies Δg calculated within this model agree well with the observed ones. The form of Bouguer anomalies within the two-layer model assumes the existence of mass excess under the southern ridge and lesser excess under the northern ridge. It is likely that a zone of rocks with lesser density exists under the axial part of the Vema Fracture Zone, which can be filled with hot mantle material ascending in the pull-apart setting.

Data on the Vema Fracture Zone discussed above suggest the existence of a high southern ridge mainly composed of the mantle-derived serpentinized peridotites, significant deconsolidation of the mantle under the transform valley, anomalously high values of the heat flux, and specific character of the free-air Δg anomalies. All these facts suggest that the Vema Fracture Zone includes an active high-temperature intrusion, which ascended along the fracture up to the level of thermal water penetration. Calculation of the thermal regime of lithospheric blocks contacting along the transform fault during intrusion of the high-temperature mantle matter to a depth of 10,000–11,000 mbsl demonstrates that properties of the studied profile, which makes up 1.5–2.0 km of the total fault-line ridge height, is affected by temperature increases at the edges of blocks owing to processes of intrusion (Dubinin 1987).

Properties of the remaining 2 km of the fault-line ridge height (Dubinin 1987; Dubinin and Ushakov 2001) can be explained by the serpentinite diapirism of the mantle. According to the calculations, the serpentinization temperature interval (~500–300°C) is located at a depth of 9,500–7,000 mbsl under the fracture valley axis. Low-density fissured material of mylonites, basalt, gabbro, and serpentinites underlies the fracture valley beneath the 1-km-thick sedimentary layer in the zone of intense deformations.

Walls of the valley and the southern fault-line ridge are also composed of these rocks. At a depth of about 8,000–9,000 m, density of rocks and, consequently, seismic wave velocity increase sharply, probably, owing to the presence of gabbro and peridotite rocks with a low degree of serpentinization. Seismic boundary at a depth of 11,000 m can represent the roof of the high-temperature mantle intrusion. The gravity field based on such geodynamic model agrees well with the observed field.

After termination of the active mantle intrusion, the thermal component of the topography and, consequently, the corresponding part of the fault-line ridge height decrease rapidly (during 10–15 Ma). At the same time, intrusions of serpentinized ultrabasic rocks continue during a longer time and can be traced in passive zones of the fracture zone.

Thus, the model of pull-apart transform faults explains the facts related to mantle intrusions, such as high values of the heat flux, manifestation of serpentinites along the slopes of ridges, and existence of a thin deconsolidated crust under the axial zone of transform valleys.

1.4 Romanche Fracture Zone

The Romanche transform fault (Fig. 5) is among the largest tectonic fractures in the Central Atlantic. In the Equatorial Atlantic, it extends from the coast of Africa to South America. The main morphological and tectonic features of the Romanche Fracture Zone were revealed for the first time in (Heezen et al. 1964a) and later supplemented in (Bonatti et al. 1977; Bonatti and Chermak 1981; Belderson et al. 1984). Active segments of the spreading axis of the Mid-Atlantic Ridge are displaced over 930 km along the fracture zone (Pushcharovsky 2005).

Abundance of nearly latitudinal linear troughs both north and south of the modern fracture valley is a specific feature of the Romanche Fracture Zone (Fig. 1.9). The troughs are covered with sediments several hundred meters thick. The modern valley of the Romanche Fracture Zone is not strictly linear. Its extension sharply changes near 19°–20° W.

The fracture zone represents a valley, which is more than 7,000 m deep east of 22° W and is less prominent west of 22° W. The width of the valley bottom is 3–10 km while the depth usually exceeds 6,500 m and reaches 7,856 in the Vema Deep (Metcalf et al. 1964; Belderson et al. 1984) at 18°30′ W. The valley is bounded by steep walls (inclination more than 30° in some places), which grade into slopes of the high fault-line ridges that accompany the valley along its entire extension on

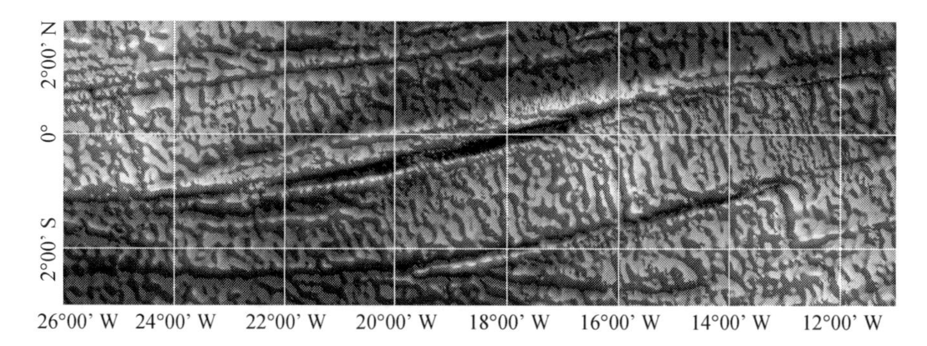

Fig. 1.9 Active zones of the Romanche (center of the image) and Chain (right lower angle) transform faults

both northern and southern sides (Fig. 1.10). Amplitudes of the topographic gradient in the Romanche transform fault frequently exceed 5,000 m. The total width of the fracture zone exceeds 100 km (Belderson et al. 1984; Bonatti and Fisher 1971). Individual summits of the fault-line ridges reach a depth of 1,000 mbsl (Fig. 5).

The fracture valley bottom accommodates a small median ridge (100–800 m), which is most prominent in the eastern part of the active zone (Bonatti and Fisher 1971; Belderson et al. 1984; Dubinin 1987). In the passive part of the fracture zone, it is partly masked by a large amount of sediments. The southern and northern fault-line ridges in the western passive part of the fracture zone are traced in bathymetry up to 31° W and 33° W, respectively. Further up to 36° W, they are reflected in the basement topography and gravity anomalies (Cochran 1973). Fault-line ridges are distinguished in the bottom topography of the eastern passive zone up to ~7° W. West of 7° W up to the steep continental slope of Ghana, the fracture zone is recognized as a large ridge and basement trough, which are buried under the sedimentary

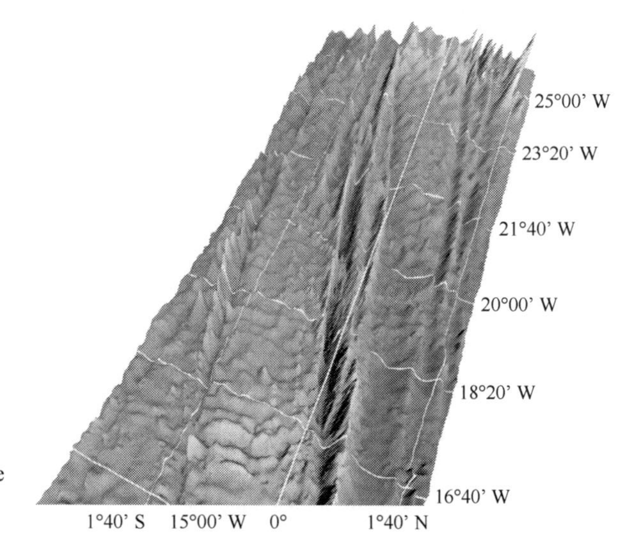

Fig. 1.10 Active zone of the Romanche Fracture Zone (view from the east)

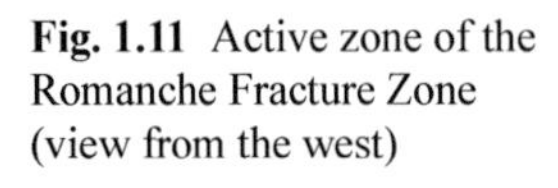
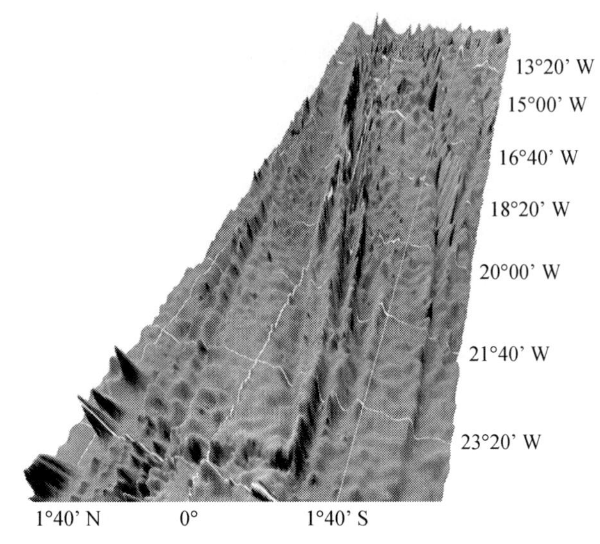

Fig. 1.11 Active zone of the Romanche Fracture Zone (view from the west)

layer (Dubinin 1987; Dubinin and Ushakov 2001) and a double belt of positive and negative Δg anomalies up to 100 km wide.

South of the Romanche transform fault, the western slope of the Mid-Atlantic Ridge is characterized by flatter surfaces (Figs. 5 and 1.11) located at a depth of 4,000–5,000 m. The sedimentary layer on the bottom of the western and eastern depressions is thin or absent, and crystalline rocks of the basement are exposed. However, thickness of the sedimentary layer in some intermontane depressions reaches 500 m and even 800–1,000 m in rare cases (Bonatti et al. 1977, 1994). Sometimes, these depressions are united in literature into the Rom 2 fracture valley (Pushcharovsky 2005).

This zone is composed of clayey silts and coarse material, which includes fragments of igneous rocks from walls of the fracture valley. Thickness of sediments in passive zones of the fracture zone increases toward the continental margins. According to the results of DSS and MRW seismic profiling, the crust in the Romanche Fracture Zone has a block structure (Bonatti et al. 1977, 1994; Dubinin 1987; and others) with a complex morphology of reflecting boundaries. Seismic layer 2 located at a depth of 4,500–5,800 m is divided by numerous faults and fractures into small blocks up to 5,000–7,000 m in size. Thickness of the layer ranges from 3,500 to 5,800 m (average 4,300 m) and reaches the maximum values in elevated zones. Seismic layer 3 is located in the depth interval 8,000–18,000 m (mean thickness 6–8 km). Its thickness decreases in depressions and increases to 10 km in elevated zones, where the boundary of the layer descends to the maximum depths.

The dredging data suggest that, as in the Vema Fracture Zone, walls of the fault-line ridges in the Romanche transform fault, are composed of basalts, gabbro, serpentinized peridotites, metabasalts, and metagabbro of the greenschist facies of metamorphism, as well as basalt breccia, mylonites, and sedimentary rocks, such as quartz sandstones, limestones, phosphorites, and siliceous shales (Bonatti and Honnorez 1976; Bonatti et al. 1994; Dubinin 1987). Serpentinized peridotites were found only at 34 of the total 59 dredging stations along the Romanche Fracture Zone. As in the Vema Fracture Zone, the southern and northern walls of the fault show petrological asymmetry not only in the lower part of the section, but also over the entire slope. For example, the southern (up to 4,000 m high) wall is almost completely composed of serpentinized peridotites. On the northern wall, they are less developed and associated with basalt and gabbro rocks. Many rocks (Bonatti et al. 1994), especially gabbro, underwent tectonic crushing up to the formation of mylonites, suggesting intense tectonic motions in the fracture zone.

Partly phosphoritized oolite biogenic limestones with an age of 5 ± 1 Ma were recovered from peaks of the northern ridge (depth 1,000 mbsl) in the eastern part of the active zone of the fault (17° W) (Bonatti and Fisher 1971; Bonatti et al. 1977). The analysis of limestones revealed that they were deposited almost at the sea surface level. Given that the sea level did not change appreciably during the last 5 Ma (Bonatti et al. 1977; Dubinin 1987) and individual peaks of the fault-line ridge are located now at a depth of about 1,000 m, submergence rate of the ridge is estimated at 0.2 mm/year (Bonatti and Fisher 1971; Bonatti et al. 1977), which agrees well with the mean submergence rate of a cooling thermal relief after the termination of intense activity of the mantle intrusion (2 km/10 Ma) (Dubinin 1987).

The Romanche Fracture Zone is located in the transition zone 0°–15° N between two geochemical provinces, which were formed at the early stages of the Atlantic Ocean opening. This fracture zone is characterized by the abundance of sodium tholeiites and alkaline basalts that are atypical of other Mid-Atlantic Ridge segments. Depleted tholeiites dominate in the northern and southern zones (Schreider et al. 2006b; Kashintsev et al. 2008). In the eastern part of the study region on the northern slope of the fracture at 20°06′ W, alkaline magmatites are especially widespread (as much as 95% of the total recovered material). At the same time, basalts are associated with their deep analogs (alkaline gabbro), suggesting the genetic unity of the entire alkaline magmatite complex. The study area of the fracture zone is also characterized by the abundance of ultrabasic rocks, especially lherzolites (Bonatti and Honnorez 1976).

The primitive nature of the mantle material in this area suggests a low degree of its manifestation. This conclusion agrees well with the development of alkaline basalts that are products of small volumes of melts. Peridotites are almost ubiquitous and especially widespread (up to 95%) on the southern slope of the fracture zone. We note that the degree of serpentinization and secondary oxidation of magnetite, which is formed during the serpentinization of ultrabasic rocks dredged from the fracture zone, promoted their development in zones of both high and low magnetization of the inverse magnetoactive layer. Thus, differently serpentinized ultrabasic rocks along with basalts can be deposited at the bottom of the Romanche Fracture Zone. This is typical for a large number of transform fractures.

Geological data suggest that vertical tectonics is widespread along with the shear component of tectonic motions within the fracture zone. Reef limestones formed 20 Ma ago within the northern block were buried to a depth of 800 m after 5 Ma owing to drastic subsidence of the crustal block (Gasperini et al. 1997), probably, according to the mechanism described in (Lonsdale, 1994).

In the Romanche Fracture Zone, positive free-air Δg anomalies reach up to 100 and 50 mGal at the fault-line ridges of old and young lithospheric blocks, respectively. Negative free-air Δg anomalies (up to −160 mGal) are confined to the trough. Topography of the fault-line ridges suggests that they are situated at the stage of thermal cooling. A geodynamic model of the structure and thermal regime of the lithosphere in the study region was contrived on the basis of interpretation of the gravity field and theoretical suppositions (Dubinin 1987). According to these data, the fracture valley zone is composed of fractured low-density rocks (most likely, fissured serpentinites with a density of 2.4–2.5 g/cm^3) subjected to shear deformations. They are underlain by rocks of higher density (probably, weakly serpentinized peridotites with a density of 3.0 g/cm^3) that overlie a pile of unaltered rocks of the upper mantle with a density of 3.3 g/cm^3. Under this thermal regime, the fault-line ridges should be dominated by partly serpentinized ultrabasic rocks mixed with basalts and gabbro. Variations in the topography of such fault-line ridges occur more slowly relative to the thermal component of topography. Isotherm depths corresponding to serpentinization temperatures (300–500°C) in the profiles are likely located beyond the zones of hydrothermal convection (Dubinin 1987).

The anomalous magnetic field has a non-uniform structure (Fig. 1.12). Positive and negative anomalies range from +100 to −200 nT. Many anomalies have an equant shape. They merge into latitudinal anomalous fields of variable sign. Correlation between the local topographic forms and anomalous magnetic field is virtually absent. Actually, there should be no such correlation, because the study region is located near the equator where inclination of the magnetizing vector of the Earth's magnetic field T is close to zero.

Under the conditions of equatorial magnetization, the following phenomenon is observed in this case: the southern and northern margins of the nearly meridional magnetism-exciting bodies would be associated with the negative values of magnetic field anomalies T. From this point of view, there are no perspectives to be gained by studying magnetic inhomogeneities of the crust based on the 2D modeling of magnetic field anomalies.

Under these conditions, magnetic inhomogeneity of the bottom can only be studied using the 3D reconstruction of parameters of inverse magnetoactive layer based on the magnetic survey data. A method of such research was developed and described in (Bulychev et al. 1997). Modification of this method for equatorial magnetic latitudes is given in detail in (Bulychev et al. 2004).

Data on anomalous magnetic field and topography of the acoustic basement provide insight into magnetic inhomogeneity of the bottom in the study region based on the concept of lithospheric plate tectonics (Fig. 1.12). Thickness of the inverse magnetoactive layer of the ocean in such calculations was assumed to be equal to 0.5 km (e.g., Sclater et al. 1976). The upper boundary of the layer coincides with the acoustic basement surface or the bottom surface if the sedimentary layer is absent or thin. Parameters of the Earth's modern magnetic field for the study region were adopted from the definitive geomagnetic reference field (DGRF) of the survey time.

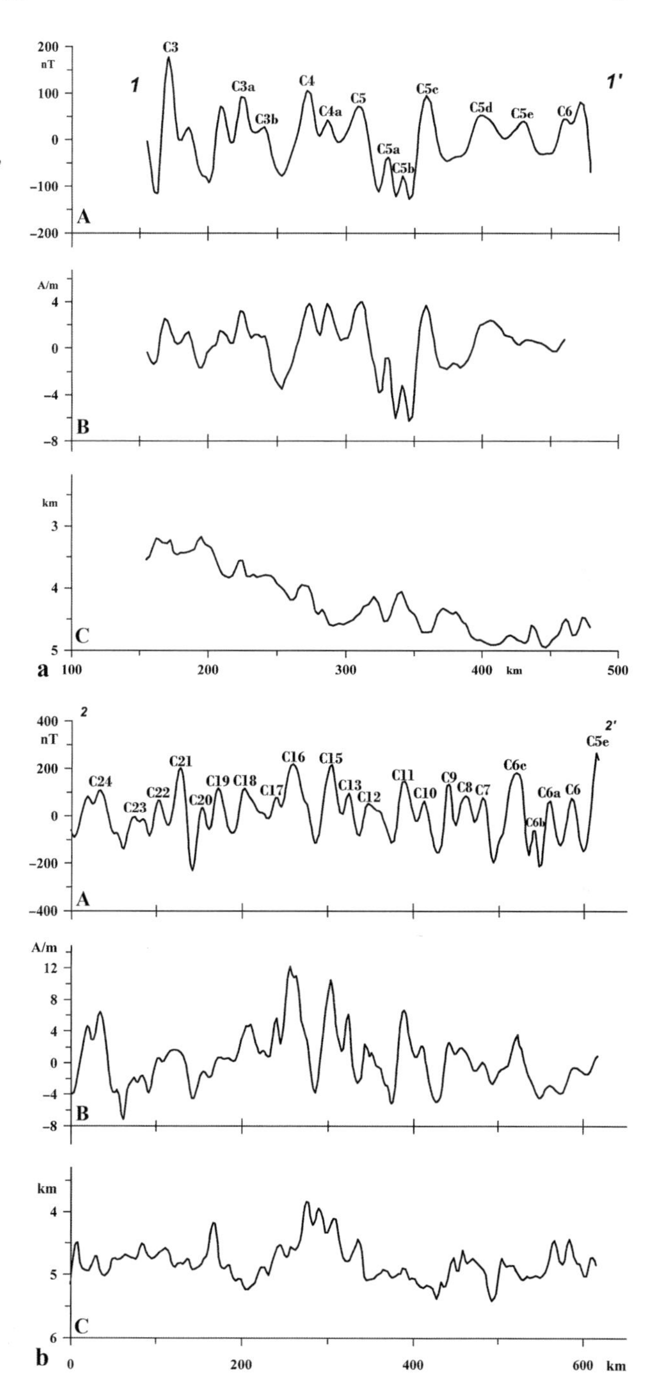

Fig. 1.12 Distribution of the pole-reduced anomalous magnetic field (A, nT), magnetization (B, A/m), and seafloor depth (C, km) along profile 1 (**a**) and profile 2 (**b**) shown in Fig. 1.13 (adopted from (Schreider et al. 2006a)). Linear magnetic anomalies are identified. Hereinafter, the nomenclature and age limits of chrons are adopted from (Schreider 2001; Gradstein et al. 2006)

Inclination of the magnetization vector was calculated as an approximation of the Earth's main magnetic field of the Earth by an axisymmetric dipole.

The method used in the calculations allows us to compute the anomalous magnetic field from the known distribution of magnetization of the inverse magnetoactive layer (Bulychev et al. 2004), which could be observed in the case of vertical magnetization of the magnetoactive rocks in the field (Fig. 1.12). Thus, the magnetic anomalies obtained in our case differ from the pole-reduced transformants of the anomalous magnetic field known from literature.

Comparison of magnetic anomalies calculated for the vertical magnetization of magnetoactive rocks of the inverse magnetoactive layer and for magnetic anomalies in the bottom spreading model allows us to identify paleomagnetic anomalies. In addition to comparative analysis of magnetic anomalies, comparative analysis of magnetization distribution in the reconstructed inverse magnetoactive layer and the bottom spreading model also facilitates the reliable identification of anomalies and determination of chron boundaries.

A zone of normally magnetized rocks with intensity exceeding 10 A/m is distinguished in the central Mid-Atlantic Ridge (Bulychev et al. 2004; Schreider et al. 2006a). The zone (about 20 km wide) is shifted to the southeast in the southern area and the total amplitude of displacement is nearly 5 km. According to the results of modeling and identification of paleomagnetic anomalies (Figs. 1.12 and 1.13), this zone was formed in the Brunhes epoch of direct polarity (chron C1n, 0–0.78 Ma).

Distribution of magnetization is non-uniform within the zone of normally magnetized rocks (Schreider et al. 2006a). Two narrow (2–5 km) zones of high magnetization (up to 15–25 A/m) distinguished in the northern area of this zone correspond in the bottom topography to the deepest part (up to 4,000 m or more) of the rift valley. According to the results of modeling, fields of highly magnetized rocks correspond

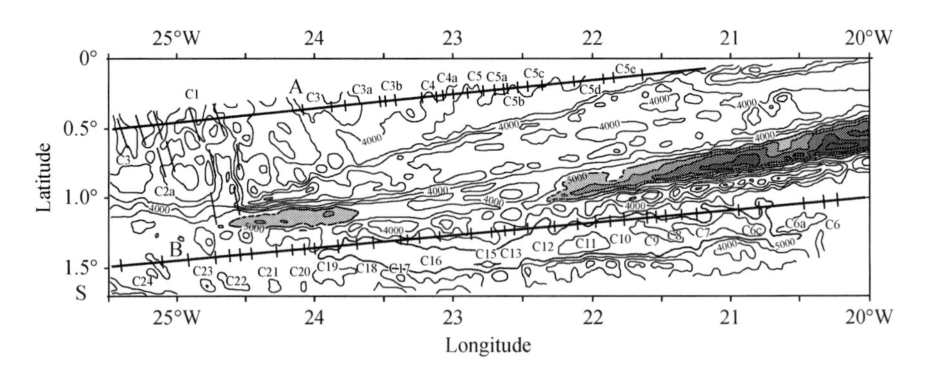

Fig. 1.13 Reconstructed bottom topography before the Brunhes transtension episode. The reconstruction is furnished with modern coordinate grids for both northern and southern walls of the Romanche fault (Schreider et al. 2006a). The figure shows spatial location of chrons with direct (alphanumerical notations above *solid lines* of the observation profiles A see Fig. 1.12a; and B, see Fig. 1.12b) and inverse polarity in the seafloor spreading model. Along with other isobaths (in km), 5-km-isobath segments used in the calculation of transtensile geodynamic parameters are shown. The 5-km-isobath segments omitted in the calculations are shown with a *dashed line*

to the neovolcanic zone of the distribution of the youngest oceanic crustal rocks. The mean horizontal size of the neovolcanic zone likely corresponds to the transverse dimension of an instantaneous spreading axis marked by the formation of new crust.

Inverse magnetized zones of the inverse magnetoactive layer are developed on the western and eastern sides of this field of direct polarity rocks. Magnetization of the rocks is generally 5–10 A/m lower relative to the rocks of direct polarity described above. The mean value is 4–10 A/m.

A highly magnetized zone (5 km wide) formed during chron C2 was distinguished east of the axial zone at the periphery of normally magnetized rocks at the eastern flank of the northern spreading cell. The next zone of normally magnetized crust (intensity up to 2 A/m) can be coeval with chron C2A. Further to the east, we can distinguish zones of directly and inversely magnetized rocks (about 5–20 km wide) extending parallel to the Mid-Atlantic Ridge. Their relative magnetization intensity is close to 5–8 A/m, which is approximately 2–3 times lower than magnetization in the central part of the ridge. These zones correspond to chrons C3–C5. The easternmost band of the normally magnetized crust corresponds to chron C5En (18.433–18.906 Ma).

A zone of normally magnetized rocks (relative intensity more than 5 A/m) is distinguished at the southern wall of the fault on the meridian of the axial spreading zone of the northern flank. The zone is about 20 km wide. Several zones of directly and inversely magnetized rocks are distinguished east of this zone. According to the modeling of paleomagnetic anomalies (Schreider et al. 2006a), they were formed during chrons C4–C6 (Figs. 1.12 and 1.13).

Determinations of spreading rate using the seafloor geochronology based on the anomalous magnetic field showed that the mean spreading rate in the eastern cell of the northern wall of the Romanche transform fault was 1.65 cm/year during the last 19 Ma.

Calculations of the instantaneous spreading rate indicate that the new oceanic crust was growing with an irregular rate that reached approximately 0.5 cm/year about 12–15 Ma ago. The rate exceeded 3.5 cm/year 5–7 Ma ago and then slowed down to the present-day rate.

Detailed modeling of the inverse magnetoactive layer shows that oceanic crust was growing relatively uniformly (mean rate 1.56 cm/year) on the southern wall of the fracture zone in the chron interval C6–C24 (19–54 Ma). The minimum (less than 1 cm/year) spreading rate is noted 43–56 Ma ago (Schreider et al. 2006a).

Detailed bottom chronology restored for the Equatorial Atlantic based on the study of flanks of the Romanche transform fault supports the concepts of spreading discussed in (Bulychev et al. 2004; Trukhin et al. 2005). The temporal distribution of instantaneous spreading rate is close to the pattern in the mid-oceanic North Weddell (America-Antarctic) Ridge (Schreider et al. 2006b).

Geological and geophysical investigation of the Romanche transform fault showed that its horizontal displacement, which reflects the mutual shear displacement of lithospheric plates relative to the Euler pole located at a point with coordinates 62° N, 36° W (e.g., Sykes 1970), has a pull-apart component (Bonatti 1973; Bonatti et al. 1994; Dubinin 1987). In this relation, modern western and eastern depressions in the internal part of the Romanche Fracture Zone are considered as morphological expression of a transtension zone characterized by the interaction of shear and extension. Transtension in one zone can be replaced by transpression (shear motion in the course of compression) in a neighboring zone (Schreider et al. 2006a). We note that a small transtension zone is also distinguished in the central Romanche zone east of 19.5° W (Bonatti et al. 1994).

The bathymetric chart based on observations in expeditions (Fig. 1.13) describes the main features of the surface of basement rocks within the Romanche transform fault and reflects the joint action of transtension components. Their surface became the basis for paleogeodynamic reconstructions. The Euler poles and rotation angles were determined by superposition of the same isobaths on the northern and southern slopes of depressions on the basis of the assumption of synchronous opening of these depressions.

If we assume that breakdown of the crust was related to transtension, configuration of the slope should reflect the configuration of the fracture, along which the breakdown was initiated. In order to check this assumption, we analyzed configurations of slopes of the western and eastern depressions.

Alignment of the segments of northern and southern depression slopes (Fig. 1.13) was made using the electronic chart of bottom topography with a step of 200 m. A minimum of root-mean-square deviation between isobath coordinates at matched points was chosen as the quantitative criterion of optimum matching. The closest result was obtained when the root-mean square minimum was normalized by the total length of the aligned segments of isobaths. The best result of alignment in the Romanche Fracture Zone was obtained for the 5-km-isobath segments in the western and eastern depressions. The digitizing interval of aligned isobath segments did not exceed 10 km.

If we locate the Euler pole of finite rotation at 2°37′ N, 27°39′ W, it is possible to align the northern and southern slopes of the depression along the 5-km isobath over a length exceeding 50 km in the western depression and 200 km in the eastern depression. The error of depth line alignment was ± 4.6 km if the matched points were located within 560–890 km from the Euler pole. The rotation angle was 1.78° ± 0.03°.

The calculated coordinates of the Euler pole and rotation angle describe the opening of the internal part of the Romanche transform fault (Fig. 1.13) as the result of transtension, but they do not give any information about the timing of this opening. In order to estimate the latter parameter, we shall analyze anomalies of the magnetic field in the internal part of the Romanche Fracture Zone.

We mentioned above that modern seismicity is a characteristic feature of the western and eastern depressions. Thus, the depressions are tectonically active structures. Based on the concept of lithospheric plate tectonics, their bottom should rest on normally magnetized rocks of the oceanic lithosphere formed in the recent Brunhes epoch. Precisely such magnetization is characteristic of the basement rock material under the bottom of depressions (Schreider et al. 2006a). At the same time, bottom rocks in the interdepression seamounts and hills are characterized by inverse magnetization. This fact can suggest that transtensile stresses accumulated here have not yet provoked fracture.

If our dating of the transtension is justified, it is possible to determine the instantaneous velocity of the relative displacement of fracture walls in the Brunhes epoch as a component of transtension rate along the accumulated shear displacement, described by the Euler pole of mutual displacement of the American and African lithospheric plates (62° N, 36° W). Our estimate (3.12 cm/year), obtained for the first time by independent calculation, refines the previous theoretical rates for the Romanche fault walls based on NUVEL-1 and NUVEL-1A models (Cande et al. 1988; Bonatti et al. 2001).

The structure of the "abandoned" rift located west of the eastern rift/transform fault junction (intersect) extends in a nearly meridional direction similarly to the main rift (Bonatti et al. 1994, 2001; Bonatti 1996). The abandoned rift is divided into two segments divided by a low-amplitude fault in the longitudinal section. The southern segment of this structure is expressed as an equant depression (Pushcharovsky 2005; and others). Basalts, gabbroids, and ultrabasic rocks were dredged from the slopes around the rises, while limestones were recovered from the top of one of the peaks. The northern segment of the "abandoned" rift is prominent in the topography as a narrow graben-shaped deflection with steep walls. Dredging of the slopes and bottom of the northern segment of the abandoned rift yielded mainly ultrabasic rocks with the subordinate gabbro and limestones. Appearance of the abandoned rift implies a spatial jump of the spreading axis.

1.5 Chain Fracture Zone

The Chain transform fault located 200 km southeast of the Romanche transform fault (Fig. 5) is responsible for the left-lateral displacement of the axes of the Mid-Atlantic Ridge over 315 km. The active part of the fault is a deep (more than 5,000 m) valley, bounded from the south and north by fault-line ridges, which slightly exceed the mean topographic level of the adjacent oceanic basement (Figs. 1.15 and 1.16).

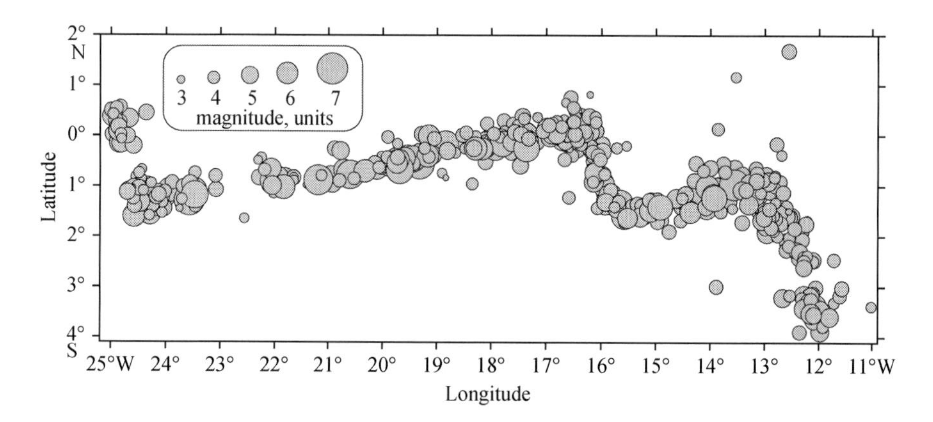

Fig. 1.14 Seismic activity of the Romanche and Chain transform faults

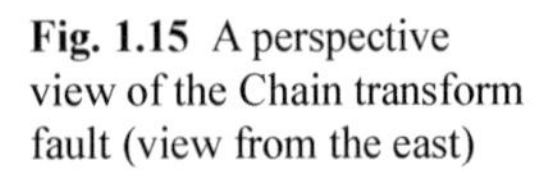

Fig. 1.15 A perspective view of the Chain transform fault (view from the east)

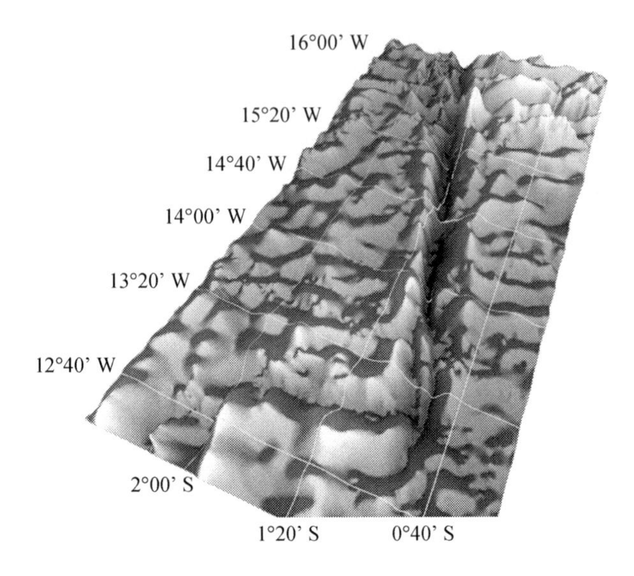

The ridges are covered with sediments. The trough is traced along the passive part of the Chain Fracture Zone and can be found at the margins of Africa and South America (Emery et al. 1975).

However, the basement depth and thickness of the sediments are different at both sides of the fault. The majority of earthquake epicenters are confined to the transform valley in the active part of the fault (Fig. 1.14). Ultrabasic rocks were dredged within the active zone of the fault (Bonatti et al. 1992).

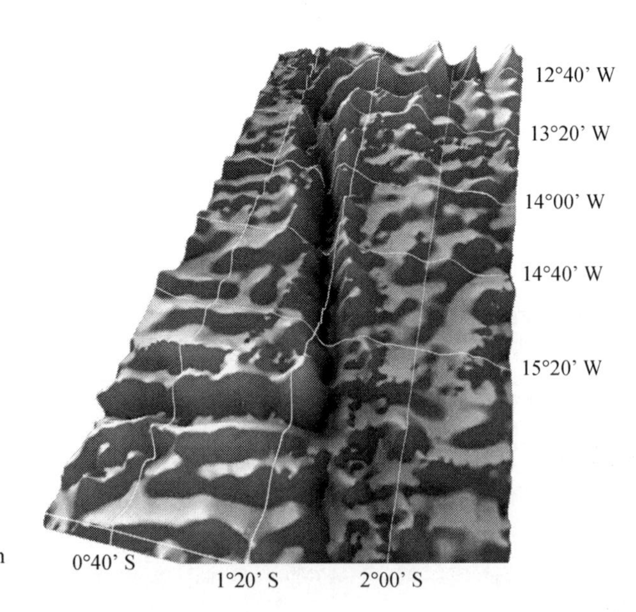

Fig. 1.16 A perspective view of the Chain transform fault (view from the west)

1.6 Vema Channel

The Mid-Atlantic Ridge and margin of South America are separated by two deep-water basins: Brazil Basin (depth 5,200–5,600 m) and Argentine Basin (depth 4,800–6,000 m). The basins, in turn, are separated by the large Rio Grande Rise with a complex structure and basement composed of the Santonian alkaline basalts (86 Ma).

The sedimentary cover includes Senonian-Cenozoic carbonate sediments, which grade into deep-water sediments upward on the profile and along the slopes (Khain 2001). The southern end of the Vema Channel is located at 30° S, 39° W at the western flank of the Rio Grande Plateau (Rio Grande Rise). The maximum depth of the channel exceeds 4,800 m, whereas the width does not exceed a few tens of kilometers. The total length of the channel exceeds 700 km (Figs. 3 and 1.17). The channel is located in an aseismic zone. Sediments are almost lacking in the channel. Therefore, the channel resembles an exogenic gully (Figs. 1.18 and 1.19) likely related to destruction of the basement rocks of the Rio Grande Plateau under the influence of the bottom current of cold waters from the Argentine Basin to the Brazil Basin. The velocity of water flow through the Vema Channel is usually 20–30 cm s^{-1}. According to Emelyanov (2008), the velocity should occasionally exceed 100 cm s^{-1}, because only such flows can erode the consolidated Quaternary and pre-Quaternary rocks. Traces of turbidites (with vegetation remains) were also found in the western part of the Romanche Fracture Zone. The Antarctic Bottom Water appeared in the Rio Grande Rise area in the Paleogene (Eocene) 38–56 Ma ago. The Antarctic

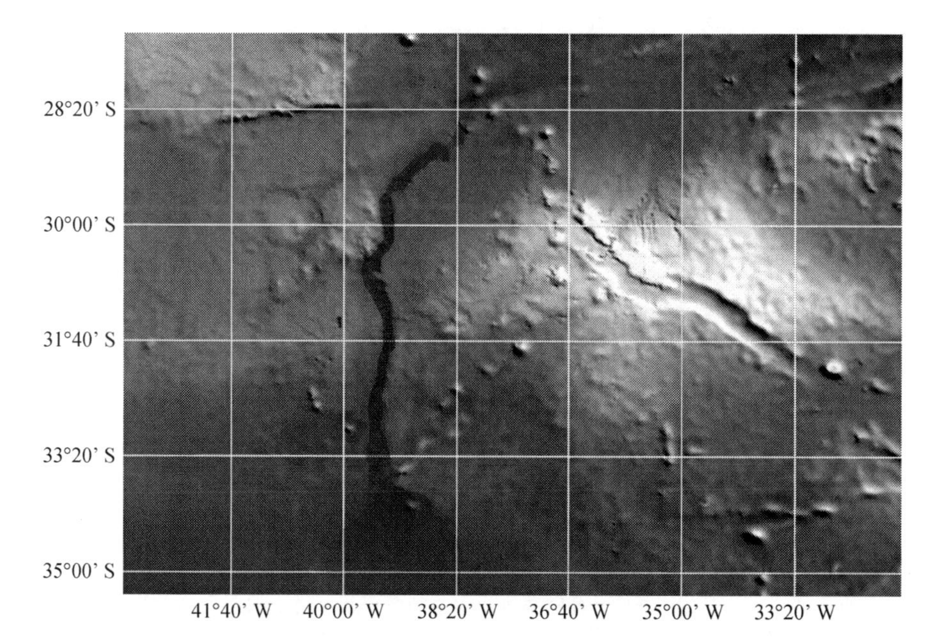

Fig. 1.17 A perspective view of the Vema Channel zone (view from above)

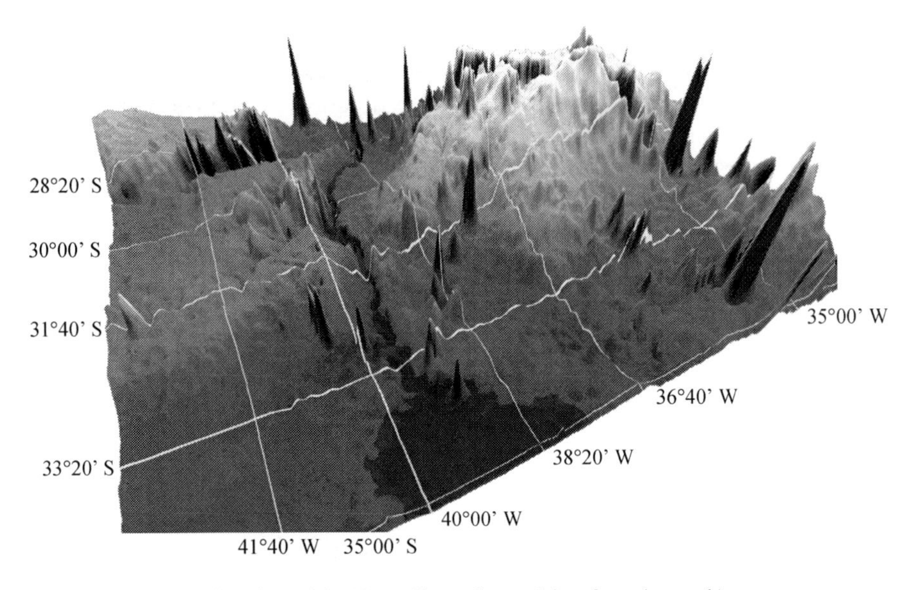

Fig. 1.18 A perspective view of the Vema Channel zone (view from the south)

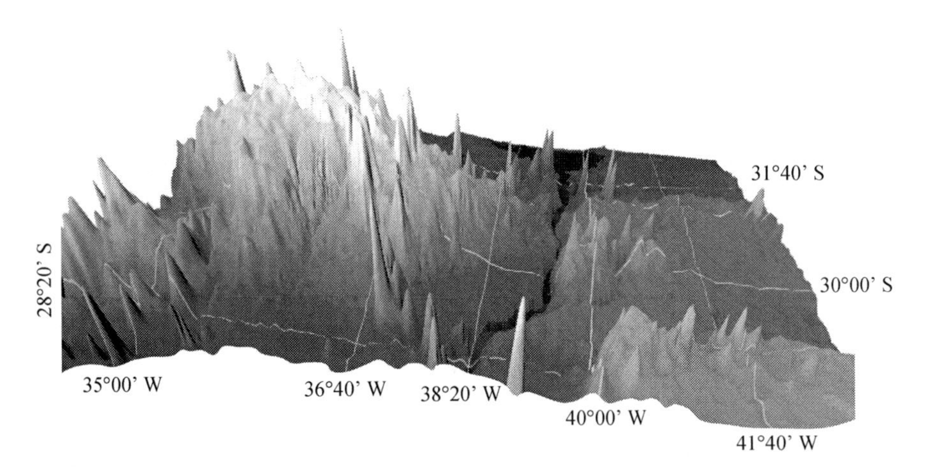

Fig. 1.19 A perspective view of the Vema Channel zone (view from the north)

Bottom Water flows were maximal before strong glaciations (especially at isotope stages 7/6 and 3/2) (Emelyanov 2008).

Thus, the analysis described above suggests that transform faults (Charlie Gibbs, Vema, Romanche, and Chain fracture zones) are characterized by several common features (the Vema Channel is an exception, because it is not a transform fault by its nature). Their main feature is the existence of a deep valley (with a relative incision down to 2,000 m or more) or a pair of conjugated valleys oriented across the extension of the Mid-Atlantic Ridge. The width of valleys is only a few tens of kilometers, whereas the length of the active part of the transform faults between the neighboring spreading axes can reach many hundreds of kilometers. The valleys are bounded by high fault-line ridges (sometimes, they can rise above the ambient bottom for a few kilometers). The active part of transform faults is characterized by modern seismicity with the average earthquake magnitude reaching 5–6. The depth of hypocenters usually does not exceed 35 km. The transform fault zones are characterized by high heat flux (frequently exceeding 2 HFU or more) and specific features of transtension.

Chapter 2
Deep Water Masses of the South and North Atlantic

2.1 General Description

The Southern Ocean and Antarctic Circumpolar Current isolate the Antarctic continent from other regions of the Earth. Thus, conditions in the study region provide the formation of a special water structure around Antarctica, whereas water structure in the northern regions is determined by interactions between waters of the North Atlantic and those of Arctic origin.

Many publications have been devoted to the formation and propagation of bottom waters of the Antarctic origin. Nevertheless, a generally accepted opinion about the evolution, pathways, and intensity of their transport is lacking even for the well-studied Atlantic Ocean. The objective of this chapter is to summarize our knowledge about the structure and propagation of waters in deep and bottom layers of the Atlantic Ocean.

Deep and bottom waters are formed as a result of thermohaline interaction in both subpolar regions of the oceans. Deep and bottom waters are formed in three sectors of the Southern Ocean corresponding to three individual oceans. In the Northern Hemisphere, they are formed only in the Atlantic Ocean. Winter cooling in the North Atlantic basins (North European Basin, Norwegian and Greenland seas, and Labrador Sea) results in the descent of water from the ocean surface to the deep layers. The density of waters of North Atlantic origin is not high enough to occupy the bottom layer over the major part of the World Ocean (Orsi et al. 1999), because density of waters of the Antarctic origin is greater. Conditions of water formation in the Arctic Ocean are similar to those in the Antarctic region, but the dense waters that fill the bottom layer of the Arctic Ocean cannot overflow the two thresholds (between Greenland and Iceland and between Iceland and the Faroe-Shetland Islands), because the depth of these sills is less than 1,000 m. Therefore, only intermediate waters from the Arctic Ocean can flow into the Atlantic. As a consequence, all waters in the bottom layer of the Atlantic have a component of the Antarctic origin.

The descent of surface and intermediate waters in the North Atlantic is only a part of the global thermohaline circulation and meridional overturning circulation (Fig. 2.1). Schematic principles of this circulation, which was called the Global

E. G. Morozov et al., *Abyssal Channels in the Atlantic Ocean*,
DOI 10.1007/978-90-481-9358-5_2, © Springer Science+Business Media B.V. 2010

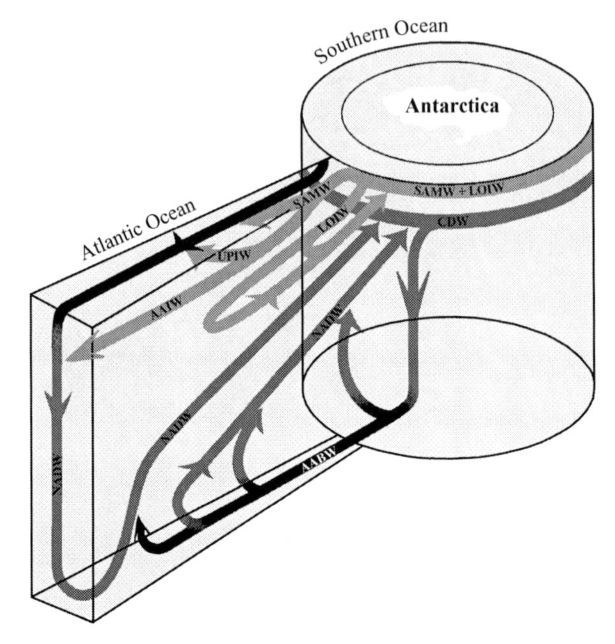

Fig. 2.1 A three dimensional flow scheme with meridional-vertical sections for the Atlantic and Southern oceans and their horizontal connections. The surface layer circulations, intermediate and Subantarctic Mode Water, deep, and near-bottom zones are shown in different gray tints. (Modified and redrawn from Schmitz 1996b). Notations: *SAMW* Subantarctic Mode Water, *CDW* Circumpolar Deep Water, *AABW* Antarctic Bottom Water, *LOIW* Lower Intermediate Water, *UPIW* Upper Intermediate Water, *AAIW* Antarctic Intermediate Water, *NADW* North Atlantic Deep Water

Conveyor Belt (Lappo 1984; Broecker 1991) (Fig. 2.2) are as follows. North Atlantic Deep Water descends as a result of winter cooling in the Norwegian, Greenland, and Labrador seas. While this water propagates toward mid-latitudes of the North Atlantic, the deep water masses formed at high latitudes are subjected to a strong influence from other water masses and intense mixing with Mediterranean Water, which is of the same density range, resulting in the formation of the main distinguishing indicator of this water identified as salinity maximum. The water mass thus formed propagates southward at depths greater than 1,500–2,000 m. In the Southern Ocean, it is advected by currents to all oceans. A compensating reverse transport of warmer waters should exist in the upper layers from the Pacific and Indian oceans to the Atlantic to maintain the quasi-stationary state of the global circulation on a secular time scale. This flow is responsible for anomalously high meridional water transport back to the northern basins of the Atlantic Ocean.

The scheme described above is simplified, because it is related only to the interaction between North Atlantic Deep water and the warm surface waters. Later, this scheme became more complex (Figs. 2.1 and 2.2). Its essential element is the meridional thermohaline circulation cell in the Antarctic (Deacon cell) described by the classical scheme (Fig. 2.3) (Deacon 1937; Speer et al. 2000). Relatively warm and saline waters propagating around the Antarctic continent, together with the Antarctic Circumpolar Current, flow to the south along the iso-surfaces of potential density and ascend from the deeper layers to the ocean surface south of the South Polar Front. Antarctic Surface water and bottom waters of the Antarctic basin are formed here as a result of freshening and oxygen enrichment. In addition, renewal of circumpolar waters occurs during these processes (Orsi et al. 1999, 2002). When

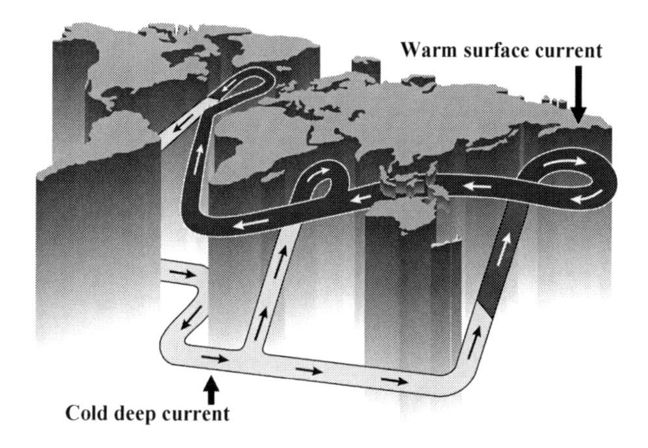

Fig. 2.2 A strongly simplified scheme of the global conveyor belt

Antarctic Surface Water reaches the South Polar frontal zone, it descends below the warmer and Subantarctic waters of lower density as Antarctic Intermediate Water.

The above description is a simplified qualitative scheme of deep and bottom waters circulation. There is no commonly accepted detailed classification of these waters. The situation in the Atlantic Ocean is most difficult and confusing. Authors of different papers actually use different terms to name one and the same water layer. Opposite examples exist when one and the same term is used to name water layers with different properties. For instance, in the majority of publications related to water masses of the Antarctic region, some authors (Patterson and Whitworth 1990; Tomczak and Godfrey 1994) distinguish Lower Circumpolar Deep Water, which originates from North Atlantic Deep Water. On the other hand, other researchers

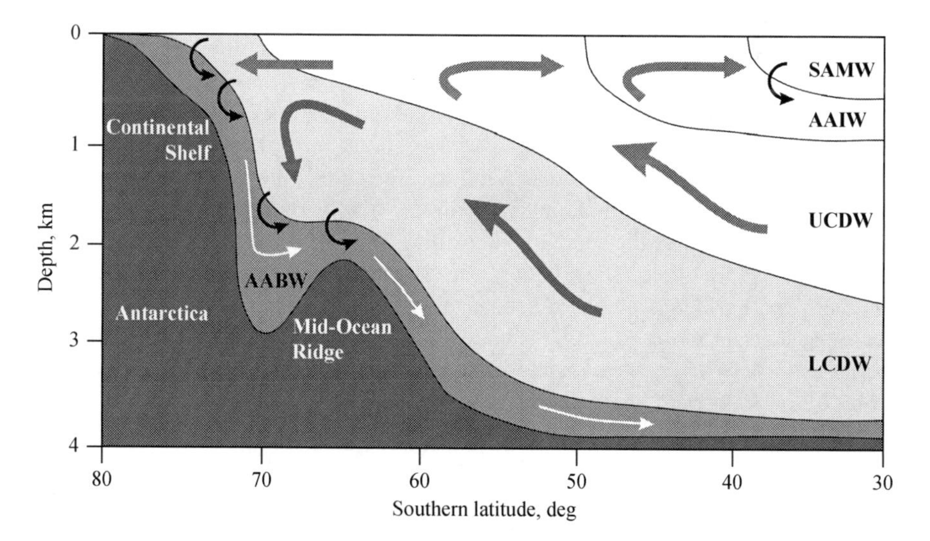

Fig. 2.3 Sketch view of the meridional overturning circulation (Deacon cell). (Modified after Speer et al. 2000)

(Larque et al. 1997; Reid et al. 1977; Saunders and King 1995; Messias et al. 1999) distinguish Lower Circumpolar Deep Water, which is a part of Antarctic Bottom Water and propagates to the region from the south. Unfortunately, lack of a single concept code in such definitions is characteristic of the major part of water masses in these regions.

Below, we give a table of different names of water masses used in different publications (Table 2.1). Since there is no generally accepted terminology for the Antarctic water masses in the Atlantic, we present two different classifications: (Wüst 1936; Reid et al. 1977) and classification for the Southern Ocean (Orsi et al. 1999).

Wüst (1936) was the first to distinguish waters of the Antarctic origin in the bottom zone of the Atlantic Ocean based on data of the German Atlantic Expedition on R/V *Meteor*. In this elaborated benchmark book, the bottom water was named Antarctic Bottom Water. It was found that this water of the so-called oceanic stratosphere is formed in the Weddell Sea. Its potential temperature (θ) is less than $-0.8°C$ in the Weddell Sea, increases to $+0.6°C$ at the equator, and reaches $+1.8°C$ near the Bermuda Islands.

According to Wüst, Antarctic Bottom Water unites all waters of the Antarctic origin in the bottom layer of the Atlantic Ocean. We note that the term 'Antarctic Bottom Water' in the sense of Wüst's classification is widely used in modern publications. Usually, its upper boundary is associated with the isotherm of potential temperature $\theta = 2°C$ (Table 2.1). Authors of some papers change this value to close values of potential temperature ranging from 1.8 to 2.1°C or use the corresponding values of potential or neutral density γ^n (see, for example, Mercier and Speer 1998; McCartney et al. 1991). Neutral density is determined from the condition that there is no work against buoyancy force during the particle motion at iso-surface γ^n (Jackett and McDougall 1997).

Mantyla and Reid (1983) found that basins north of the southern polar latitudes are filled with waters not directly from the sources of water formation at the Antarctic slope, but from the lower layers of circumpolar waters that circulate around Antarctica together with the Antarctic Circumpolar Current. These layers occupy an intermediate position between dense southern waters near the bottom and overlying more saline waters of North Atlantic origin. In its turn, the Antarctic Circumpolar Current zone separates the southern subtropical zone of the World Ocean from the subpolar cyclonic gyres. Thus, the former zone acts as a buffer zone between the southern and northern waters. This means that the lower layers of circumpolar waters appear far from the Antarctic slope over a large part of the Antarctic continent perimeter. These layers are elongated distributed sources of bottom water in basins north of the Antarctic Circumpolar Current zone. Thus, two components of bottom waters of the southern origin can be distinguished at least in the Atlantic Ocean: circumpolar water from the region of the Antarctic Circumpolar Current and denser water transported from the subpolar cyclonic gyre south of the Antarctic Circumpolar Current (Mantyla and Reid 1983; Schmitz 1996a, b). Gordon (1971, 1972) used the term 'Antarctic Bottom Water' only in relation to the second component determining this component as water layer with potential temperature $\theta < 0.0°C$.

Table 2.1 Terminology of water masses used in different publications

Author	Bottom water masses	Comments Citations from authors
Classification of Wüst (1936); θ < 2°C		
Wüst (1936)	AABW	
Sverdrup et al. (1942)	AABW	
Hogg et al. (1982)	AABW	
Friedrichs and Hall (1983)	AABW	
Harvey and Arhan (1988)	AABW	
Broecker (1991)	AABW	
Speer and McCartney (1992)	AABW	
McCartney and Curry (1993)	AABW	
Rhein et al. (1996)	AABW	
Hall et al. (1997)	AABW	
Stramma and Rhein (2001)	AABW	
Schott et al. (2003)	AABW	
Lavin et al. (2003)	AABW	
Classification of Reid et al. (1977); LCPW < 2°C; WSDW < 0.2°		
Reid et al. (1977)	Lower Circumpolar Water, Weddell Sea Deep Water	
Mantyla and Reid (1983)	AABW = LCPW + WSDW	
Reid (1989)	Circumpolar, Weddell Sea Upper Layers	
Speer and Zenk (1993)	AABW = LCDW + WSDW	
Tsuchiya et al. (1994)	LCPW, WSDW	
Zenk and Hogg (1996)	AABW = LCDW + WSDW	
Larque et al. (1997)	WSDW LCDW	
Rhein et al. (1998)	AABW = LCDW + WSDW	
Arhan et al. (1998)	AABW = LCPW + WSDW	
Arhan et al. (1999)	WSDW, LCDW, SPDW	
Stramma and England (1999)	AABW = LCDW + WSDW	
Messias et al. (1999)	LCDW, AABW	
Wienders et al. (2000)	AABW = LCPW + WSDW	
Memery et al. (2000)	AABW = LCPW + WSDW	
Sandoval and Weatherly (2001)	AABW = LCPW + WSDW	
Vanicek and Siedler (2002)	AABW = LCDW + WSDW	
McDonagh et al. (2002)	AABW = LCPW + WSDW	
Zenk and Morozov (2007)	AABW = CDW + WSDW	
Original opinions		
Fu (1981)	Only LCPW	
Roemmich (1983)	Only LCPW	
Smythe-Wright and Boswell (1998)	AABW primary derived from WSDW	
Onken (1995)		LCDW consists of WSDW and LCDW transported through Drake Passage
Johnson (2008)	AABW = WSBW	

Table 2.1 (continued)

Author	Bottom water masses	Comments Citations from authors
Southern Ocean classification; AABW $\gamma^n > 28.27$; ACCbw $28.27 > \gamma^n > 28.18$; LCDW $28.18 > \gamma^n > 28.00$		
Mantyla and Reid (1983)	LCDW = NADW + AABW	
Patterson and Whitworth (1990)	LCDW = NADW + AABW	
Orsi et al. (1999)	LCDW + ACCbw + WSDW	
Orsi et al. (1999)	AABW = WSDW + WSBW	
Hoppema et al. (2001)	WDW = LCDW; WSDW + WSBW	
Naveira Garabato et al. (2002b)	AABW = WSDW + WSBW	
Klatt et al. (2002)	WDW = LCDW; WSDW + WSBW	
Klatt et al. (2005)	WDW = LCDW; WSDW + WSBW	
Barre et al. (2008)	AABW = WSDW + WSBW	
Southern Ocean classification applied to the South Atlantic		
Coles et al. (1996)	LCDW, AABW	
van Aken (2007)	LCDW, AABW	

According to (Foster and Carmack 1976), Antarctic Bottom Water in the Atlantic Ocean is a mixture of the Weddell Sea Bottom Water (WSBW) and overlying waters of circumpolar origin. Weddell Sea Bottom Water is the water related to the mixing of circumpolar waters with Antarctic Shelf Water. This is the water that descends along the Antarctic slope to the ocean bottom. When analyzing water structure in the Southwest Atlantic, Reid et al. (1977) emphasized that the water under consideration is actually not bottom water in the Weddell Sea in the above approach for distinguishing Antarctic Bottom Water. They suggested substitution of this term by Weddell Sea Deep Water. This is the water that leaves the Weddell Gyre and spreads in the bottom layer of the Atlantic Ocean. They assigned this term only to the water that cannot overflow the threshold in the Drake Passage with the Antarctic Circumpolar Current ($\theta < 0.2$°C). Later, Orsi et al. (1999) developed the approach of (Gordon 1971, 1972; Foster and Carmack 1976) and suggested a return to the term 'Antarctic Bottom Water' that would join not only Weddell Sea Deep Water (WSDW) and Weddell Sea Bottom Water (WSBW), but also the bottom waters formed in the Indian and Pacific sectors of the Southern Ocean, strictly confining their classification to specific density ranges. According to (Orsi et al. 1999), Antarctic Bottom Water is any water in the Southern Ocean (excluding very cold shelf waters), which is dense enough not to circulate with the easterly Antarctic Circumpolar Current and cannot overflow the ridges in the Drake Passage. It should be mentioned that the dominating part of the Atlantic Antarctic Bottom Water is formed in the Weddell Sea and only a small part of this water is transported to the Atlantic from the Indian Ocean sector of the Southern Ocean (Orsi et al. 1999).

We note that the approaches in (Wüst 1936; Orsi et al. 1999) are completely different from each other in application of the term 'Antarctic Bottom Water.' According to Wüst (1936), Antarctic Bottom Water unites the Antarctic Bottom Water defined in (Orsi et al. 1999) and the lower part of circumpolar waters. The approach in (Reid et al. 1977) agrees with the classification in (Orsi et al. 1999), although the term 'Antarctic Bottom Water' is rejected. Two principally different approaches also exist in the classification of the overlying circumpolar waters. We shall discuss this issue below.

North Atlantic Deep Water with a relatively high salinity flows to the northern periphery of the Antarctic Circumpolar Current zone and then propagates to the east. In the course of circumpolar motion this water mixes with deep waters of lower density from the Indian and Pacific oceans and with denser Antarctic waters. According to (Sverdrup et al. 1942), this thick layer, which is a mixture of waters from different sources, is an independent water mass. They called it Circumpolar Deep Water, (CDW). This layer is divided into two independent layers, specific characteristics of which reflect the properties of the parental deep waters (Gordon 1967). Upper Circumpolar Deep Water (UCDW) has a minimum oxygen concentration, while maximum contents of phosphates and nitrates are confined to its upper layer. These properties originate from the Indian and Pacific oceans. The salinity maximum in Lower Circumpolar Deep Water (LCDW) originates from the North Atlantic (Patterson and Whitworth 1990). Later, Orsi et al. (1999) distinguished the lowest layer of circumpolar waters flowing over the threshold in the Drake Passage to the east. This layer is not renewed isopycnically by North Atlantic waters. They defined this water as independent water (Bottom Water of the Antarctic Circumpolar Current). However, in this book we shall keep to the other term 'Circumpolar Bottom Water' (Koshlyakov and Tarakanov 2003b). According to the classification in (Orsi et al. 1999), this water occupies the near-bottom position over a large part of the Atlantic Ocean but also of the Indian and Pacific oceans. Although the paper by Orsi et al. (1999) was published more than ten years ago, classification of this water for the analysis of the ocean water structure has not been accepted universally, probably, because Circumpolar Bottom Water lacks the characteristic extrema of physical or chemical properties in the entire circumpolar zone. Thus, the approach in (Gordon 1967) with supplements by (Orsi et al. 1999) is a three-layer classification of circumpolar waters.

North Atlantic Deep Water is wedging into the Circumpolar Deep Water layer in the Atlantic at the northern periphery of the Antarctic Circumpolar Current. Since North Atlantic Deep Water is characterized by both high salinity and high oxygen concentration, this water divides Circumpolar Deep Water into two layers characterized, in turn, by low oxygen concentration in this region. In (Reid et al. 1977), these layers are designated as Lower and Upper Circumpolar waters. According to the classification of circumpolar waters in (Reid et al. 1977), Upper Circumpolar Water is actually the same water as Upper Circumpolar Deep Water, because they have the same main distinctive properties and density range characteristics. However, Lower Circumpolar Water is not identical to Lower Circumpolar Deep Water. The main distinguishing property of Lower Circumpolar Deep Water is the salinity maximum of the North Atlantic origin, whereas the main feature of Lower Circumpolar Water is the oxygen minimum formed between two maxima. Their specific density ranges do not coincide. These considerations are valid both for the two-layer classification by (Gordon 1967) and for the three-layer classification of circumpolar waters with supplements by (Orsi et al. 1999). Thus, Lower Circumpolar Water is an object of the bottom water structure different from Lower Circumpolar Deep Water.

Ambiguity in use of the term 'Antarctic Bottom Water' and disagreement between classifications of circumpolar waters in the Atlantic Ocean and the Antarctic Circumpolar Current caused confusion in denotation of waters of the bottom layer in the Atlantic beyond the Antarctic Circumpolar Current zone. Some authors continued to use the term Antarctic Bottom Water related to the entire bottom layer (Speer and McCartney 1992), while other researchers restricted the above term 'Antarctic Bottom Water' to Lower Circumpolar Water (Fu 1981; Roemmich 1983) (Table 2.1). However, some scientists follow the terminology in (Reid et al. 1977) and divide this layer into Weddell Sea Deep Water and Lower Circumpolar Water (Mantyla and Reid 1983). Some of the authors, for example (Coles et al. 1996; van Aken 2007) use the Southern Ocean classification, applying it to the southern parts of the South Atlantic, with simultaneous identification of Lower Circumpolar Deep Water and Antarctic Bottom Water. Moreover, the term 'Lower Circumpolar Water' introduced by Reid et al. (1977) was changed to Lower Circumpolar Deep Water, which we think is principally less helpful.

Many authors (van Aken 2000; Stephens and Marshall 2000 among others) have introduced other terms for the strongly transformed Antarctic waters in the bottom layer of the eastern Atlantic, where waters of Antarctic origin propagate through fractures and channels in the Mid-Atlantic and Walvis ridges. We shall discuss this issue later. Such approaches should be considered beyond the classifications described above.

In addition to the disagreement in terminology, the situation is aggravated by the fact that authors of various publications use different methods of water mass identification. They assume isotherms, isobars, or isopycnals of different values as boundaries of water masses, which sometimes differ strongly from each other. Thus, limits of the spreading and estimates of the transport of different water masses are not the same. Our book is aimed at a comparison of different approaches.

Thus, we can formulate the objective of this study as an analysis of the available information about the waters of the Atlantic bottom layer and its critical analysis to elucidate the water structure in the bottom layer and the corresponding transport estimates.

2.2 Antarctic Intermediate Water

Antarctic Intermediate Water (AAIW) occupies the water layer above North Atlantic Deep Water over a significant part of the Atlantic Ocean and plays the role of a mixing agent with the latter. Our prime interest within the context of this book is in the lower boundary of Antarctic Intermediate Water. Here, we give a short description of properties of Antarctic Intermediate Water needed for consideration of properties of the lower water masses.

Antarctic Intermediate Water reaches the surface between the South Polar Front in the south and the Subantarctic Front in the north. This water is found in the layer up to 300–800 m at lower latitudes and approximately between 700 and 1,300 m in the zone north of the Subantarctic Front. Here, this layer is characterized by a north-extending low-salinity water tongue, which is seen in any hydrographic section from the Southern Hemisphere (Reid 1965; Kuksa 1983; Patterson and Whitworth 1990; Schmid et al. 2000).

The scheme of Antarctic Intermediate Water propagation in the South Atlantic is shown in Fig. 2.4.

Several hypotheses are known about the sources of Antarctic Intermediate Water formation. According to (Sloyan and Rintoul 2001a), Antarctic Intermediate Water is formed between the South Polar Front and the Subantarctic Zone as a result of the transformation of Antarctic Surface Water transported to this zone by pure drift current. This transformation is caused by a strong excess of precipitation

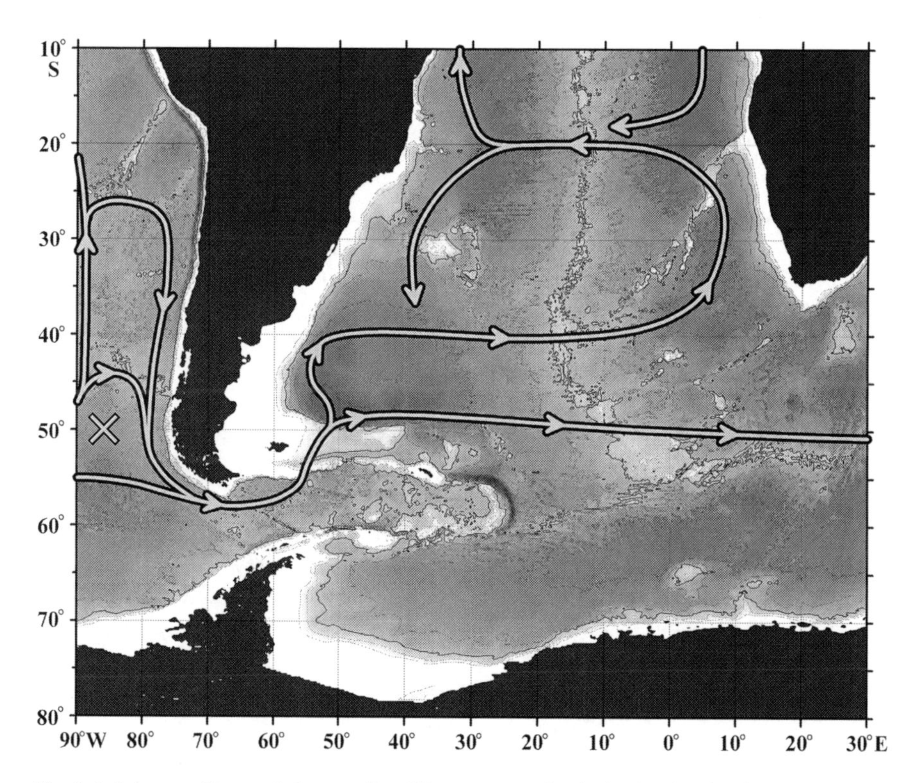

Fig. 2.4 Scheme of Antarctic Intermediate Water propagation in the South Atlantic. The scheme is combined from the results reported in Davis (1998) and Larque et al. (1997). The *cross* indicates the domain of waters of the coolest and freshest type, which is also the only place of Antarctic Intermediate Water formation at the surface. (McCartney 1977)

over evaporation in this region of the ocean. Spreading of low-saline water from the upper layer of the ocean in the region south of the Subantarctic Front to the intermediate layer north of this front is mainly related to the separation of cyclonic meanders of the Subantarctic Front elongated to the north. The next step in Antarctic Intermediate Water formation is the decay of quasi-geostrophic cyclonic eddies of the front. As they decay, the water of lower salinity follows the descent of the domes of isopycnal surfaces in the eddies and descends from the upper layer of the ocean to the intermediate layer. Mesoscale isopycnal mixing of this water with the surrounding waters of the ocean occurs simultaneously (Golivets and Koshlyakov 2004).

McCartney (1977) suggested another hypothesis about non-circumpolar formation of Antarctic Intermediate Water. According to this hypothesis, the entire volume of Antarctic Intermediate Water is formed as a result of deep winter convection north of the Subantarctic Front in the southeastern part of the Pacific Ocean. When transported to the Atlantic Ocean through the Drake Passage this water is transformed in the Drake Passage and western part of the Scotia Sea. In this case, Antarctic Intermediate Water consists only of the coldest, freshest, and densest type of Subantarctic Mode Water.

Martineau (1953) put forward a hypothesis about the third source of Antarctic Intermediate Water near the South American coast extending from 45° S to 25° S. Waters of the Brazil and Falkland currents are mixed up in this region. Waters formed as a result of mixing descend below lighter waters of the Brazil Current and join Antarctic Intermediate Water formed near the South Polar Front.

Circulation of the low-salinity Antarctic Intermediate Water in the South Atlantic and the associated dynamical processes were studied in (Schmid et al. 2000) based on recent and historical hydrographic profiles, Lagrangian and Eulerian current measurements, as well as wind stress observations. The circulation pattern inferred for Antarctic Intermediate Water clearly supports the hypothesis of an anticyclonic basinwide recirculation of the intermediate water in the subtropics.

Antarctic Intermediate Water is transported to the northern part of the South Atlantic in a system of gyres generally directed opposite to the surface current system (Larque et al. 1997). Weakening of salinity minimum in the Antarctic Intermediate Water core, after spreading of this water from their sources of formation, is caused mainly by the diapycnal mixing ('diapycnal mixing' means mixing across isopycnal surfaces) of Antarctic Intermediate Water with more saline waters. However, at present, no commonly accepted point of view exists about the degree of Antarctic Intermediate Water spreading in the ocean. According to (Kuksa 1983), these waters propagate up to tropical latitudes of the North Atlantic, while Zubov (1947) believed that they spread only to the equator in the eastern basin. If we assume that the northern boundary of Antarctic Intermediate Water spreading is not salinity anomaly but disappearance of the extreme concentrations of nutrients, then Antarctic Intermediate Water can be traced up to the Canary archipelago (Perez et al. 2001) or even to the Arctic basin (Tsuchiya 1989). It is possible that Antarctic Intermediate Water descends to the deeper layer in the North Atlantic (Broecker et al. 1976).

2.3 Upper Circumpolar Water and Upper Circumpolar Deep Water

As was mentioned above, two terms are used in the literature: Upper Circumpolar Water and Upper Circumpolar Deep Water.

Upper Circumpolar Deep Water (UCDW) is a product of isopycnal and diapycnal mixing of Indian Ocean Deep Water and Pacific Deep Water, which spread into the Antarctic Circumpolar Current zone from the north, with Antarctic Surface Water, Antarctic Intermediate Water, and Lower Circumpolar Deep Water. The scheme of Upper Circumpolar Deep Water circulation together with its parental Pacific and Indian Deep Waters is shown in Fig. 2.5. Pacific Deep Water underlying Antarctic Intermediate Water propagates from the northern Pacific to its southern part, where

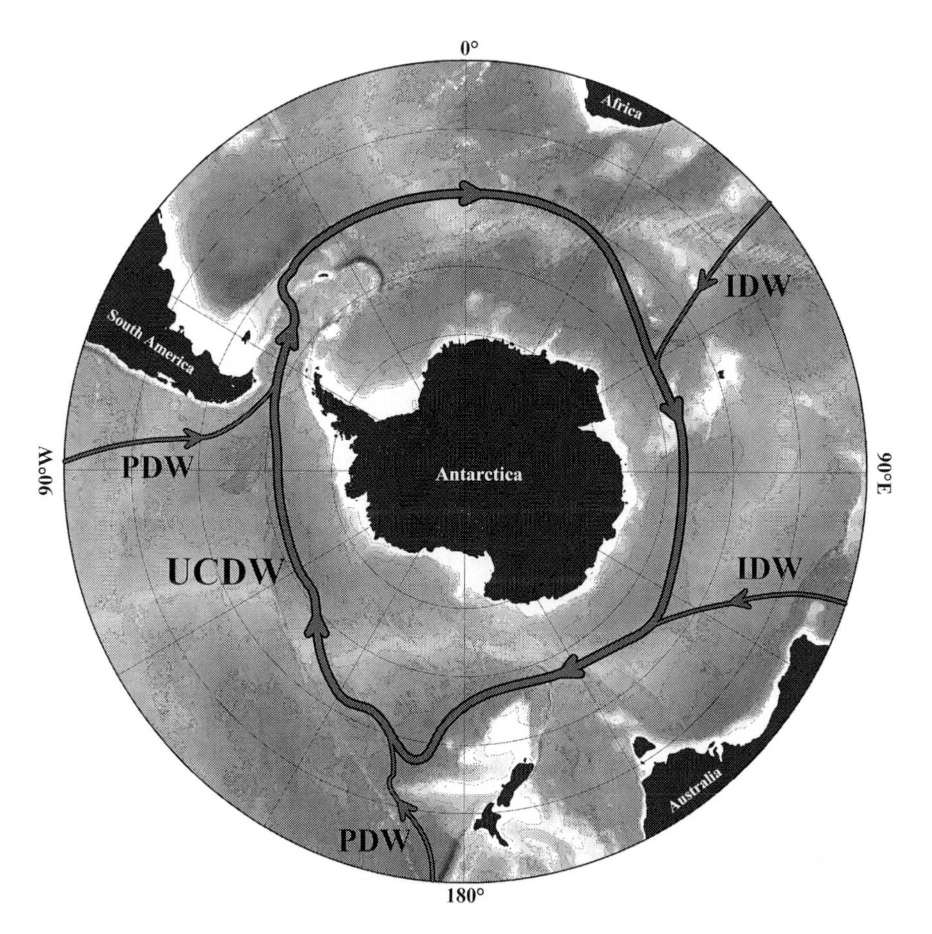

Fig. 2.5 Scheme of Upper Circumpolar Water spreading in the Southern Ocean (*thick lines*). *Thin lines* show the Indian Ocean Deep Water inflow to the region of Antarctic Circumpolar Current and Pacific Deep Water inflow to the Antarctic Circumpolar Current region

it is formed as a result of mixing of Lower Circumpolar Deep Water and Circumpolar Bottom Water with North Pacific Intermediate Water and Subarctic Surface Water (Koshlyakov and Tarakanov 2004; Tomczak and Godfrey 1994). Indian Ocean Deep Water, in turn, is related to mixing of Lower Circumpolar Deep Water and Circumpolar Bottom Water. Together with Antarctic Intermediate Water and Red Sea Water, the former waters spread in the bottom layer from the south to the Subtropical zone of the Indian Ocean in its western and eastern deep basins (Neiman et al. 1997; Tomczak and Godfrey 1994). Transport of Indian Ocean Deep Water and Pacific Deep Water to the Antarctic Circumpolar Current zone at 1,500–3,000 m and their transformation to Upper Circumpolar Deep Water occur predominantly at the western and eastern margins of the Indian and Pacific oceans (Neiman et al. 1997; Orsi et al. 1995; Sloyan and Rintoul 2001b; Tomczak and Godfrey 1994). According to (Sloyan and Rintoul 2001b), transport of Upper Circumpolar Water with the Antarctic Circumpolar Current is 60 Sv at the Tasmania meridian and increases to 69 Sv in the Drake Passage, mainly owing to the additional transport of Pacific Deep Water from the north. Lower values of oxygen concentration and higher content of nutrients in Indian Ocean Deep Water and Pacific Deep Water are caused by oxidizing processes combined with the old age of these waters. Like Indian Ocean Deep Water and Pacific Deep Water, Upper Circumpolar Deep Water is characterized by a minimum of oxygen concentration and maxima of phosphates and nitrates. In the zone south of the South Polar Front, which is at the same time the southern boundary of Antarctic Intermediate Water, Upper Circumpolar Deep Water comes into contact with Antarctic Surface Water. In this region, Upper Circumpolar Deep Water is characterized by a temperature maximum at the intermediate depth, related to low temperature of the overlying Antarctic Surface Water. The term 'Upper Circumpolar Deep Water' is used by the authors of papers devoted to the Southern Ocean (Patterson and Whitworth 1990; Naveira Garabato et al. 2002a; Well et al. 2003), and the Southern Atlantic (Arhan et al. 1998; Ruth et al. 2000; van Aken 2007; and others).

Oxygen minimum and maximum of nutrients are distinguishing properties of Upper Circumpolar Water (UCPW) in the Atlantic Ocean (Larque et al. 1997; Saunders and King 1995; Vanicek and Siedler 2002). Another distinguishing property of this layer in the Atlantic north of the Antarctic Circumpolar Current is a temperature minimum between two temperature maxima in North Atlantic Deep Water from below and Antarctic Intermediate Water from above. This feature is observed only in the Brazil and Argentine basins (Vanicek and Siedler 2002). Frequently, this layer is not distinguished as an independent water mass in the South Atlantic and is included into the overlying Antarctic Intermediate Water.

In the South Atlantic, this water spreads in the western basin, gradually changing propagation to the southeastern direction (Larque et al. 1997). These authors do not exclude that a small portion of Upper Circumpolar Water propagates in the system of the Deep Western Boundary Current. According to (Memery et al. 2000; Demidov 2003), it is not likely that it spreads north of 20° S. Nevertheless, Messias et al. (1999) distinguish Upper Circumpolar Deep Water (in our context: Upper Circumpolar Water) near the equator. The term 'upper branch of Circumpolar Water' for the South Atlantic was used in (Reid et al. 1977) for the upper layer of circumpolar waters dividing North Atlantic Deep Water into the upper and lower parts, i.e., Upper Circumpolar Water (UCPW) and Lower Circumpolar Water (LCPW). This term was later used in (Tsuchiya et al. 1994; Arhan et al. 1998; Outdot et al. 1998; Memery et al. 2000). We shall keep to the approach traditionally accepted in the literature and use the term 'Upper Circumpolar Deep Water' for the Antarctic waters of the ocean and use the term 'Upper Circumpolar Water' for waters of the South Atlantic north of the Antarctic Circumpolar Current.

2.4 North Atlantic Deep Water

North Atlantic Deep Water (NADW) is distinguished from the overlying and underlying Antarctic waters by higher salinity, temperature, and oxygen concentration, as well by low concentration of nutrients (Wüst 1936; Larque et al. 1997; Vanicek and Siedler 2002). North Atlantic Deep Water occupies a layer between 1,200 and 4,000 m (Baum 2004). A scheme of North Atlantic Deep Water propagation in the North Atlantic is shown in Fig. 2.6.

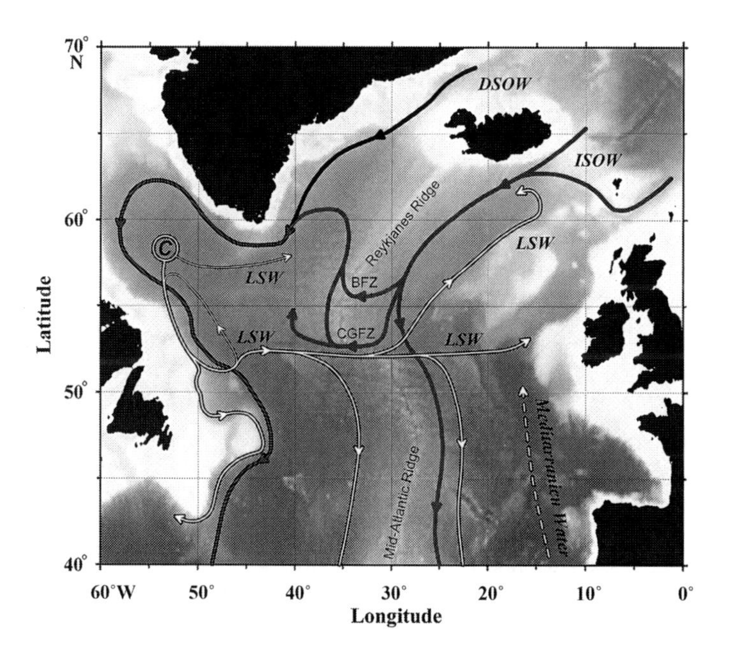

Fig. 2.6 Circulation layers in the North Atlantic. Currents in the intermediate layers are shown by *white lines*. The Mediterranean water pathway is shown with a *dashed line*. Near-bottom flows are shown with *black* and *gray lines*. *C* denotes region of convection. Other notations: *BFZ* Bight Fracture Zone, *CGFZ* Charlie Gibbs Fracture Zone, *LSW* Labrador Sea Water, *DSOW* Denmark Strait Overflow Water, *ISOW* Iceland Scotland Overflow Water

North Atlantic Deep Water is formed according to the following scheme (Koltermann et al. 1999). The densest waters of the Northern Hemisphere are located in bottom layers of the Arctic Ocean, but they cannot propagate to the Atlantic owing to limited deep depths (600–850 m) of the straits and thresholds separating the Arctic Ocean from the Atlantic Ocean. Waters located above the bottom waters flow to the Atlantic through the Denmark Strait (Denmark Strait Overflow Water, DSOW) (Macrander et al. 2007) and Iceland Scotland thresholds (Iceland-Scotland Overflow Water, ISOW) (Olsen et al. 2008; van Aken and de Boer 1995). Formation of North Atlantic Deep Water is related not only to the inflow of waters from the Arctic Basin, but also to deep convection in the Labrador Sea and Irminger Sea (Pickart et al. 2003). Among them, Labrador Sea Water (LSW) has a lower density and waters from the Denmark Strait are the densest ones. Iceland-Scotland Overflow Water flows to the Irminger Basin through the Charlie Gibbs Fracture Zone (Worthington and Volkman 1965; Harvey 1980). Beyond the fracture zone, this water is sometimes called Charlie Gibbs Fracture Zone Water (Mamayev 1992; Koltermann et al. 1999).

The region near 60° N is the southernmost region of final formation of North Atlantic Deep Water (Mamayev 1992). On the other hand, North Atlantic Deep Water is transformed strongly by the influence of Mediterranean Water far to the south of 60° N (Keeling and Peng 1995). According to (Zubov 1947), the 30° N region is also a source of deep water formation.

North Atlantic Deep Water propagates from the source region to the South Atlantic in the Deep Western Boundary Current system (Mamayev 1992). Branching from this current is observed in the South Atlantic near the equator (which is the main flow of North Atlantic Deep Water into the eastern basin). Two more branches are observed in the 5° and 15° S region (they are found only in the western basin) (Larque et al. 1997). In the 40° S region, the direction of North Atlantic Deep Water propagation changes to a quasi-latitudinal one (Larque et al. 1997).

In the South Atlantic, North Atlantic Deep Water is divided into three layers: Upper (UNADW), Middle (MNADW), and Lower (LNADW) (Molinari et al. 1990; Tsuchiya et al. 1994). Sometimes, it is divided into two layers: upper and lower (in this case, Middle North Atlantic Deep Water is related to the lower layer) (Fu 1981; Macdonald 1993; Roemmich 1983).

Labrador Sea Water is the water of lower density among North Atlantic Deep Water components. It is formed as a result of winter convection in the ice-free central and southern parts of the Labrador Sea and possibly in the Irminger Basin (Pickart et al. 2003; Bacon et al. 2003).

Iceland-Scotland Overflow Water is formed when water from the Norwegian Sea overflows the thresholds east of Iceland: Iceland Faroe Ridge and Faroe Shetland gateways (Dickson and Brown 1994; van Aken and de Boer 1995; Lankhorst and Zenk 2006) with depths of 480 and 840 m, respectively. Since the warmer and more saline intermediate waters are mixed and entrained when they descend to greater depths, salinity of Iceland-Scotland Overflow Water is greater than that in the western North Atlantic. Iceland-Scotland Overflow Water propagates along the eastern slope of the Reykjanes Ridge. Then, it flows successively to the west through the deep Charlie Gibbs Fracture Zone and to the north along the western slope of the Reykjanes Ridge. After reaching the northern boundary of the Labrador Basin, it flows along the coast of America (Worthington 1970). In the western basin of the North Atlantic, Iceland-Scotland Overflow Water is identified by local salinity, silicate maximum, and oxygen minimum (van Aken and de Boer 1995). Recent investigations show that the densest waters in the straits are related to mixing of deep and intermediate waters of the Norwegian Sea, respectively), but the ratio of their contribution is variable (Fogelqvist et al. 2003).

Denmark Strait Overflow Water, the densest and fresher water component, is formed during the flow of intermediate waters of the Arctic basin over the threshold in the Strait of Denmark with a depth of 620 m. Spatial and temporal structure of the Denmark Strait Overflow was revealed by acoustic observations (Macrander et al. 2007) as well as field measurements and models (Käse and Oschlies 2000; Jungclaus et al. 2001). Intermediate waters of the Arctic Basin are formed as a result of winter convection in the Greenland Sea (Swift et al. 1980; McCartney 1992).

The contribution of the densest waters does not exceed 10% of the total volume of Denmark Strait Overflow Water (Swift et al. 1980). After overflowing the threshold, Denmark Strait Overflow Water propagates along the continental slope of Greenland to the south (Girton et al. 2000), then flows along the western boundary of the Labrador Sea, and eventually continues its motion in the general direction to the southwest (Worthington 1970). It can easily be identified in the Irminger Basin and Labrador Sea by the bottom oxygen maximum and minimum of temperature, salinity, and silicates. Denmark Strait Overflow Water propagates at great depths and cannot spread into the eastern part of the Atlantic even through the Charlie Gibbs Fracture Zone (Mantyla and Reid 1983; Dickson et al. 1990).

A significant contribution of Mediterranean Sea Water to the North Atlantic Deep Water is discussed in many publications (Wüst 1936; Keeling and Peng 1995; Schmitz 1996a; Iorga and Lozier 1999). Thus, we can consider that North Atlantic Deep Water is formed as a result of mixing between Labrador Sea Water, Iceland-Scotland Overflow Water, Denmark Strait Overflow Water, and Mediterranean Sea Water with a possible influence of Antarctic Bottom Water shown in (Schmitz and McCartney 1995).

According to the summary publication (Schmitz 1996a), quantitative estimates of deep water transport (density range $\sigma_0 = 27.5 - 27.8$) are as follows: 3 Sv through the Strait of Denmark, 2 Sv east of Iceland, and 8 Sv is contributed by waters of the intermediate layer. According to (Schmitz 1996a), the amount of deep waters transported from the Newfoundland Bank to a latitude of 30° S with the Deep Western Boundary Current is 14–18 Sv.

High salinity is a distinguishing indicator of the upper layer of North Atlantic Deep Water. Local maxima of oxygen concentration indicate the middle and lower layers (Tsuchiya et al. 1994). According to (Molinari et al. 1992), the minimum chlorofluorocarbon (CFC) concentration can be used for distinguishing Middle North Atlantic Deep Water from Lower and Upper North Atlantic Deep Waters, which have local CFC maxima. South of 25° S in the western basin, it is not reasonable to divide North Atlantic Deep Water into components, because the Middle

North Atlantic Deep Water layer does not exist there and properties of the Upper and Lower North Atlantic Deep Water are very similar (Demidov 2003).

Vanicek and Siedler (2002) divide North Atlantic Deep Water into four components. They distinguish: Upper NADW, Labrador Sea Water, old Lower NADW, and overflowing Lower NADW and indicate that Upper North Atlantic Deep Water approximately corresponds to shallow Upper NADW according to the classification in (Rhein et al. 1996). The components are also characterized by maxima of salinity, temperature, concentration of CFC and minimum of nutrients concentration. Labrador Sea Water, the core of which is associated with the oxygen maximum, and old Lower NADW are a part of the Middle North Atlantic Deep Water. They are characterized by minimal concentration of oxygen and CFC, in contrast to the overflowing Lower North Atlantic Deep Water. The North Atlantic Deep Water component can also be distinguished by the minimal concentration of the terrigeneous helium (Vanicek and Siedler 2002).

There is no commonly accepted opinion about the origin of North Atlantic Deep Water components. For example, according to (Arhan et al. 1998; Larque et al. 1997), Upper North Atlantic Deep Water originates mainly from Mediterranean Water. According to (Pickart 1992; Andrie et al. 1998), Upper North Atlantic Deep Water originates from waters of the southern Labrador Sea. Ambar (1983) believes that Upper North Atlantic Deep Water is formed from these two water masses. This controversy is explained by the following fact; distinguishing indicators of this water, such as salinity maximum and minimum of silicate concentration, are typical for Mediterranean Water, while high oxygen concentration is a characteristic feature of Labrador Sea Water.

Some authors (Messias et al. 1999; Tsuchiya 1989) associate Middle North Atlantic Deep Water with the maximum oxygen concentration, while others (Molinari et al. 1990; Andrie et al. 1998) believe that a minimum of CFC concentration is the best indicator of these waters, which does not coincide with the maximum of oxygen. This fact explains a wide spectrum of opinions about their origin. For example, Arhan et al. (1998) consider that Middle North Atlantic Deep Water is formed in the Labrador Sea. In addition, Andrie et al. (1998) suggest that these waters are mixed with circumpolar waters and Iceland-Scotland Overflow Water. Ambar (1983) considers that Middle North Atlantic Deep Water originates directly from Iceland-Scotland Overflow Water.

Lower North Atlantic Deep Water originates from Denmark Strait Overflow Water (Wüst 1936; Andrie et al. 1998; Arhan et al. 1998; Tsuchiya et al. 1994). Ambar (1983) indicates that Lower North Atlantic Deep Water is formed from Denmark Strait Overflow Water and modified Antarctic Bottom Water propagating along the eastern slope of the Mid-Atlantic Ridge. Speer and McCartney (1992) assume that Lower North Atlantic Deep Water is formed from the mixture of Iceland-Scotland Deep Water and Denmark Strait Overflow Water. A minimum of nutrients is also a characteristic property of Lower North Atlantic Deep Water formed from the waters of the Strait of Denmark transported to the Equatorial Atlantic by the Deep Western Boundary Current (Speer and McCartney 1991; Tsuchiya et al. 1992).

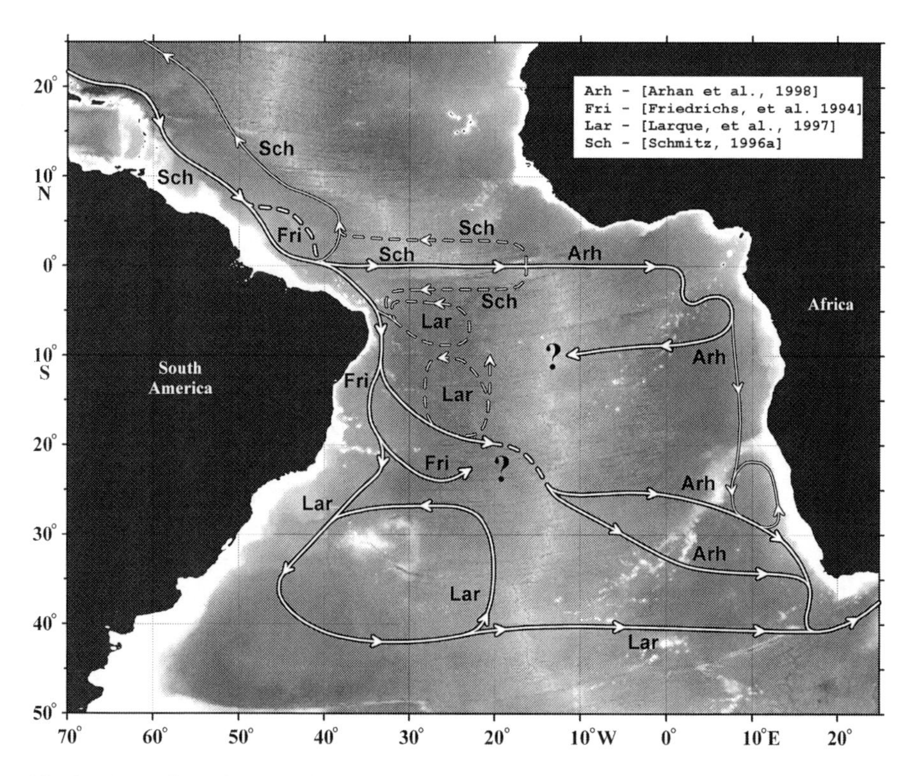

Fig. 2.7 Spreading of North Atlantic Deep Water in the South Atlantic based on the opinion of different authors indicated in the legend. *Question marks* denote questionable pathways of further NADW propagation. *Dashed lines* show uncertain pathways

South of 25° S, division of North Atlantic Deep Water into components is not justified due to the absence of Middle North Atlantic Deep Water and strong similarity between Upper and Lower North Atlantic Deep Waters (Demidov 2003). In the Southwest Atlantic, salinity maximum south of 25° S is continued from the layer of Lower North Atlantic Deep Water (Tsuchiya et al. 1994).

A scheme of North Atlantic Deep Water spreading in the South Atlantic is shown in Fig. 2.7. This scheme is based on publications shown in the figure. Distinguishing properties of North Atlantic Deep Water are given in Table 2.2. Different opinions about the formation of North Atlantic Deep Water components are summarized in Table 2.3.

Table 2.2 Difference between extrema of different NADW components

Wüst (1936); Molinari et al. (1990)			Rhein (1995); Vanicek and Siedler (2002)		
	Max	Min		Max	Min
UNADW	S, CFC		SUNADW	T, S, CFC	Si, P
MNADW	O_2	CFC	LSW	O_2	
			LNADW-old		O_2, CFC
LNADW	O_2, CFC		OLNADW	O_2, CFC	

Table 2.3 Different opinions about the origin of NADW components

	Wüst (1936)	Baum (2004)	Tsuchiya et al. (1992)	Rhein et al. (1995)	Schmitz (1996a)	Andrie et al. (1998)
UNADW	MIW	MIW	MIW + LSW	uLSW	LSW + MW	LSW
MNADW	LSW	LSW	ISOW + EBW + DSOW	LSW + CGFZW	NSOW + AABW	CW + LSW + ISOW
LNADW	DSOW	NSOW	AABW + DSOW	DSOW	NSOW + AABW	DSOW

2.5 Lower Circumpolar Water and Lower Circumpolar Deep Water, Circumpolar Bottom Water, Southeast Pacific Deep Water, and Warm Deep Water

We noted in the beginning of the chapter that Lower Circumpolar Deep Water and Lower Circumpolar Water are different objects in the Atlantic water structure.

North Atlantic Deep Water is the main component of Lower Circumpolar Deep Water (Patterson and Whitworth 1990). Intense, mainly isopycnal mixing of North Atlantic Deep Water with deep waters of the Antarctic Circumpolar Current occurs at the northern periphery of the Antarctic Circumpolar Current in the Atlantic sector of the Southern Ocean (Stramma and England 1999; Patterson and Whitworth 1990). The resultant new water mass was called Lower Circumpolar Deep Water (Gordon 1967; Patterson and Whitworth 1990). It is the opinion of Mantyla and Reid (1983) that Lower Circumpolar Deep Water is a mixture of Antarctic Bottom Water and North Atlantic Deep Water. Unlike North Atlantic Deep Water, it is characterized by smoother extrema found in North Atlantic Deep Water (salinity maximum and minima of the concentrations of phosphates and nitrates). During the further easterly transport of Lower Circumpolar Deep Water by the Antarctic Circumpolar Current, these extrema become weaker, but they are still clearly manifested along the entire pathway of these waters around Antarctica up to the next renewal of Lower Circumpolar Deep Water by North Atlantic Deep Water in the Southwestern Atlantic.

The main distinguishing property of Lower Circumpolar Water at the northern periphery of the Antarctic Circumpolar Current in the Atlantic Ocean is an oxygen minimum caused by the existence of higher oxygen concentrations in the overlying and underlying layers in Weddell Sea Deep Water and North Atlantic Deep Water (Reid et al. 1977). This feature allows us to distinguish Lower Circumpolar Water only in the Argentine Basin. Already in the Brazil Basin, waters of this density range have intermediate characteristics between the bottom water and North Atlantic Deep Water.

If we use the two-layer classification of circumpolar waters (Gordon 1971, 1972), Lower Circumpolar Deep Water would include Lower Circumpolar Water. However, if we use the three-layer classification with supplements by (Orsi et al. 1999), which is likely more reasonable, density ranges of these waters would only partially overlap. Hereinafter, we shall keep to the three-layer classification of circumpolar

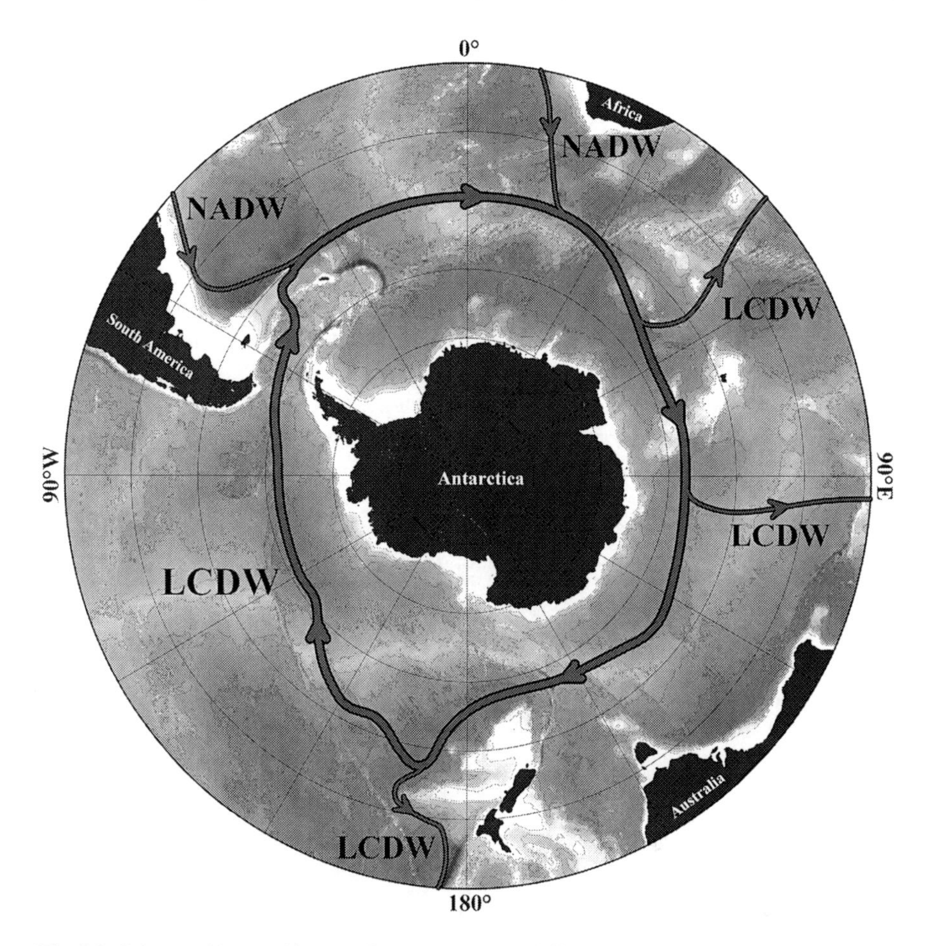

Fig. 2.8 Scheme of Lower Circumpolar Deep Water and Circumpolar Deep Water propagation in the Southern Ocean (*thick lines*). *Thin lines* show inflow of North Atlantic Deep Water to the Antarctic Circumpolar Current region and outflow of Lower Circumpolar Deep Water to the Indian and Pacific oceans

waters in the Antarctic Circumpolar Current region. A scheme of Lower Circumpolar Deep Water and Circumpolar Deep Water propagation in the Southern Ocean is shown in Fig. 2.8.

Circumpolar Bottom Water (CBW) occupies a layer between Antarctic Bottom Water (in the definition by (Orsi et al. 1999)) and Lower Circumpolar Deep Water in the Southern Ocean. During the motion with the Antarctic Circumpolar Current, Lower Circumpolar Deep Water is partly mixed with the underlying colder and denser Antarctic Bottom Water. Circumpolar Bottom Water is actually formed as a result of this mixing and also due to isopycnal inflow of waters from Antarctica (Orsi et al. 1999, 2002). This water can overflow the underwater thresholds in the Drake Passage during the motion from west to east because its density is low. However, its density is high enough to occupy the isopycnal interval of North Atlantic

Deep Water renewing Lower Circumpolar Deep Water in the South Atlantic. Because of such specific properties of this water, relative to Antarctic Bottom Water and Lower Circumpolar Deep Water, it should be distinguished as an independent water mass, as was stated for the first time in (Orsi et al. 1999).

Circumpolar Bottom Water (Fig. 2.9) is the most uniform water layer in the Antarctic Circumpolar Current. It is subjected to least variability of properties in the course of its propagation in the circumpolar circle compared with the overlying and underlying layers (Orsi et al. 1999). Circumpolar Bottom Water rather than waters of the Weddell and Ross seas propagate to the equator in the Pacific and Indian oceans. According to (Orsi et al. 1999), the neutral density range for Circumpolar Bottom Water is $28.18 < \gamma^n < 28.27$. Later, it was accepted that isopycnal $\gamma^n = 28.26$ is the boundary between Weddell Sea Deep Water, Water which is the upper component of Antarctic Bottom Water, and circumpolar water (Naveira Garabato 2002a).

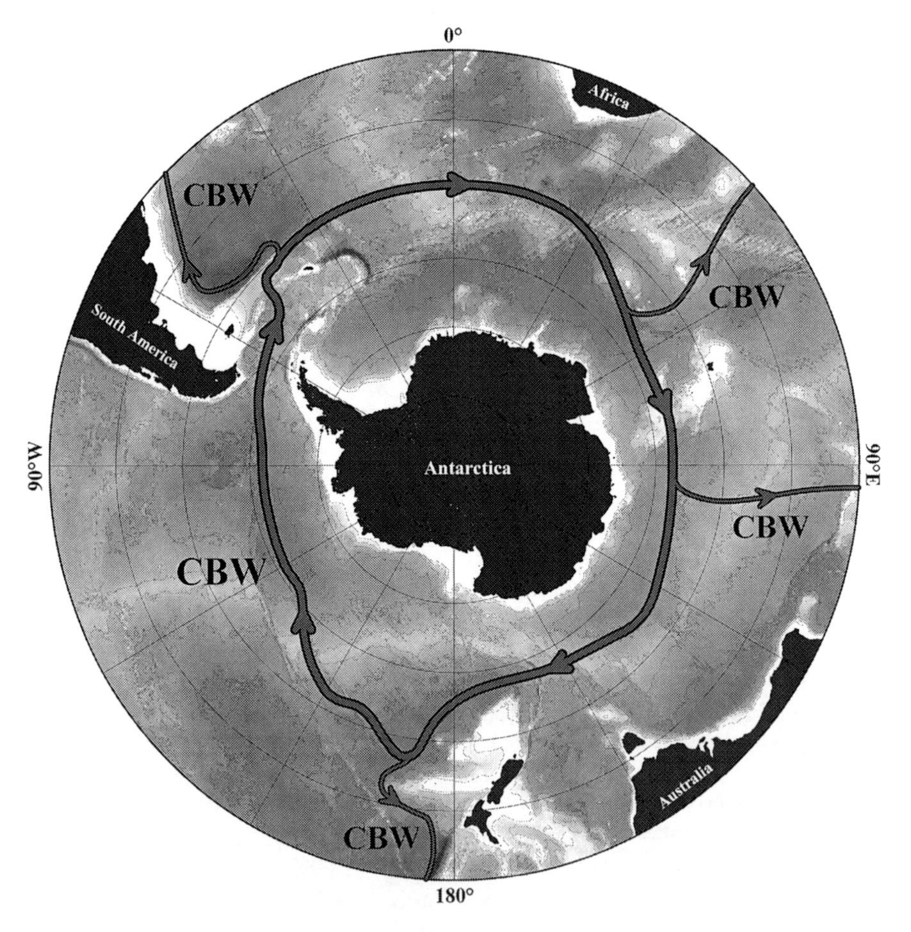

Fig. 2.9 Scheme of Circumpolar Bottom Water propagation in the Southern Ocean (*thick line*). *Thin lines* show outflow of Circumpolar Bottom Water to the north

Koshlyakov and Tarakanov (2003a) showed that this value of neutral density gives the best indicator for this boundary in the Pacific sector of the Southern Ocean.

Gordon (1966) was first to demonstrate the existence of Pacific origin waters in the Atlantic Ocean (in the Argentine Basin). Later, Sievers and Nowlin (1984) distinguished Southeastern Pacific Deep Water (formed in the southeastern sector of the Pacific Ocean) in the Drake Passage as a mixture of overlying Lower Circumpolar Deep Water with bottom waters of the Ross Sea. According to (Orsi et al. 1999), Circumpolar Bottom Water in the Drake Passage is actually equivalent to Southeastern Pacific Deep Water. A high age of the mixing waters in the Pacific Ocean explains high silicate concentration in Southeastern Pacific Deep Water in the Drake Passage compared to a relatively low content in the younger Antarctic Bottom Water of the Weddell Sea.

Koshlyakov and Tarakanov (2003b) analyzed the properties of deep waters in the Pacific Ocean and showed that Circumpolar Bottom Water in the southeastern part of the Pacific Ocean is a product of diapycnal mixing of the overlying Lower Circumpolar Deep Water and the underlying Antarctic Bottom Water. Antarctic Bottom Water in this region is represented by the Weddell Sea Deep Water that passed the route with the Antarctic Circumpolar Current from the Weddell Sea to the east that mixes on this pathway with other modifications of Antarctic Bottom Water. As already mentioned above, formation of Antarctic Bottom Water in the classic scheme of Deacon (1937) is accompanied by isopycnal renewal of Circumpolar Water. According to estimates in (Orsi et al. 2002) based on the analysis of CFC balance in the World Ocean, the total volume of renewed circumpolar waters (Circumpolar Bottom Water and Lower Circumpolar Deep Water) in the Southern Ocean is 9.4 Sv, which is approximately equal to the known estimates of the volume of Antarctic Bottom Water formed in the entire Southern Ocean. For example, the estimate reported in (Orsi et al. 1999) is 10 Sv. It is likely that formation of Antarctic Bottom Water and renewal of Circumpolar Deep Water are confined to specific regions around Antarctica. The northwestern periphery of the Weddell Gyre (60°–35° W) is one of the regions where this process is intense (Whitworth et al. 1994). The other region is George V Coast (120°–160° E) (Foster 1995; Rintoul 1998; Rintoul and Bullister 1999).

Traditionally, the relatively warm and saline layer in the Weddell Gyre, which occupies an intermediate position between Antarctic Bottom Water and Antarctic Surface Water, is called Warm Deep Water (see, for example, (Orsi et al. 1993)). Salinity maximum in this water indicates its circumpolar origin, precisely from Lower Circumpolar Deep Water. According to (Orsi et al. 1993; Gordon 1998), Lower Circumpolar Deep Water propagates to the Weddell Sea approximately at 25° E. It flows along the southern arc and later along the western arc of the Weddell Gyre. Gradual cooling and freshening of its core are observed during this propagation. If we use classification in (Orsi et al. 1999), this water joins Circumpolar Bottom Water and lower layers of Lower Circumpolar Deep Water. Thus, renewal of Circumpolar Deep Water at the northwestern periphery of the Weddell Gyre is related to inflow of the cooler and fresher Warm Deep Water from the Weddell Sea.

Hence, several water masses formed or renewed as a result of convection in the Weddell Sea and Antarctic Circumpolar Current owing to their interaction with the

North Atlantic Deep Water are distinguished in the Antarctic bottom zone. We shall consider peculiarities of their northward propagation in the Atlantic Ocean.

2.6 Antarctic Bottom Water

Definition of Antarctic Bottom Water suggested by Wüst (1936) is generally accepted for water mass analysis in the Atlantic Ocean. Wüst defines Antarctic Bottom water as water with potential temperature less than 2°C.

Definitions of Antarctic Bottom Water are different for research in the Southern Ocean and Antarctic region. According to the definition of Antarctic Bottom Water given in (Whitworth et al. 1998; Orsi et al. 1999), the limiting density value of neutral density is $\gamma^n = 28.27$. This property does not allow Antarctic Bottom Water to overflow the meridional ridges in the Drake Passage in the eastern direction together with the Antarctic Circumpolar Current (Reid et al. 1977; Orsi et al. 1999). Thus, the term 'Antarctic Bottom Water' covers all bottom waters with density $\gamma^n > 28.27$ (Whitworth et al. 1998; Orsi et al. 1999). This density approximately corresponds to the water with potential temperatures lower than 0.2°C. An additional criterion was suggested in (Whitworth et al. 1998) to distinguish Antarctic Bottom Water from Antarctic Shelf Water: potential temperature $\theta > -1.7$°C.

Since a large part of our book is dedicated to the review of literature, we shall keep to both definitions of Antarctic Bottom Water, and try to follow the definitions accepted for each individual Atlantic region. Each definition will be specially emphasized if not clear from the text.

According to the modern concepts (using the definition by (Orsi et al. 1999)), Antarctic Bottom Water is formed over the Antarctic slope as a result of mixing of the cold and heavy Antarctic Shelf Water with the lighter, warmer, and more saline Circumpolar Deep Water. In its turn, Antarctic Shelf Water is formed in the autumn-winter season over the Antarctic shelf due to cooling of the relatively fresh Antarctic Surface Water to nearly freezing point temperature and salinification due to ice formation (Fig. 2.10). Formation of Antarctic Shelf Water, its further downwelling along the Antarctic slope to the ocean abyssal, and formation of Antarctic Bottom Water occur only in the case of a combination of specific conditions: existence of permanent large polynya supported by strong offshore winds, leading to the withdrawal of heat to the atmosphere; sufficient width of the shelf that provides significant time for interaction with the atmosphere; and existence of crossing canyons that allow outflow of the heaviest forms of Antarctic Shelf Water from the shelf. Cooling and freshening at the bottom of shelf glaciers also takes place during the formation (cabbeling) of Antarctic Shelf Water. This is especially characteristic of the western part of the Ross Sea, where the temperature under the ice column can reach extremely low values $\theta < -2.0$°C (Patterson and Whitworth 1990). If the salinity of the mixed waters is relatively high and the temperature is low, the resulting water mass reaches the ocean floor, thus forming Antarctic Bottom Water. Because of these limitations, formation of Antarctic Bottom Water is confined to a

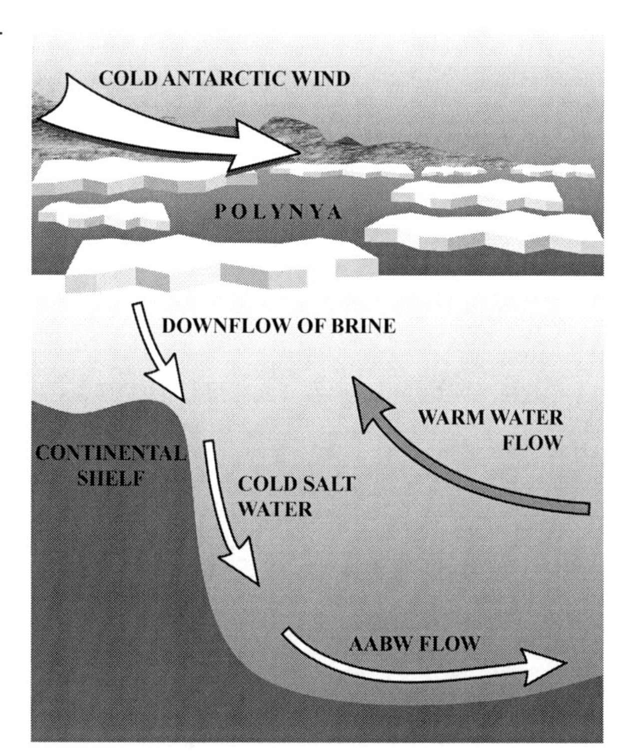

Fig. 2.10 Scheme of Antarctic Bottom Water formation

few regions along the perimeter of the Antarctic slope. At present, one can identify several regions (Baines and Condie 1998; Orsi et al. 1999; Whitworth et al. 1998): southern and western parts of the Weddell Sea; Commonwealth Sea in the Indian Ocean sector of Antarctica; George V Coast (or Adelie Land) at the boundary between the Indian Ocean and Pacific sectors; western and eastern parts of the Ross Sea (Fig. 2.11). The Weddell Sea is a region marked by the formation of 60–65% of Antarctic Bottom Water (Orsi et al. 1999; Rintoul 1998; Carmack 1977). There are a few other regions of lesser importance for Antarctic Bottom Water formation (Whitworth et al. 1998). Waters formed in these sources are also called Antarctic Bottom Water, but their properties can strongly differ from those of waters from the Weddell and Ross seas and other sources in the Southern Ocean.

Hypotheses about the formation of cold bottom waters in the Antarctic region as a result of mixing of cold shelf waters with warmer deep waters were first put forward by Brennecke (1921) and Mosby (1934). Later, Fofonoff (1956) indicated the possibility of cabbeling during mixing of these waters that changes the water density. Gordon (1978) and Whitehead (1989) suggested another mechanism of Antarctic Bottom Water formation: deep convective mixing in polynyas within mesoscale eddies in the open ocean intensified by strong cooling in the surface layer.

In the early 1970s, such convection was observed in the Southern Ocean in the Maud Rise region (Greenwich meridian). However, satellite observations of ice cover in the ocean demonstrated that necessary conditions for this mechanism ex-

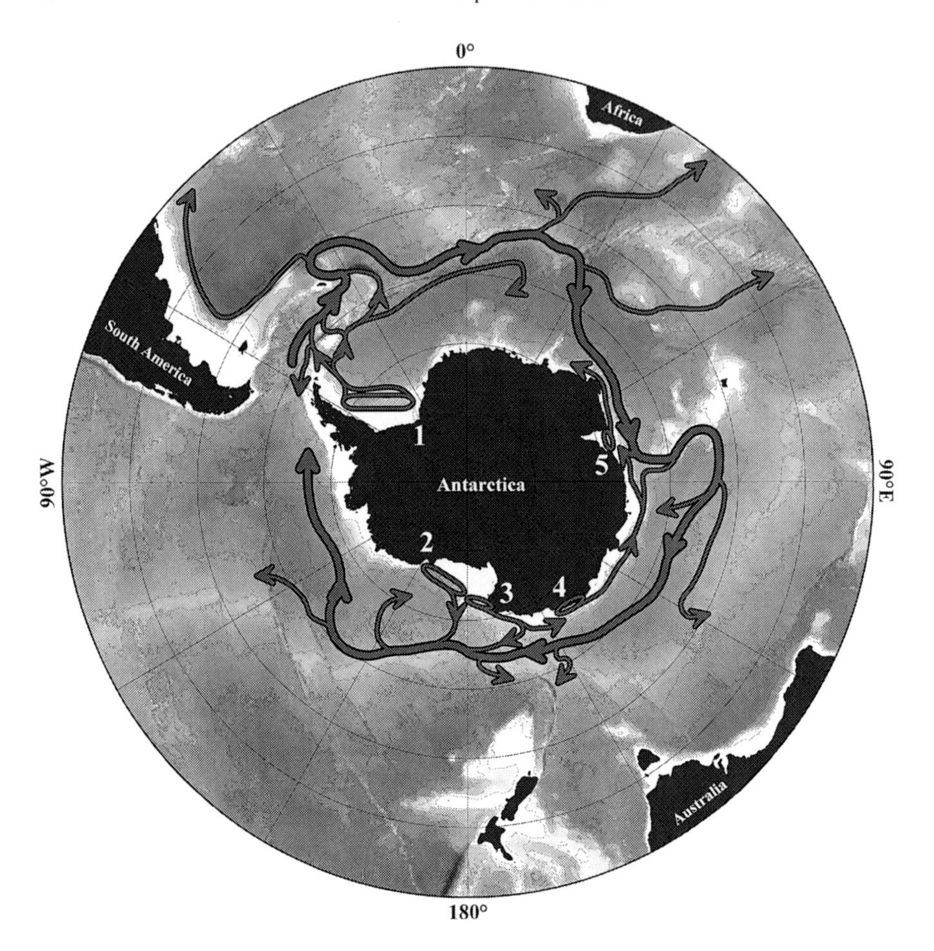

Fig. 2.11 Scheme of Antarctic Bottom Water propagation in the Southern Hemisphere according to (Orsi et al. 1999). *Thick lines* denote quasi-circumpolar flow of Antarctic Bottom Water. *Thin lines* correspond to outflow of this water from the sources and pathways for further Antarctic Bottom Water propagation from the Southern Ocean. Numerals indicate main regions of AABW formation. (*1*) Weddell Sea; (*2*) East part of the Ross Sea; (*3*) West part of the Ross Sea; (*4*) Adelie Land; (*5*) Commonwealth Sea

isted only during 11% of the time of such observations, which could provide the formation of approximately 2.4 Sv of deep waters (Gordon 1978). The obtained mean estimate of the rate of deep water formation in the Weddell Sea due to deep convection is more than one order of magnitude smaller than estimates of the rate of Antarctic Bottom Water formation over the Antarctic slope. However, Whitworth et al. (1998) demonstrated that this process should be considered as potentially possible over the major part of the perimeter of the Antarctic slope.

Antarctic Bottom Water formation is not uniform in time because of the seasonal character of Antarctic Shelf Water formation and sporadic down-flow of Antarctic Shelf Water down the Antarctic slope in the form of plumes and individual portions

(Baines and Condie 1998) Some interseasonal and interannual differences in the characteristics (including density) of the waters flowing from the shelf are quite natural.

According to the formation mechanism in all regions of the Antarctic Bottom Water spreading, it is characterized by low temperature values. In the major part of regions of Antarctic Bottom Water formation mentioned here, Antarctic Shelf Water is also characterized by lower salinity compared to the overlying Circumpolar Deep Water. Therefore, Antarctic Bottom Water corresponding to different types of shelf waters also differs by lower salinity compared to the overlying waters. Antarctic Shelf Water and, correspondingly, Antarctic Bottom Water from the western part of the Ross Sea are exceptions. According to (Patterson and Whitworth 1990), this peculiarity is caused by quasi-isolated location of the cyclonic gyre near the Wilkes Coast in the northwestern part of the Ross Sea shelf, which, in turn, leads to extremely strong salination of the local Antarctic Shelf Water (salinity reaches 35.0 psu). Antarctic Bottom Water within the Antarctic Circumpolar Current is distinguished by increased content of oxygen. Concentrations of oxygen and nutrients depend on the age of the bottom waters and trajectory of their flow. In some cases, these features allow us to distinguish Antarctic Bottom Water formed in different Antarctic regions. (Patterson and Whitworth 1990; Koshlyakov and Tarakanov 1999, 2003a). Potential temperature of Antarctic Bottom Water in the Weddell Sea can be as low as −1.5°C but even lower temperatures are recorded; salinity is approximately 34.65 psu, and silicate concentration is nearly 100 μmol/kg. The type of Antarctic Bottom Water from the Ross Sea has a slightly higher temperature (−0.2°C) and salinity (34.75 psu), while salinity is slightly smaller (34.4–34.7 psu) near the George V Coast (Adelie Land). Mean values of Antarctic Bottom Water in the region of 60° S are equal to −0.4°C and 34.66 psu, respectively (Mamayev 1992), while the silicate concentration is approximately 125 μmol/kg (Ostlund et al. 1987). Propagation of Antarctic Bottom Water around Antarctica is shown in Fig. 2.11.

Spreading of Antarctic Bottom Water in the Atlantic Ocean can be seen from the temperature at the bottom (Morris et al. 2001). Figure 2.12 shows topography of the Atlantic Ocean and distribution of potential temperature at the bottom, which characterizes spreading of Antarctic Bottom Water. Potential temperature strongly increases after passage of the narrow abyssal channels (Vema Channel, Vema Fracture Zone, Romanche and Chain fracture zones).

Orographic peculiarities have a determining influence on propagation of abyssal waters of the Antarctic origin. The deepest part of the Atlantic Ocean is a series of alternating depressions and elevations of the ocean bottom in the western and eastern parts of the ocean divided by the Mid-Atlantic Ridge. The publications by (Wüst 1936; Hogg and Zenk 1997; Saunders 1994; Koltermann et al. 1999; Friedrichs and Hall 1993; Harvey and Arhan 1988) demonstrate the significant role of fractures in the Mid-Atlantic Ridge, such as Romanche (0°), Vema (11° N), and Charlie Gibbs (53° N), in propagation of the deep and especially bottom waters. In the eastern parts of the ocean, their propagation is limited by the Walvis Ridge crossing the entire eastern basin. Although Connary and Ewing (1974) indicate that a small por-

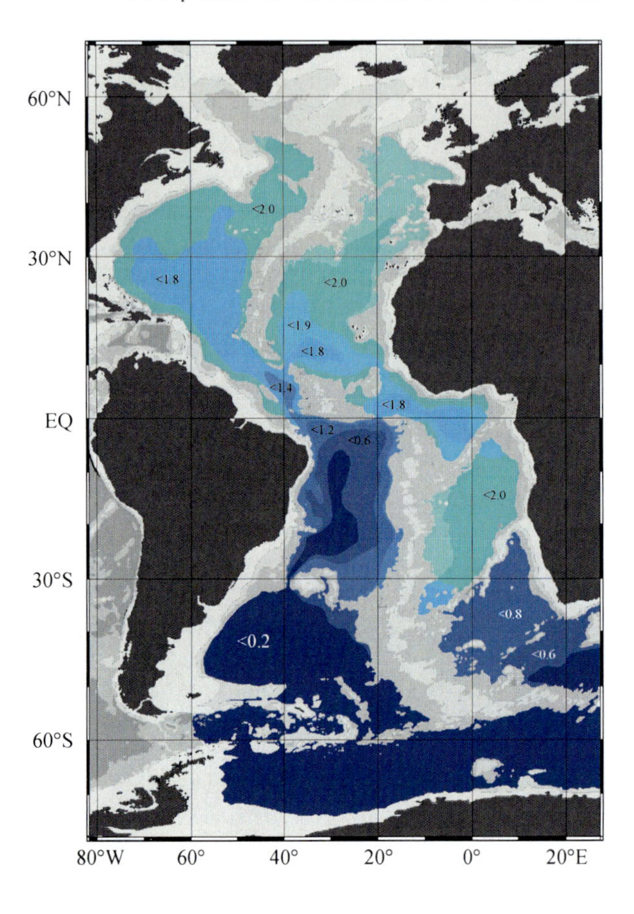

Fig. 2.12 Distribution of potential temperature (°C) at the bottom in the Atlantic Ocean that gives an idea of Antarctic Bottom Water spreading in the Atlantic Ocean

tion of Antarctic Bottom Water (in the classification of (Wüst 1936)) propagates to the Angola Basin through a channel in the Walvis Ridge, the main role in the water exchange between the western and eastern parts of the ocean belongs to the deep fractures in the Mid-Atlantic Ridge.

Munk and Wunsch (1998) estimated that without mixing, Antarctic Bottom Water would fill the entire ocean within a few thousand years and turn it into a stagnant pool of cold water. Only a thin upper layer of warm water would remain at the surface. Internal tide is the main source of ocean mixing, especially at hot spots over slopes of bottom topography. Impressive laboratory experiments in (Whitehead and Wang 2008) demonstrate this.

Chapter 3
Source Regions, Abyssal Pathways, and Bottom Flow Channels (for Waters of the Antarctic Origin)

3.1 General Description

Generally, propagation of Antarctic waters in the bottom layer of the Atlantic Ocean is confined to depressions in the bottom topography. The general flow of these waters can be presented as follows (Fig. 3.1).

Antarctic Bottom Water (in the concept developed in Orsi et al. (1999)) propagates from the Weddell Sea to the north through the passages in the South Scotia Ridge and through the South Sandwich Trench and South Sandwich Abyssal Plain. In the Scotia Sea, part of Antarctic Bottom Water flows to the west to the Drake Passage. The remaining part of Antarctic Bottom Water propagates through the Georgia and Northeast Georgia passages to the Georgia Basin. Here, it occupies the bottom layer together with the Antarctic Bottom Water that passed through the South Sandwich Trench. The further northward propagation of Antarctic Bottom Water to the Argentine Basin occurs through the Falkland Gap in the Falkland Ridge (Whitworth et al. 1991). A part of this flow propagates along the southern and western margins of the Argentine Basin. Another part of this flow is entrained into the Antarctic Circumpolar Current and propagates to the east (Whitworth et al. 1991). Part of this easterly current in its turn flows around the Zapiola Ridge as an anticyclonic gyre (Smythe-Wright and Boswell 1998).

Circumpolar waters in the South Atlantic propagate to the east with the Antarctic Circumpolar Current. At the northeastern periphery of the Weddell Gyre, part of these waters is entrained into the cyclonic circulation of the Weddell Gyre. From the Scotia Sea, circumpolar waters in the Southwest Atlantic penetrate to the north into the Argentine Basin over the Falkland Plateau. Here, together with Antarctic Bottom Water, they form the bottom structural zone, waters of which actually propagate further to the north in the Atlantic.

After the Argentine Basin, waters of the Antarctic origin (Antarctic Bottom Water together with circumpolar waters) are transported to the Brazil Basin. It is traditionally considered that this propagation occurs in three places: through the Vema Channel, Hunter Channel, and over the Santos Plateau. Further in the north, the flow of Antarctic water splits. A part of the flow is transported to the eastern basin through the Romanche and Chain fracture zones, influencing the waters of the bottom layer

E. G. Morozov et al., *Abyssal Channels in the Atlantic Ocean*,
DOI 10.1007/978-90-481-9358-5_3, © Springer Science+Business Media B.V. 2010

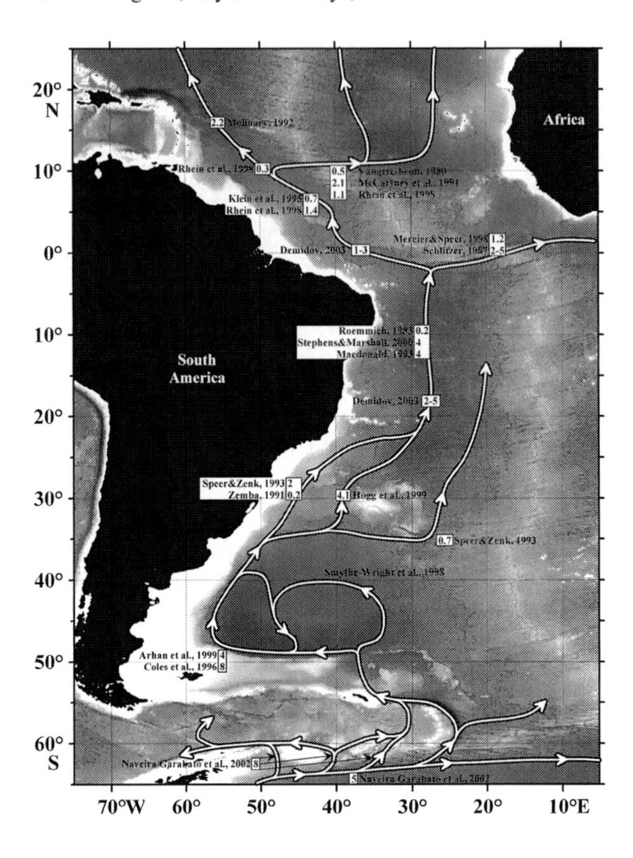

Fig. 3.1 Spreading of bottom water of the Antarctic origin. The scheme is based on literature data

in the Southeast Atlantic. The other part flows through the nameless Equatorial Channel and Guyana Basin, propagating further to the Northeast Atlantic through the Vema Fracture Zone and to the North American Basin in the west, where it is entrained into cyclonic gyre within its northward spreading zone, reaching the Newfoundland Bank. Below, we shall describe the structure and circulation of bottom waters of the Antarctic origin with quantitative estimates in each of the basins.

Figure 3.2 shows a principle scheme of water mass distribution in the South Atlantic and the location of their cores distinguished from the extreme values of any physical or chemical characteristic. Figure 3.3 shows contour lines of different characteristics, which are associated in literature with the boundaries of water masses of Antarctic origin in the bottom layer of the Atlantic. The figure is based on the data of quasi-meridional section A17 occupied in 2003 from R/V *Akademik Sergey Vavilov*. The section crossed the Brazil and Argentine basins from north to south. Figure 3.4 shows the corresponding regions of these waters spreading. The scheme is based on climatic WOCE dataset. We do not show all possible boundaries used in literature so as not to overload the figure.

Group of contour lines $\sigma_2=37.11$ (brown line in Fig. 3.3), $\theta=2.0°C$ (red line), $\gamma^n=28.11$ (orange line), $\sigma_4=45.87$ (yellow line) corresponds to the upper boundary of Lower Circumpolar Water (Wüst 1936; Reid et al. 1977; Ganachaud 2003; Demidov et al. 2007a). One can see that contour lines $\gamma^n=28.11$, $\sigma_4=45.87$ are very close

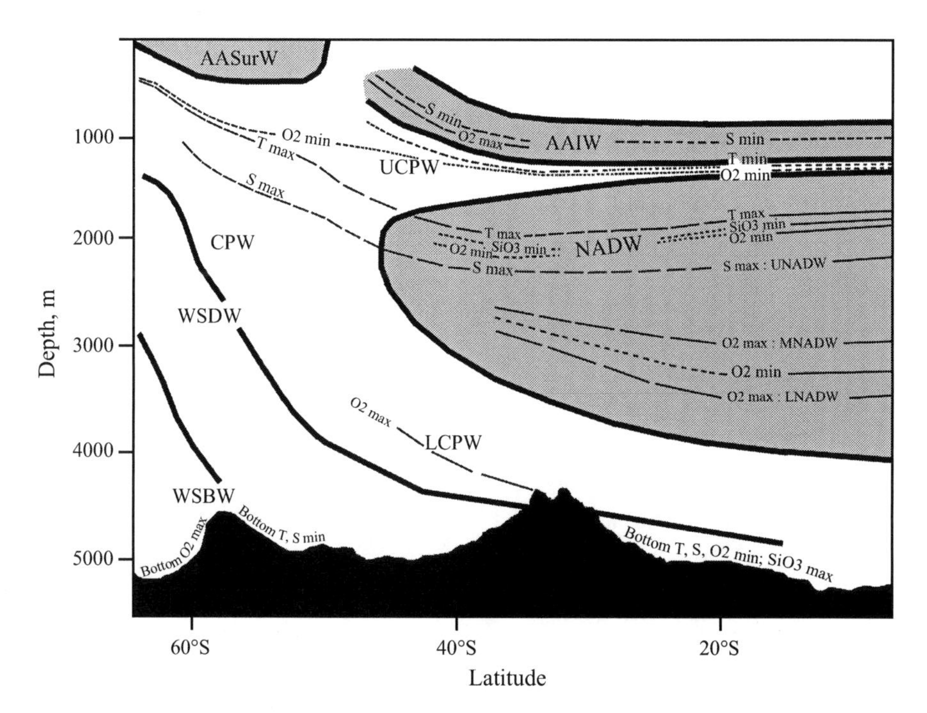

Fig. 3.2 Schematic representation of water mass distribution and frontal position along a full-depth meridional section through the South Atlantic based on Fig. 3 from Peterson and Whitworth (1989)

to each other (Fig. 3.3) and the spreading regions bounded by these lines practically coincide (Fig. 3.4). However it is not the same with other characteristics. Isotherm θ=2.0°C practically coincides with γ^n=28.11 and σ_4=45.87 in the Brazil Basin and northern part of the Argentine Basin (Fig. 3.3). The regions of their spreading are also very close excluding the northern part of the Northeast Atlantic, in which the isotherm reaches only 37° N (Fig. 3.4). The isotherm and isopycnals diverge strongly in the southern part of the Argentine Basin. This divergence is clearly seen at the Deep Front (34° S, Fig. 3.3), which according to McDonagh et al. (2002), is related to the change in the dominating propagation of North Atlantic Deep Water from the meridional to zonal directions. (Hogg and Thurnherr (2005) studied this peculiarity of circulation in detail). The depth and spreading region of isopycnals σ_2=37.11 differs strongly from the other three characteristics. In the Brazil Basin, this isopycnal is located deeper than θ=2.0°C, γ^n=28.11, and σ_4=45.87 approximately by 300–500 m (Fig. 3.3). Unlike the similar selection of Lower Circumpolar Water boundary, its spreading is strictly limited to the region east of the Mid-Atlantic Ridge. It spreads only in the Cape Basin and basins located in the equatorial and tropical parts of the Atlantic Ocean (Fig. 3.4). We emphasize that north of the Deep Front, this isopycnal is close to contour line γ^n=28.16 (Fig. 3.3). The latter is assumed the best approximation of the boundary between Lower Circumpolar Deep Water and Circumpolar Bottom Water on the basis of the analysis of temperature and salinity gradients in the southern part of the Pacific Ocean (Koshlyakov and

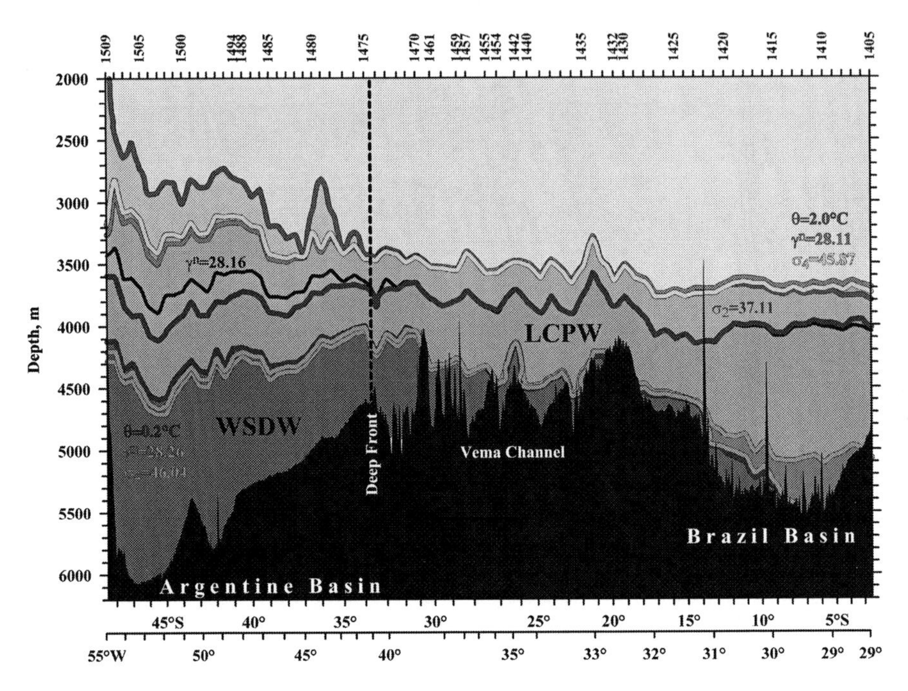

Fig. 3.3 Location of water mass isopycnals related to the boundaries between them determined by different characteristics used in various publications

Tarakanov 2003b). Thus, it is our opinion that $\sigma_2=37.11$ is the best choice of the boundary that corresponds to the circumpolar water mass classification of Antarctic waters (Orsi et al. 1999). The authors of Orsi et al. (1999) identified this boundary with $\gamma^n=28.18$. In the literature, the upper boundary of Antarctic waters in the bottom layer of the Atlantic Ocean is also identified with isobaths (3,200 and 4,000 m) and isotherms $\theta=2.07$°C, $\theta=1.9$°C, $\theta=1.8$°C.

The group of contour lines $\theta=0.2$°C (blue line in Fig. 3.6), $\gamma^n=28.26$ (light blue line), $\sigma_4=46.04$ (green line) corresponds to the upper boundary of Weddell Sea Deep Water (Reid et al. 1977; Koshlyakov and Tarakanov 2003a; Naveira Garabato et al. 2002b). The boundaries determined from these contour lines are quite close to each other, although in the Brazil Basin the spreading region of water with $\theta=0.2$°C and cooler is smaller than the spreading regions bounded by other characteristics (Fig. 3.4). This boundary is also identified with contour lines $\gamma^n=28.27$ (Orsi et al. 1999) and $\sigma_4=46.06$ (Arhan et al. 1999).

Each of the groups described above determines the upper boundary of Antarctic waters according to two approaches. The first approach (Wüst 1936) (isotherm close to $\theta=2.0$°C) unites all bottom waters in the Atlantic, which have Antarctic rather than North Atlantic origin under the general term Antarctic Bottom Water. The second approach divides these waters into the waters of circumpolar and Antarctic origin (Reid et al. 1977). In this case, the term Antarctic Bottom Water unites all Antarctic waters in the Southern Ocean with a density high enough not to overflow the threshold

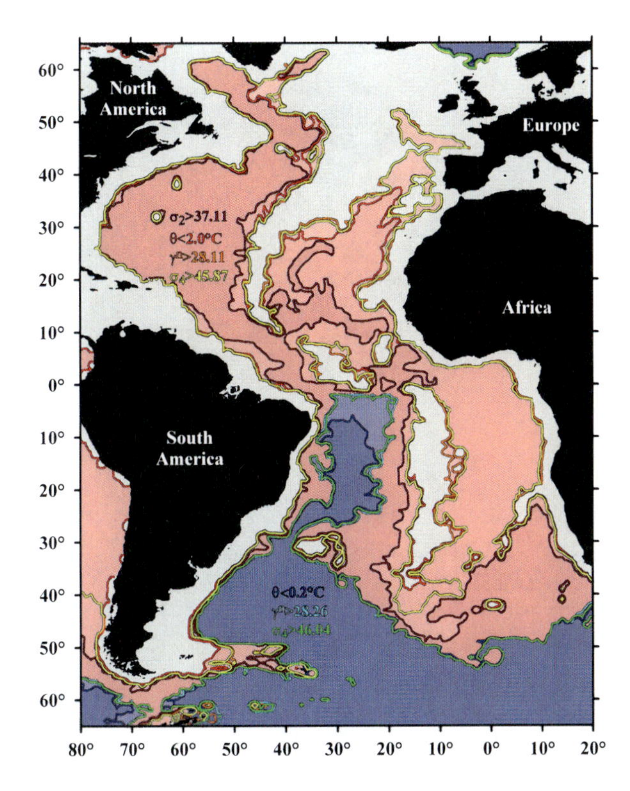

Fig. 3.4 Spreading regions of Antarctic waters in the bottom layer of the Atlantic according to different selection of boundaries (see Fig. 3.3)

in the Drake Passage (approach developed in Orsi et al. (1999)) (isotherms close to $\theta=0.2$°C). Naturally, the estimates of transport and regions of spreading of these waters are principally different. Even more, differences can be notable even within one approach to classification of Antarctic waters. This is especially related to the approach of Wüst (1936), because the authors of different publications applying this approach use a wide range of characteristics and their values to determine the upper boundary of Antarctic Bottom Water. Thus, ambiguity in classification, different (but sometimes the same) terminology, and different boundaries complicates comparison of quantitative characteristics of waters in publications by different authors. We shall make an attempt to solve this problem applying mainly the data of our measurements.

3.2 Weddell Sea and Weddell Gyre

The Cyclonic Weddell Gyre spreading to the bottom occupies a vast region between Antarctica in the south to the South Scotia Ridge and the American-Antarctic and Southwest Indian ridges in the north (Fig. 3.5). The Antarctic Peninsula is the western boundary of the gyre. Its eastern boundary is associated with the propagation of circumpolar waters from the north, which is confined to a quite wide gap (in the region of Prince Edward Fracture Zone) in the Southwest Indian Ridge at 25°–32° E

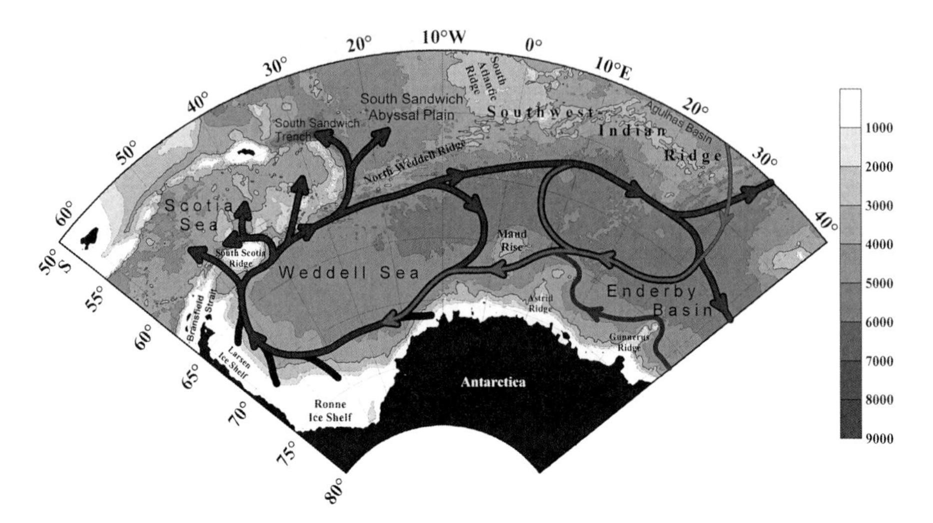

Fig. 3.5 Cyclonic Weddell Gyre. Deep and bottom water circulation in the Weddell Sea based on Orsi et al. (1993). The *thick line* shows circulation of deep and bottom waters. The *black line* shows circulation of Weddell Sea Bottom Water. The *dark gray line* shows circulation of Weddell Sea Deep Water. The *light gray line* shows the main inflow of Lower Circumpolar Deep Water. Inflow of Antarctic Bottom Water along the Antarctic continental slope from the Indian Ocean is also shown with a *gray line*

(Orsi et al. 1993). The measurements evidence penetration of this gyre further to the east approximately to 53° E (Park et al. 2001). Northern branching of the gyre observed in the South Sandwich Islands region is associated with the corresponding deflection of the Antarctic Circumpolar Current from the zonal direction during the flow of this current around the arc of the South Sandwich Arc (Naveira Garabato et al. 2002a). The Weddell Gyre at the surface is a unique cyclonic system elongated from southwest to northeast. In the deep layers, its central part splits into two independent sub-gyres divided at 0°–20° W depending on the depth (Orsi et al. 1993).

As was mentioned above, Antarctic Bottom Water in the region described here is divided into Weddell Sea Bottom Water and the overlying Weddell Sea Deep Water. Weddell Sea Bottom Water is a layer, in which Antarctic Bottom Water is predominantly renewed as a result of mixing of Antarctic Shelf Water and Circumpolar Deep Water over the Antarctic Slope. Weddell Sea Deep Water is a product of mixing of Weddell Sea Bottom Water with the surrounding waters: Antarctic Shelf Water and Warm Deep Water. Some publications indicate that Weddell Sea Deep Water is formed as a result of direct mixing of Antarctic Surface Water and modified Warm Deep Water (Orsi et al. 1993; Fahrbach et al. 1995; Weppernig et al. 1996). Sources of the Antarctic Bottom Water in the Indian Ocean sector of the Southern Ocean form a contribution to Weddell Sea Deep Water (Meredith et al. 2000; Hoppema et al. 2001; Klatt et al. 2002). At the northwestern periphery of the Weddell Gyre, the Weddell Sea Deep Water of lowest density penetrates to the Bransfield Strait (Gordon et al. 2000; von Gyldenfeldt et al. 2002) and the Scotia Sea (Whitworth et al. 1994). Heavier forms of Weddell Sea Deep Water outflow from the Weddell Sea through the passages in the South Scotia Ridge (Phillip, Orkney, Bruce, and

Discovery passages, and maybe Hoyer Passage) to the Scotia Sea and Georgia Basin through the South Sandwich Trench (Nowlin and Zenk 1988; Locarnini et al. 1993; Orsi et al. 1999; Naveira Garabato et al. 2002a; Schodlock et al. 2002). The heaviest forms of Weddell Sea Deep Water flow through the latter of these passages, leaving the Weddell Gyre in the Southwest Atlantic. The flow of Antarctic Bottom Water to the Agulhas Basin is confined to a topographic depression in the Southwest Indian Ridge approximately at 25°–30° E (Reid 1989; Orsi et al. 1993). The densest forms of Weddell Sea Deep Water reached the gap between the Crozet Islands and Kerguelen Ridge at 55° E, through which Antarctic Bottom Water penetrates to the Crozet Basin in the Indian Ocean (not shown in Fig. 3.5) (Haine et al. 1998). Propagation of Weddell Sea Bottom Water is confined to the Weddell Gyre including the South Sandwich Abyssal Plain (Naveira Garabato et al. 2002b; Orsi et al. 1999). We also note that upper layers of Weddell Sea Deep Water propagate to the east and other sectors of the Southern Ocean, thus forming a quasi-circumpolar bottom circulation, which includes Antarctic Bottom Water from different Antarctic basins (Fig. 2.11), and forms a single water mass (Orsi et al. 1999).

As compared to the Weddell Sea Deep Water, the younger Weddell Sea Bottom Water is characterized by lower temperature, salinity, and silicate concentration, as well as higher oxygen concentration. Generally, temperature and salinity of Weddell Sea Deep Water are intermediate between Weddell Sea Bottom Water and Warm Deep Water (Orsi et al. 1993). At the same time, several maxima and minima in silicate concentration are found in the Weddell Sea Deep Water layer. This peculiarity of Weddell Sea Deep Water is related to its mixing at the northwestern periphery of the Weddell Gyre with circumpolar silicate-rich waters, solutions of sediment-containing silicates, and transport of Antarctic Bottom Water from the Indian Ocean sector of the Southern Ocean, which is characterized by lower concentration of silicates.

The isotherm $\theta=-0.7°C$ is most frequently accepted as the upper boundary of Weddell Sea Bottom Water (Fahrbach et al. 1995; Weppernig et al. 1996; Gordon 1998; Gordon et al. 2001; Fahrbach et al. 2001; Harms et al. 2001; Naveira Garabato et al. 2002b), which approximately corresponds to isopycnal $\gamma^n=28.40$ or the isotherm $\theta=-0.8°C$ (Foldvik et al. 1985; 1993; Fahrbach et al. 1994; Muench and Gordon 1995; Foldvik et al. 2004; Mensch et al. 1998; Yaremchuk et al. 1998). Other methods of determination of this boundary are applied rarely (Carmack and Foster 1975; Foster and Carmack 1976; Weiss et al. 1979; Mensch et al. 1997) (Table 3.1). Following (Gordon 1971, 1972), a potential temperature isotherm of 0°C is most frequently accepted as the upper boundary of Weddell Sea Deep Water and Antarctic Bottom Water, in general (Carmack 1977; Fahrbach et al. 1994; Meredith et al. 2001; Yaremchuk et al. 1998; Naveira Garabato et al. 2002b). In Orsi et al. (1999), $\gamma^n=28.27$ is taken as the upper boundary of Antarctic Bottom Water. Sievers and Nowlin (1984) associated this boundary in the Drake Passage with the maximum of hydrostatic stability (close to $\theta=0.2°C$), which approximately corresponds to $\gamma^n=28.26$. This value of neutral density was taken in Naveira Garbato et al. (2002b) as the upper boundary of Antarctic Bottom Water. The same value was also accepted for this boundary in the Pacific sector of the Southern Ocean in Koshlyakov and Tarakanov (2003a) based on the analysis of hydrophysical and hydrochemical characteristics over the section in the entire Pacific sector of the Southern Ocean.

Table 3.1 Estimates of the formation rates Antarctic Bottom Water and Weddell Sea Bottom Water in the Weddell Sea. (Adapted from Naveira Garabato et al. 2002b) (1 Sv = 10^6 m^3/s)

Source and method of estimating	Water mass and boundaries	Transport, Sv
Transport of the Deep Western Boundary Current in the Weddell Sea		
Carmack and Foster (1975)	WSBW ($-1.4 < \theta < -1.2$°C)	2–5
Foster and Carmack (1976)	WSBW ($\theta = -1.3$°C)	3.6
Foldvik et al. (1985)	WSBW ($\theta < -0.8$°C)	2–5
Gordon et al. (1993)	WSBW ($\theta < -0.8$°C)	3
Fahrbach et al. (1995)	WSBW ($\theta < -0.7$°C)	1–4
Muench and Gordon (1995)	WSBW ($\theta < -0.8$°C)	2.5–3
Gordon (1998)	WSBW ($\theta < -0.7$°C)	4–4.8
Fahrbach et al. (2001)	WSBW ($\theta < -0.7$°C)	1.3 ± 0.4
Gordon et al. (2001)	WSBW ($\theta < -0.7$°C)	5
Foldvik et al. (2004)	WSBW ($\theta < -0.8$°C)	4.3 ± 1.4
Transport in the Weddell Sea Gyre		
Fahrbach et al. (1994)	AABW ($\theta < 0.0$°C)	3.3–5.6
	WSBW ($\theta < -0.8$°C)	2.6–2.8
Balance of shelf waters		
Gill (1973)	AABW ($\theta = -0.6$°C)	6–9
Carmack (1977)	AABW ($\theta < 0.0$°C)	5–10*
Application of tracers		
Weiss et al. (1979)	AABW ($\theta < 0.4$°C)	8, 4.5
	WSBW ($\theta < -0.9$°C)	5, 3
Weppernig et al. (1996)	WSBW ($\theta < -0.7$°C)	5
Mensch et al. (1997)	AABW ($\theta = -0.5$°C)	11
	WSBW ($\theta < -1.0$°C)	3.5
Mensch et al. (1998)	WSBW ($\theta < -0.8$°C)	5
Broecker et al. (1998)	AABW ($PO_4 = 1.95$ mMol/kg)	15
Orsi et al. (1999)	AABW ($\gamma^n > 28.27$)	8.1
Meredith et al. (2001)	AABW ($\theta < 1.0$°C)	<6.6, 3.7 ± 1.6
Mass balance		
Orsi (1999)	AABW ($\gamma^n > 28.27$)	10*
Freshwater balance		
Harms et al. (2001)	WSBW ($\theta < -0.7$°C, S > 34.64)	2.6
Numerical models		
Hellmer and Beckmann (2001)	AABW ($\sigma_2 > 37.16$)	11
Inverse models		
Yaremchuk et al. (1998)	WSDW ($-0.8 < \theta < 0.0$°C)	2.6 ± 1.3
	WSBW ($\theta < -0.8$°C)	2.5 ± 1.9
Sloyan and Rintoul (2001b)	AABW ($\gamma^n > 28.30$)	11 ± 1
Naveira Garabato et al. (2002a)	AABW ($\gamma^n > 28.26$)	9.7 ± 3.7
	($\theta < 0.0$°C)	10.0 ± 3.7
	($\gamma^n > 28.27$)	9.8 ± 3.7
	($\sigma_2 > 37.16$)	9.8 ± 3.7
	($\gamma^n > 28.30$)	10.2 ± 3.8
	WSBW ($\gamma^n > 28.40$)	3.9 ± 0.8
	($\theta < -0.7$°C)	4.5 ± 0.9

Note: The estimates marked with asterisk correspond to the circumpolar rate of AABW formation.

The rate of Antarctic Bottom Water formation is determined by researchers on the basis of different methods (Table 3.1). Estimates of the rate of Weddell Sea Bottom Water formation range from 1 to 5 Sv. The rate of Antarctic Bottom Water formation in the Weddell Sea ranges from 3 to 8 Sv. The rate of Antarctic Bottom Water formation in the Southern Ocean as a whole ranges from 5 to 10 Sv.

3.3 Agulhas and Cape Basins

Cyclonic circulation in the bottom layer of these basins was found quite long ago. It existed even in the previous ages, which was revealed on the basis of sedimentary rocks (Tucholke and Embley 1984). The bottom circulation based on the analysis in Reid (1989) is shown in Fig. 3.6. Weddell Sea Deep Water propagating to the southeastern part of the Agulhas Basin from the south through fractures (Prince Edward Fracture Zone) in the Southwest Indian Ridge approximately at 25°–32° E flows mainly to the Mozambique Basin in the Indian Ocean (Reid 1989). Comparing the data of hydrophysical sections in the Southeast Atlantic with the classifica-

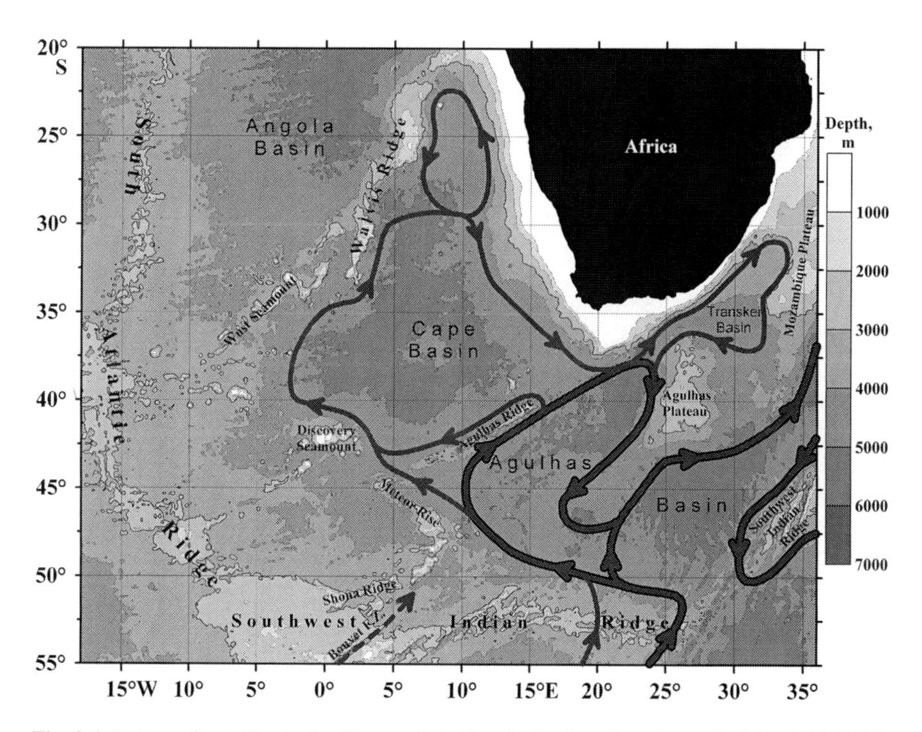

Fig. 3.6 Bottom circulation in the Cape and Agulhas basins based on the analysis in Reid (1989) and Tucholke and Embley (1984) modified by the authors. The *thick gray lines* show circulation of dense water in the entire bottom layer. The *thin gray lines* show circulation of waters of lower density only in the upper part of the bottom layer. The *dashed gray lines* show possible pathways of Antarctic Bottom Water inflow through the Bouvet Fracture Zone from the Weddell Gyre

tion in Orsi et al. (1999), we can state that there is only a small inflow of Weddell Sea Deep Water into the southwestern part of the Agulhas Basin, and Circumpolar Bottom Water occupies the bottom part of the Cape Basin. Thus, taking into account the existence of cyclonic circulation, we can state that the lowest layer of circumpolar waters flows into the Cape Basin not with the jets of the Antarctic Circumpolar Current crossing the Mid-Atlantic Ridge, but from the south through the fractures in the Southwest Indian Ridge. Anticyclonic circulation in the northern part of the Cape Basin (Fig. 3.6) is shown according to Arhan et al. (2003). The authors of this article estimate intensity of bottom water circulation in the Cape Basin at 8 ± 3 Sv taking the upper boundary as σ_4=46.0 (θ < 1.5°C, S < 34.80 psu). They also estimate weak anticyclonic circulation in the northern part of the basin at 2 Sv.

3.4 Drake Passage, Scotia Sea, and Georgia Basin

3.4.1 General Description and Bottom Topography

The southwestern part of the Atlantic Ocean, which includes the Scotia Sea, South Sandwich Abyssal Plain, and Georgia Basin, is the first basin in the water pathway of the bottom layer to the North Atlantic. Basins of this region are separated from each other and from the adjacent basins by chains of islands and high submarine ridges: e.g., the ridge in the Shackleton Fracture Zone and its continuation to the continental slope of South America (hereinafter, the Shackleton Ridge), Falkland Ridge and Southwest Indian Ridge, and other smaller ridges, as well as the arc of the Southern Antilles, which includes the North Scotia Ridge, South Georgia, South Sandwich and South Orkney Islands, and South Scotia Ridge. A few relatively narrow passages through these ridges and between islands play a key role in water exchange of the deep and abyssal waters between the South Atlantic basins and with surrounding waters of the World Ocean. Moreover, jets of the Antarctic Circumpolar Current are confined to some of these passages.

Together with the Shackleton Ridge, the South Shetland Islands arc forms a series of islands and submarine ridges surrounding the Scotia Sea. It is significant that the western part of the sea is much deeper than the eastern part. The characteristic depths are 4,000 m or more in the western part while in the eastern part they are generally less than 3,500 m (the depths are given according to the database in Smith and Sandwell (1997)). A threshold is located at a depth of 3,200 m in the eastern Scotia Sea. This threshold separates the western part of the sea from basins east of the Scotia Sea (Fig. 3.7). In the north, the threshold depth over the North Scotia Ridge is approximately 2,900 m in the Shag Rocks Passage (Zenk 1981; Walkden et al. 2008). The Georgia Passage (the threshold depth is 3,200 m) and Shag Rocks Passage is the place of the location of the southern boundary of the Antarctic Circumpolar Current and, correspondingly, the South Polar Front. The Shackleton Ridge, which limits the Scotia Sea from the west, crosses the entire Antarctic Circumpolar Current between the continental slopes of South America and Antarctica. It is separated from

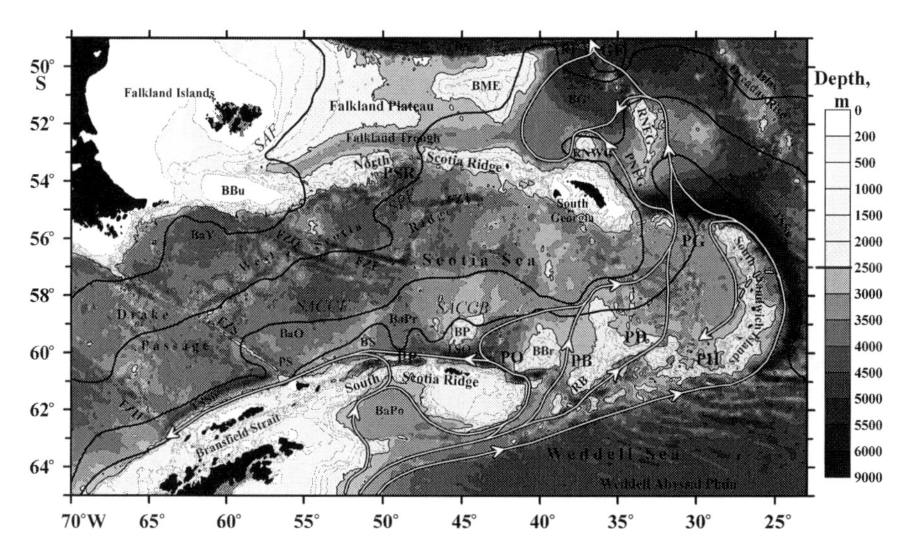

Fig. 3.7 Bottom topography in the Drake Passage and Scotia Sea. Fronts of the Antarctic Circumpolar Current (*ACC*): Subantarctic Front (*SAF*); South Polar Front (*SPF*); Southern Front of the ACC (*SACCF*); and Southern Boundary of the ACC (*SACCB*). *Arrows* show AABW spreading in this region according to Naveira Garabato et al. (2002a). Decoding of abbreviations is given in Table 3.2. (Modified and redrawn from Tarakanov 2009)

the Antarctic slope by a passage with a threshold depth of 3,300 m (hereinafter, the Shackleton Passage). In the southern part of the ridge, the depth is less than 2,500 m and the minimum depth is 750 m. The depth in the northern part is generally 2,000–3,200 m. The maximum depth is found in the passage between the continental foot-

Table 3.2 Decoding of geographical abbreviations in Fig. 3.7

BBu	Burdwood Bank	PB	Bruce Passage
BBr	Bruce Bank	PD	Discovery Passage
BME	Maurice Ewing Bank	PG	Georgia Passage
BP	Pirie Bank	PH	Hoyer Passage
BS	Shackleton Bank	PNEG	North East Georgia Passage
		PO	Orkney Passage
TSO	South Orkney Trough	PP	Phillip Passage
		PS	Shackleton Passage
TSSa	South Sandwich Trench		
TSSh	South Shetland Trench	FZE	Endurance Fracture
		FZH	Hero Fracture
BaA	Argentine Basin	FZQ	Quest Fracture
BaG	Georgia Basin	FZS	Shackleton Fracture
BaO	Ona Basin	FZT	Tehuelche Fracture
BaPo	Powell Basin		
BaPr	Protector Basin	RB	Bruce Ridge
BaY	Yaghan Basin	RF	Falkland Ridge
		RNEG	North East Georgia Ridge
GF	Falkland Gap	RNWG	North West Georgia Ridge

hills of South America (about 3,600 m). Water exchange from the southern part of the Scotia Sea with the Weddell Sea occurs through a few passages.

3.4.2 Deep and Bottom Water Masses and Previous Concepts of Circulation

For a long time, the South Sandwich Trench was considered the only transit pathway for Weddell Sea Deep Water to the northern regions of the Atlantic Ocean. After passing this trench, the bottom waters propagate to the Georgia Basin and are then transported to the Argentine Basin through the Falkland Gap in the Falkland Ridge (Fig. 3.7). Although penetration of Weddell Sea Deep Water to the Scotia Sea through the passages in the South Scotia Ridge has been known for a long time (Gordon 1966, 1967), the role of the Scotia Sea in the transport of Weddell Sea Deep Water to the north was underestimated. It was considered that the Scotia Sea is shallower than the adjacent basins of the Atlantic (Weddell Sea, South Sandwich Trench, and Georgia Basin) and, therefore, cannot serve as a transit pathway for Weddell Sea Deep Water. The essential role of the Scotia Sea in this process was reported first in Locarnini et al. (1993). Possibility of the northerly transport of Weddell Sea Deep Water from the Weddell Sea through the South Sandwich Abyssal Plain was shown in Orsi et al. (1999). South Sandwich Abyssal Plain is shown in Fig. 3.5.

According to present-day concepts, Weddell Sea waters propagate to the Scotia Sea through four passages in the South Scotia Ridge: Phillip, Orkney, Bruce, and Discovery. The Orkney Passage is the deepest one (3,500 m) and, therefore, considered the main passage (Naveira Garabato 2002a). Additional inflow of these waters to the eastern part of the sea can also occur through the Georgia Passage between the South Sandwich Islands and South Georgia (Meredith et al. 2008), as well as through passages between individual islands of the South Sandwich Islands (approximately 2,300 m deep) from the South Sandwich Trench.

After penetrating the passages in the South Scotia Ridge, Weddell Sea Deep Water is divided into two flows in the southern part of the Scotia Sea (Naveira Garabato 2002a) (Fig. 3.7). The western flow, which includes the waters from the Phillip Passage and partly Orkney Passage, propagates along the slope of the South Scotia Ridge and then along the slope of Antarctica. The major part of this flow is deflected to the north by the topographic features and then turned to the east with the Antarctic Circumpolar Current. It fills almost the entire abyssal part of the Scotia Sea (Locarnini et al. 1993). The eastern flow propagating from the Orkney Passage includes Weddell Sea Deep Water from the other passages in the South Scotia Ridge (Naveira Garabato 2002a).

Weddell Sea Deep Water is likely transported from the Scotia Sea through the following passages mentioned above (Locarnini et al. 1993): Shag-Rocks Passage (Wittstock and Zenk 1983; Walkden et al. 2008), Shackleton Passage (Nowlin and Zenk 1988), and Georgia Passage. In the first case, this flow (with the upper boundary of Weddell Sea Deep Water determined as $\theta=0.2°C$) from the Scotia Sea was considered insignificant or zero. Lack of Weddell Sea Deep Water in this passage was recently shown on the section occupied along the crest of the North Scotia Ridge (Smith et al. 2010). In the second case, Nowlin and Zenk (1988) showed that Weddell Sea Deep Water propagates through the Shackleton Passage to the South Shetland Trench in the southern part of the Drake Passage with the Antarctic Slope Current, which continues in the Pacific Ocean at least up to 120° W (Hollister and Heezen 1967). As to the third passage, transport of Weddell Sea Deep Water through this passage is currently considered comparable with the transport through the South Sandwich Trench and South Sandwich Abyssal Plain (Naveira Garabato et al. 2002b).

The Weddell Sea Deep Water transport through the passages in the South Scotia Ridge is estimated at 4.7 ± 0.7 Sv (Naveira Garabato et al. 2002b) and 5.9 ± 1.5 Sv (Naveira Garabato et al. 2003). A part of this water (0.4–0.8 Sv) is transported westward to the Drake Passage by the Antarctic Slope Current (Locarnini et al. 1993). Total transport of Antarctic Bottom Water to the north through the South Sandwich Trench and South Sandwich Abyssal Plain is estimated at 5.0 ± 4.3 Sv (Naveira Garabato et al. 2002b), including 1.1 ± 3.5 Sv transported to this region in the density range of Weddell Sea Deep Water and 3.9 ± 0.8 Sv transported initially in the layer of Weddell Sea Bottom Water and then under conditions of intense diapycnal mixing to Weddell Sea Deep Water. Northward transport through the South Sandwich Trench and South Sandwich Abyssal Plain is estimated as the value of Antarctic Bottom Water transport from the Indian Ocean sector of the Southern Ocean (Naveira Garabato et al. 2002b). According to Meredith et al. (2000), this inflow takes place in the density range of Weddell Sea Deep Water equal to 2.7 ± 0.9 Sv (Hoppema et al. 2001), which agrees with the results of calculations (2.5 Sv) reported in Schodlock et al. (2002). Rintoul (1998) argued that not more than 25% of the circumpolar production of Antarctic Bottom Water occurs in the Indian Ocean, although it is not clear what fraction of this is exported to the Weddell Gyre.

The arc formed by the South Sandwich Islands and South Georgia (Fig. 3.7) is an orographic barrier for the densest forms of Weddell Sea Deep Water. Therefore, only its upper layers can overflow through the Georgia Passage (Arhan et al. 1999). In the basins north of the Scotia Sea, properties of the upper and lower layers of Weddell Sea Deep Water, which propagated by different pathways through the Scotia Sea and South Sandwich Trench, respectively, are notably different. Recently, Weddell Sea Deep Water was divided on the basis of these properties into the Lower and Upper Weddell Sea Deep Waters (Arhan et al. 1999; Naveira Garabato 2002b; Meredith et al. 2008). The neutral density value $\gamma^n = 28.31$ is considered the threshold value (in the climatic sense) for overflowing the threshold in the Georgia Passage (Arhan et al. 1999). This value is assumed as the boundary between Upper and Lower Weddell Sea Deep Waters (Arhan et al. 1999; Meredith et al. 2008).

It is shown in Meredith et al. (2001, 2008) that traces of the more saline and warmer Weddell Sea Deep Water with density $\gamma^n > 28.31$ are found on the quasi-meridional section in the eastern part of the Scotia Sea, which penetrates the Georgia Passage. This inflow is irregular. It was detected only at one of the three sections (Meredith et al. 2008). According to Meredith et al. (2008), such irregularity is related to the variability of wind stress over the Weddell Gyre. Baroclinic properties of waters in the cyclonic Weddell Gyre increase in the periods of high wind stress. Thus, inclination of isopycnals to the gyre periphery increases, and penetration of the densest layers of Weddell Sea Deep Water through the passages in the South Scotia Ridge is hampered. At the same time, a flow of the denser Weddell Sea Deep Water, i.e., Lower Weddell Sea Deep Water from the South Sandwich Trench through the Georgia Passage to the Scotia Sea, is formed. On the contrary, when the wind stress weakens, basins of the southern part of the Scotia Sea are filled with the densest water, and inflow of Lower Weddell Sea Deep Water from the South Sandwich Trench terminates.

Circulation in the Scotia Sea and Drake Passage in the isopycnal density range of Circumpolar Deep Water is considered as easterly water transport by jets of the Antarctic Circumpolar Current. In the northern part of the basin, this transport includes the flow of a relatively fresh and cold Pacific Deep Water along the continental slope of South America. This water plays a specific role in smoothing the salinity maximum in the Lower Circumpolar Deep Water layer (Well et al. 2003). Cold and fresh waters from the Weddell Sea flow through the South Scotia Ridge into the southern part of the Scotia Sea and then get mixed with Circumpolar Deep Water. It is clear that strong eddy generation at the fronts of the Antarctic Circumpolar Current plays a significant role in this mixing, but the mechanism is not clear yet. We also note that part of Weddell Sea waters in the Scotia Sea is entrained into the Slope Antarctic Current directed to the west.

The southern part of the Scotia Sea is located south of the southern boundary of the Antarctic Circumpolar Current, thus it is included into the northwestern periphery of the Weddell Gyre (Orsi et al. 1999). It is noteworthy that waters of the circumpolar origin in the Weddell Gyre are called Warm Deep Water. Thus, Warm Deep Water, which includes the density ranges of Circumpolar Bottom Water and Lower Circumpolar Deep Water, spreads in the southern part of the Scotia Sea south of the south Antarctic Circumpolar Current boundary. Meridional minima of salinity and temperature are found in the Scotia Sea in this density range due to the easterly transport of Antarctic Shelf Water from the Bransfield Strait to the east of the Antarctic Peninsula (Whitworth et al. 1994). These minima become smoother with increasing distance in the eastern direction and almost disappear in the eastern part of the Scotia Sea.

3.4.3 Analysis of Recent Data

In this section we analyze the circulation and properties of abyssal waters in the Drake Passage and Scotia Sea based on the climatic dataset (Gouretski and Koltermann 2004), data of WOCE hydrographic sections, database WODB2005, and Russian expeditions in 2003–2007. Information about the data is given in Table 3.3 and their location scheme is shown in Fig. 3.8.

3.4.3.1 Data Description and Methods of Processing

The climatic dataset contains data on temperature, salinity, and oxygen, as well as data on other hydrochemical characteristics in the nodes of half-degree grid. At

Table 3.3 Hydrographic sections in the Drake Passage and Scotia Sea (Fig. 3.8)

Section (Fig. 3.8)	Notation (Fig. 3.8)	Time	Ship	Country
SR01a	▣	September 1992	R/V "Polarstern"	Germany
SR01g	◇	November 1994	R/V "Vidal Gormaz"	
SR01m	+	December 1997 – January 1998	RRS "James Clark Ross"	USA
A16s	▼	February 1989	R/V "Melville"	USA
A21_99	■	March-April 1999	RRS "James Clark Ross"	Great Britain
S04_99				
A23_99				
Falk_99				
Drk75a	×	February 1975	R/V "Atlantis II"	USA
Drk75b				
Drk75c				
Falk_80	◆	August–September 1980	R/V "Atlantis II"	USA
Drk80	✶	October 1980	R/V "Atlantis II"	USA
AJAX	▲	February 1984	R/V "Atlantis II"	USA
Drk93	✙	May 1993	R/V "Nathaniel B. Palmer"	USA
Drk95	▶	December 1995	R/V "Hesperides"	Spain
Drk96	◀	January 1996	R/V "Hesperides"	Spain
Dvtl97e	★	August 1997	R/V "Nathaniel B. Palmer"	USA
Dvtl97w				
Dvtl98	△	January 1998	R/V "Hesperides"	Spain
Drk03	●	December 2003	R/V "Akademik Sergey Vavilov"	Russia
Drk05a	●	November 2005	R/V "Akademik Ioffe"	Russia
Drk05b	⊞	November 2005	R/V "Akademik Ioffe"	Russia
Drk07	○	November 2007	R/V "Akademik Ioffe"	Russia

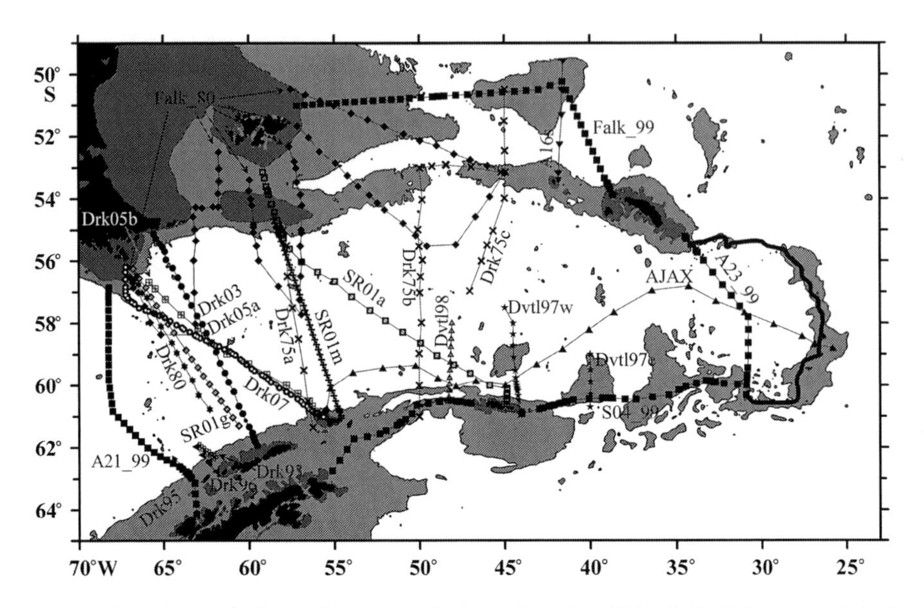

Fig. 3.8 Locations of hydrographic sections in the study region (Table 3.3). *Light gray and dark gray* colors show the regions shallower than 2,500 and 200 m, respectively. The *solid bold line* shows the eastern boundary of the study region in the Scotia Sea. (Modified and redrawn from Tarakanov 2009)

stations of the Russian sections, currents were measured using the Lowered Acoustic Doppler Current Profiler (LADCP). The data were processed using the approach in Visbeck (2002). The sector studied here is bounded by sections A21_99, S04_99, A23_99, and Falk_99 occupied in March and April 1999 under the Antarctic Large-Scale Box Analysis and the Role of the Scotia Sea (ALBATROSS) program. The other boundary is the arc of South Sandwich Islands (Fig. 3.8, Table 3.3). Figures 3.9–3.15 show the distribution of potential temperature, salinity, and neutral density γ^n, as well as boundaries between water masses over a few sections in this region of the ocean. Special attention is focused on the analysis of four Russian sections occupied near the Shackleton Fracture Zone (Figs. 3.10 and 3.12; locations of sections are shown in Fig. 3.8). Sections Drk03 and Drk05a were occupied along the same line from Tierra del Fuego to Elephant Island and they crossed the fracture in the central part of the Drake Passage. Sections Drk05b and Drk07 were occupied along the Shackleton Fracture Zone. The line of Drk07 passed along the crest, while the line of Drk05b passed northeast of the Shackleton Ridge. Figure 3.16a–c show velocity vectors, which are based on the data of current measurements by a LADCP averaged over the Circumpolar Bottom Water layer, and charts of absolute dynamic ocean level, which are available from http://www.aviso.oceanobs.com (NRT-MADT product).

In the series of papers (Koshlyakov and Tarakanov 1999, 2003a,b, 2004) dedicated to the study of water mass structure in the Pacific sector of the Southern Ocean, the authors described original methods for determining the boundaries be-

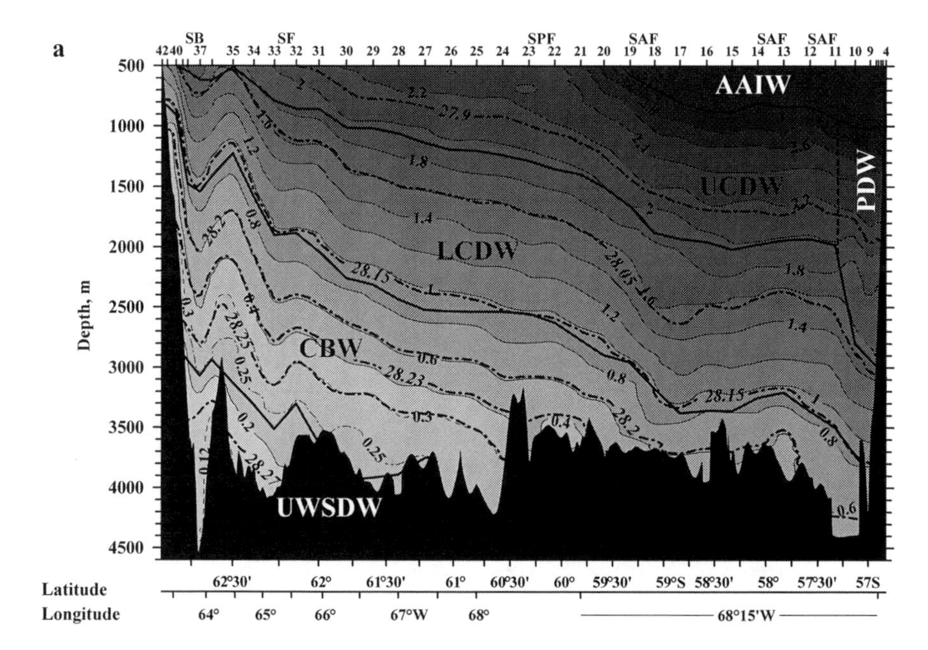

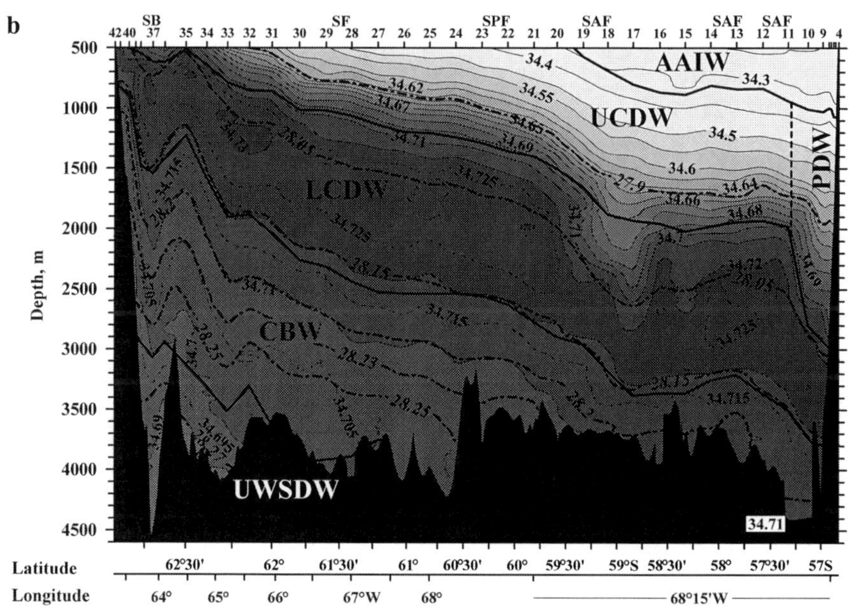

Fig. 3.9 Potential temperature, °C (**a**) and salinity, psu (**b**) over section A21_1999 (Fig. 3.8). *Solid bold lines* show the boundaries between water masses. *Dot-and-dash lines* show contours of neutral density γ^n. Numbers of stations are shown along the upper axis. Notations: (*SF*) Southern front of the ACC, (*SB*) Southern boundary of the ACC, (*SPF*) South Polar Front, (*SAF*) Subantarctic Front, (*PDW*) Pacific Deep Water

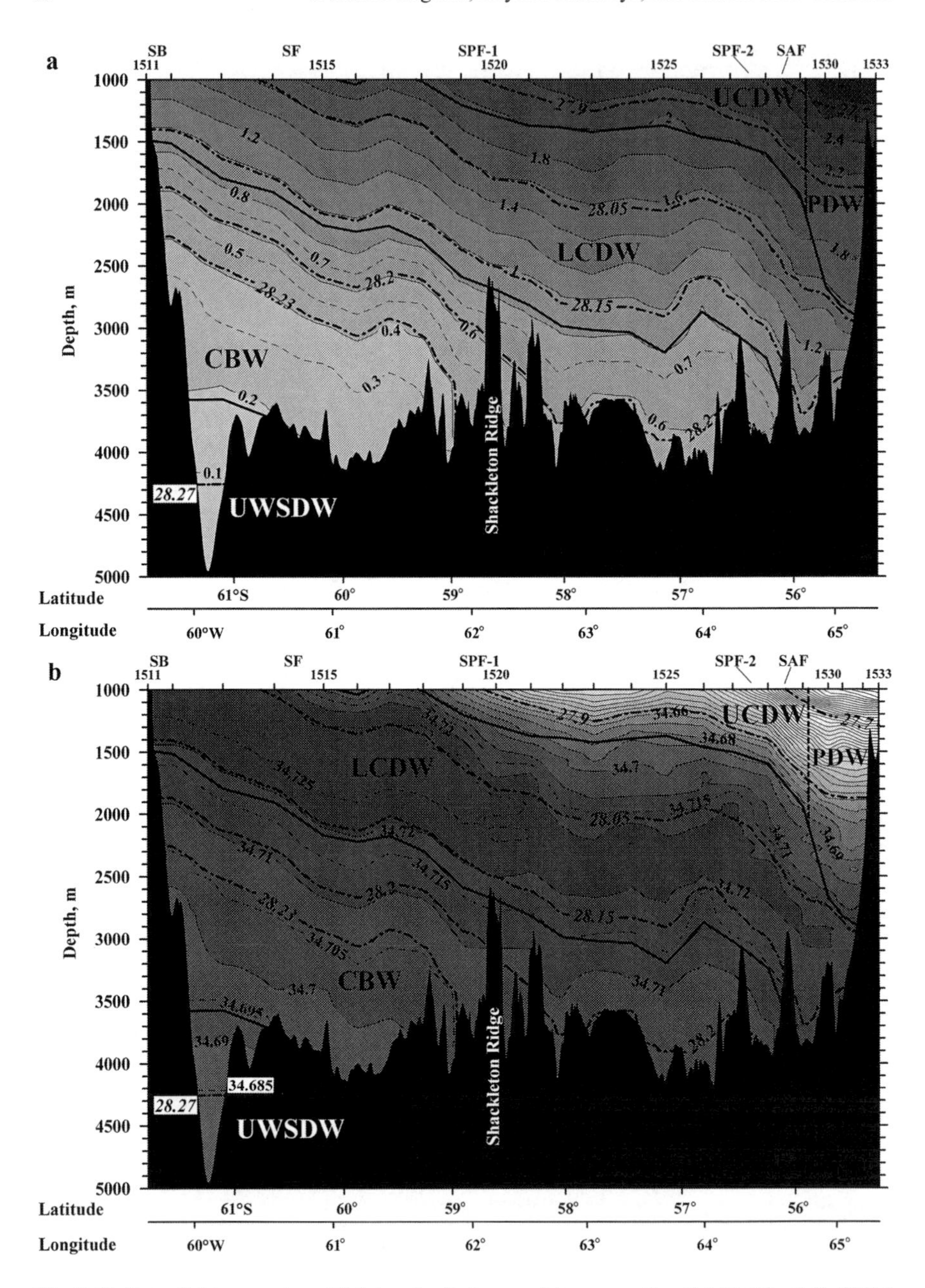

Fig. 3.10 Potential temperature, °C (**a**) and salinity, psu (**b**) over section Drk03 (Fig. 3.8). SPF-1 and SPF-2 are southern and northern branches of the South Polar Front, respectively. See Fig. 3.9 for other notations. (Modified and redrawn from Tarakanov 2010)

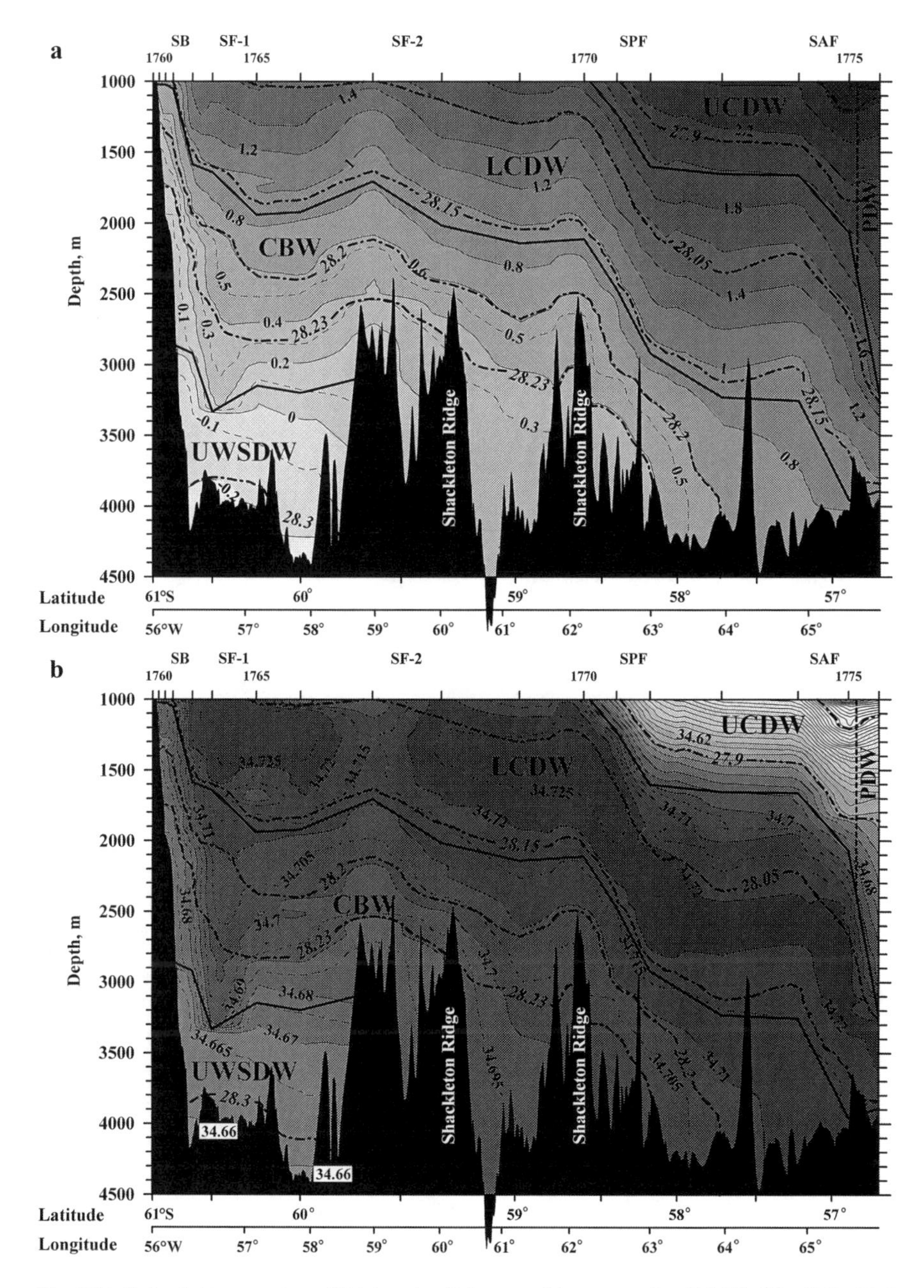

Fig. 3.11 Potential temperature, °C (a) and salinity, psu (b) over section Drk05b (Fig. 3.8). SF-1 and SF-2 are southern and northern ACC branches, respectively, in the Southern Front zone of the ACC. See Fig. 3.9 for other notations. (Modified and redrawn from Tarakanov 2010)

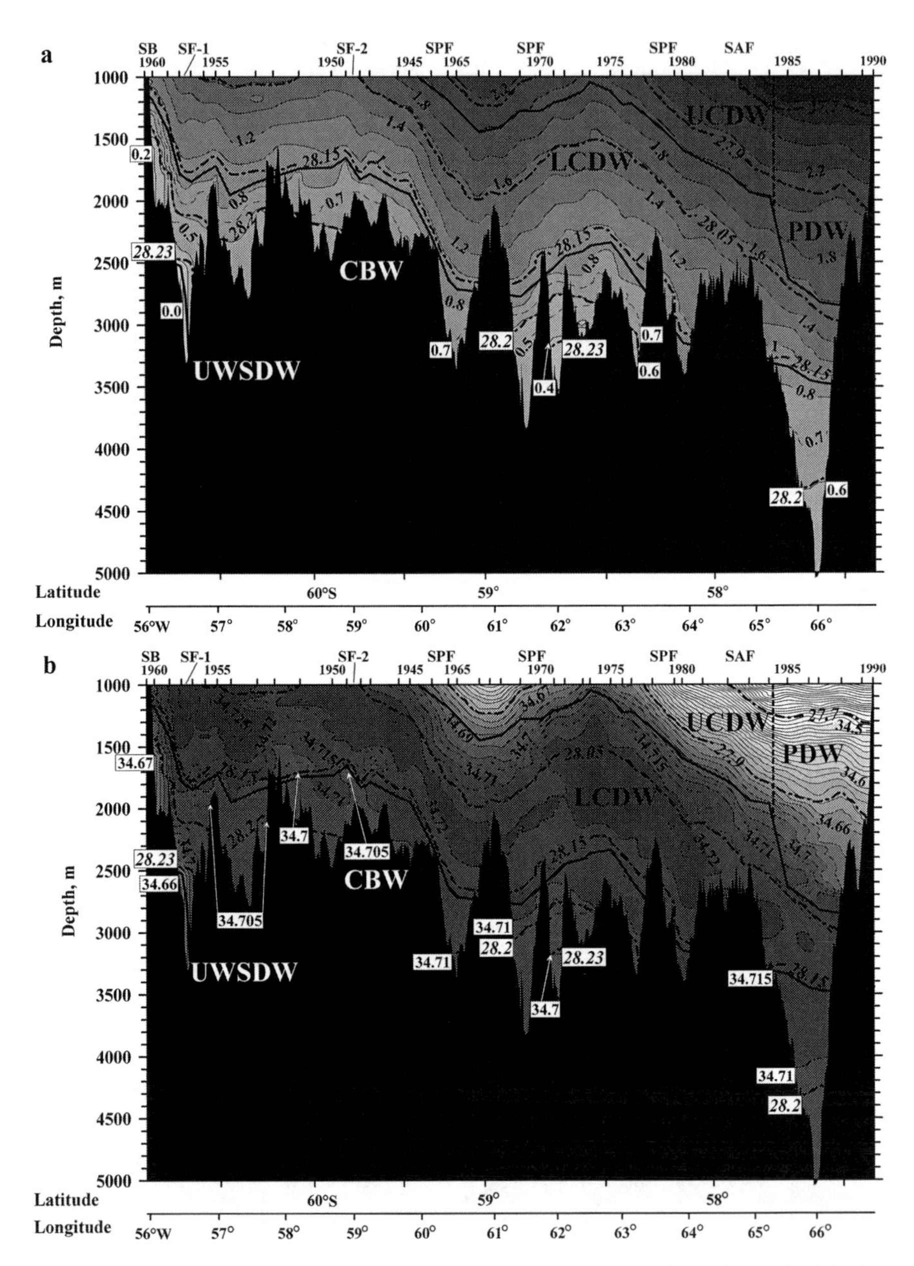

Fig. 3.12 Potential temperature, °C (**a**) and salinity, psu (**b**) over section Drk07 (Fig. 3.8). See Fig. 3.9 for other notations. (Modified and redrawn from Tarakanov 2010)

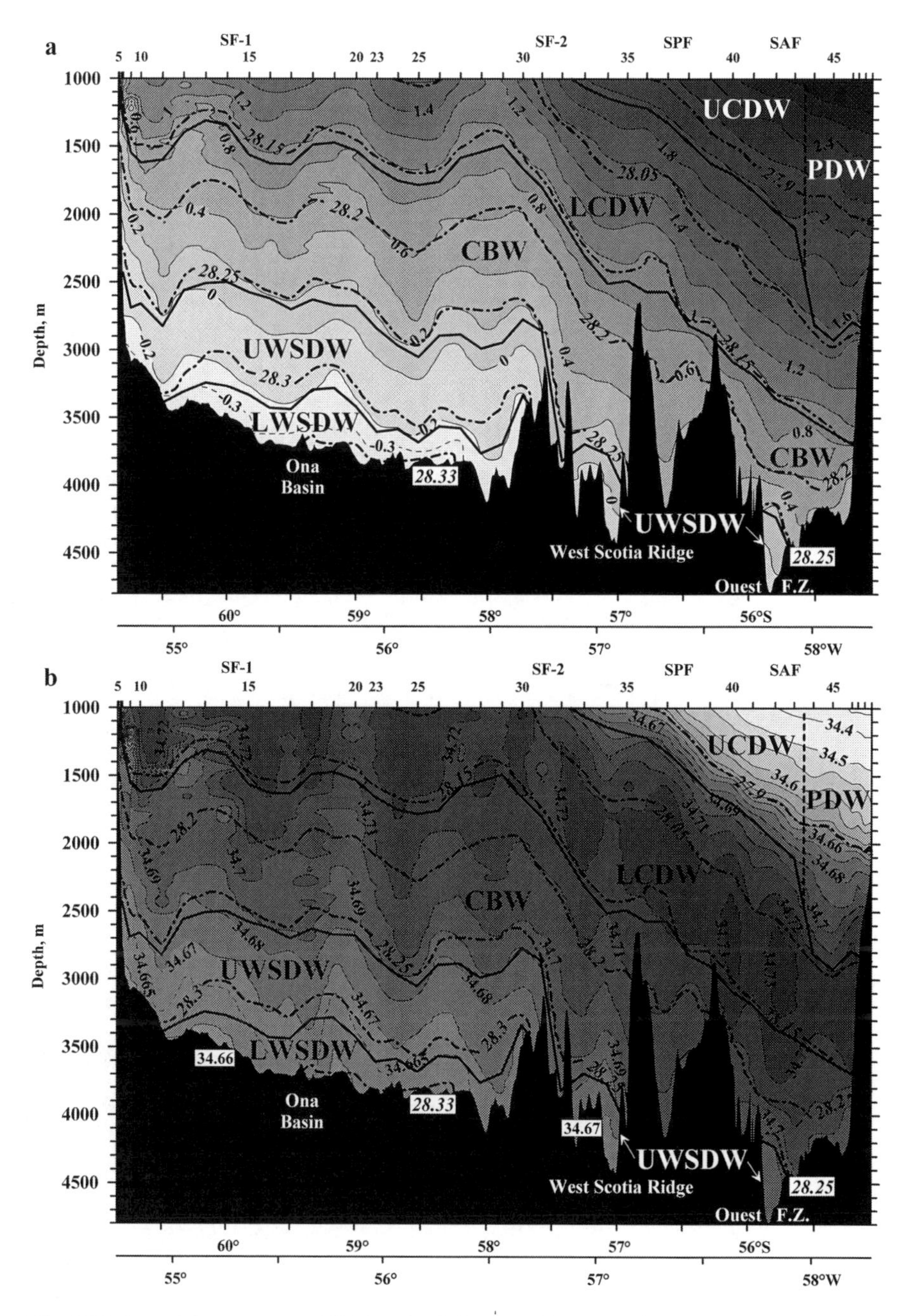

Fig. 3.13 Potential temperature, °C (**a**) and salinity, psu (**b**) over section SR01m (Fig. 3.8). SF-1 and SF-2 are southern and northern ACC branches, respectively, in the Southern Front zone of the ACC. See Fig. 3.9 for other notations

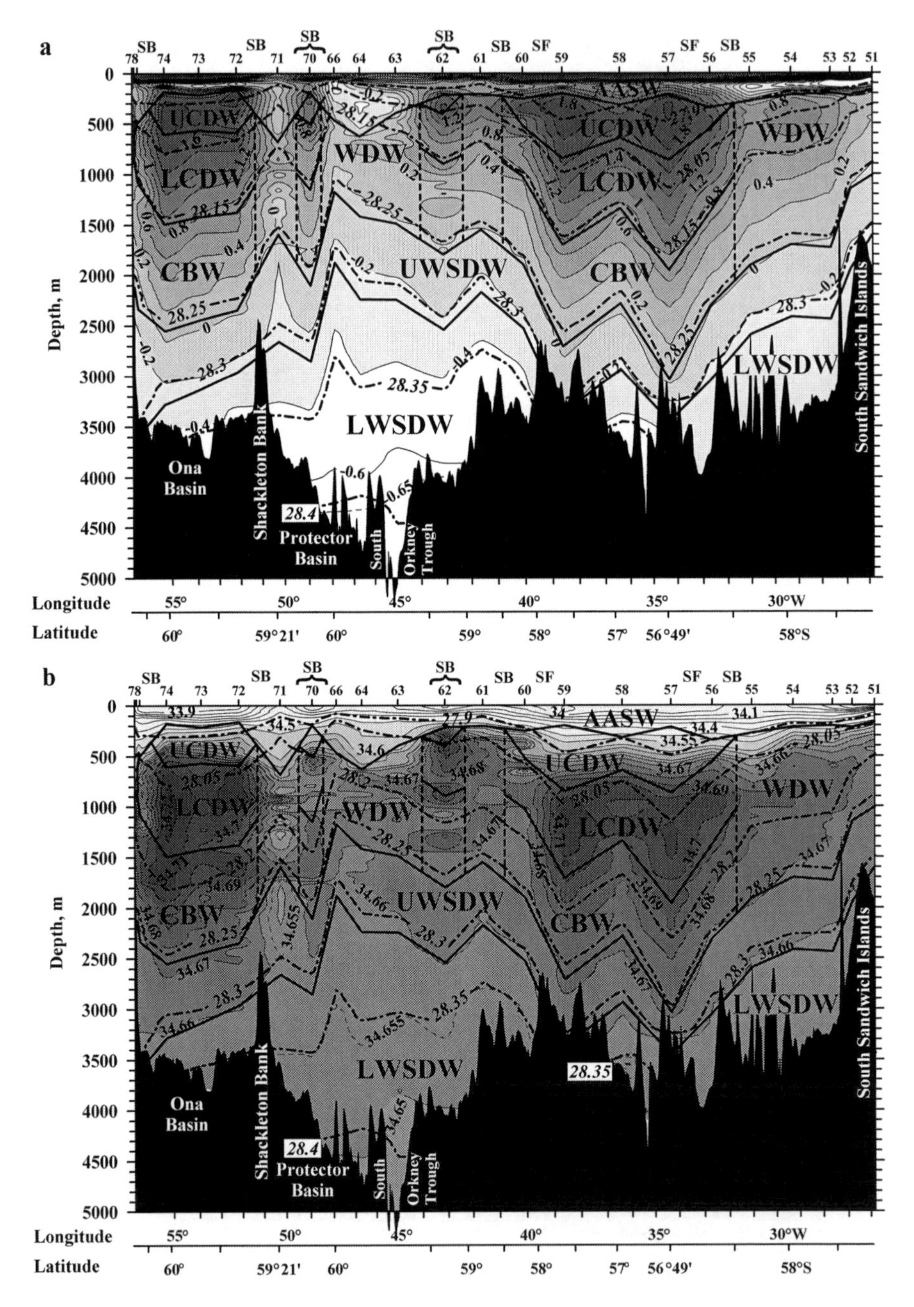

Fig. 3.14 Potential temperature, °C (**a**) and salinity, psu (**b**) over section AJAX (Fig. 3.8). See Fig. 3.9 for other notations. (Modified and redrawn from Tarakanov 2009)

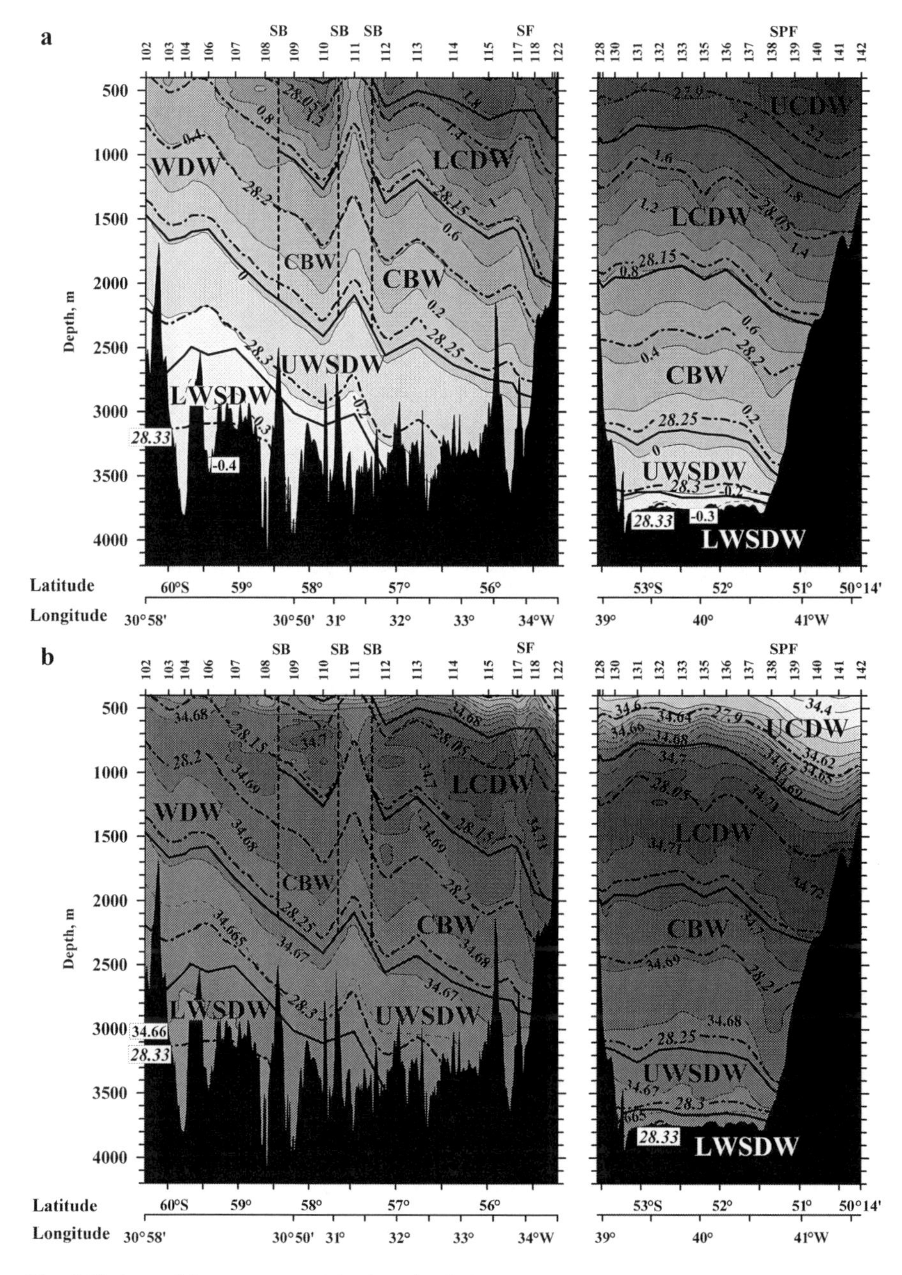

Fig. 3.15 Potential temperature, °C (**a**) and salinity, psu (**b**) over section A23_1999 and quasi-meridional part of section Falk_99 (Fig. 3.8). See Fig. 3.9 for other notations

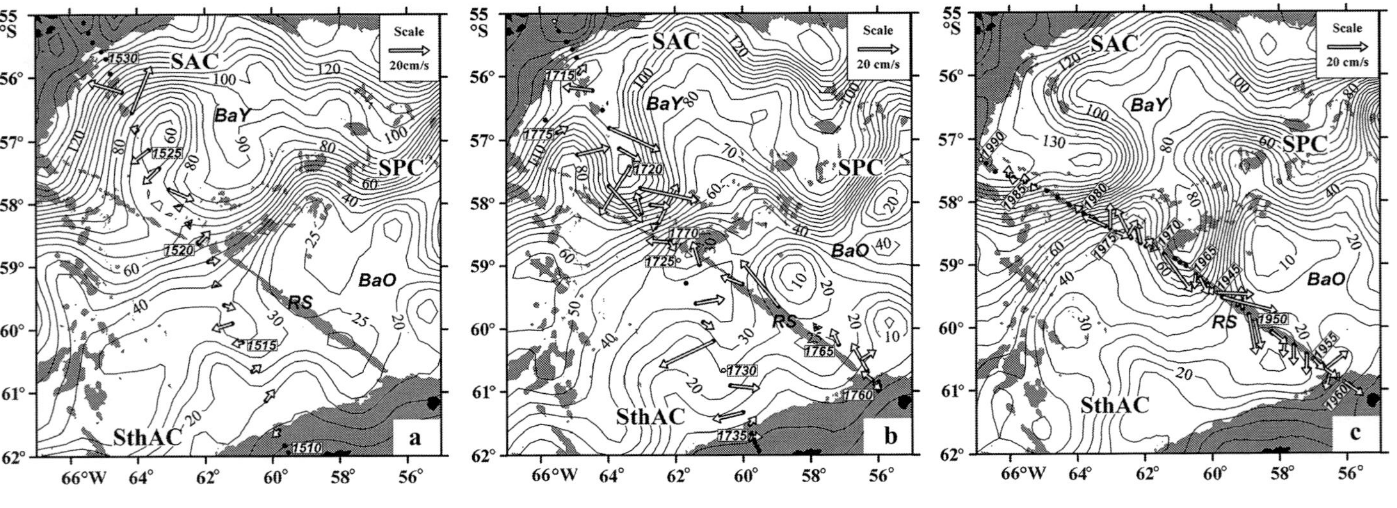

Fig. 3.16 Chart of absolute dynamic topography (cm) on December 13, 2003 (**a**), November 12, 2005 (**b**) and November 15, 2007 (c). *Arrows* show velocity vectors average over the layer of Circumpolar Bottom Water based on LADCP data for section Drk03 (**a**), Drk05a and Drk05b (**b**), and Drk07 (c). Notations: (*BaO*) Ona Basin, (*BaY*) Yaghan Basin, *RS* Shackleton Ridge. Abbreviations SthAC, SPC, and SAC denote South Antarctic Current, South Polar Current, and Subantarctic Current, respectively that are related to the corresponding fronts of the Antarctic Circumpolar Current. (Modified and redrawn from Tarakanov 2010)

tween water masses. They correlate the boundaries with the maximum of the vertical gradient of any physical or chemical characteristic. Detailed analysis of sections in the Scotia Sea demonstrated that it is impossible to apply any specific method for determining boundaries of Circumpolar Bottom Water and Weddell Sea Deep Water even within a small basin. This is possibly related to strong mesoscale variability and intense isopycnal and diapycnal mixing of Antarctic Circumpolar Current waters with the Weddell Sea waters in the Scotia Sea. Therefore, the upper boundary of Circumpolar Bottom Water, as well as the Circumpolar Bottom Water/Upper Weddell Sea Deep Water and Upper Weddell Sea Deep Water/Lower Weddell Sea Deep Water boundaries were determined over the major part of the sections based on contour lines of neutral density γ^n equal to 28.16, 28.26, and 28.31, respectively. We note that the first and second values of neutral density were the best to approximate the boundaries of Circumpolar Bottom Water in the Pacific sector of the Southern Ocean (Koshlyakov and Tarakanov 2003b). Only for sections west of the Shackleton Ridge, we used the method that relates the upper boundary of Circumpolar Bottom Water with the minimum of function

$$\Phi = \left(\frac{\partial\rho}{\partial S}\right)_{T,p}\frac{dS}{d\sigma_l} - 0.5 \equiv -\left(\left(\frac{\partial\rho}{\partial T}\right)_{S,p}\frac{d\theta}{d\sigma_l} - 0.5\right),$$

where σ_l is the local potential density:

$$\sigma_l = \sigma_0(0) + \int_0^z \left(\frac{d\sigma_{z'}}{dz'}\right)dz' \equiv \sigma_0(0) + \int_0^z \left(\left(\frac{\partial\rho}{\partial T}\right)_{S,p}\frac{d\theta}{dz'} + \left(\frac{\partial\rho}{\partial S}\right)_{T,p}\frac{dS}{dz'}\right)dz',$$

σ_z is potential density at reference level z'. Axis z is directed downward.

The upper boundary of Warm Deep Water at the southern periphery of the Scotia Sea was determined from the location of maximum of Φ. This water contacts at this boundary from above with Antarctic Surface Water and waters of the Bransfield Strait. These methods for determining boundaries of Circumpolar Bottom Water and Warm Deep Water are very close to the methods used in Koshlyakov and Tarakanov (2003b, 2004) based on the locations of maxima of the absolute values of salinity and potential temperature gradients. The methods also allow us to reduce the procedure of determining the boundaries between water masses to the analysis of one function Φ. The spatial location of boundaries in the region between the sections was determined using the algorithm described in Koshlyakov and Tarakanov (2003a). The boundaries of waters shown in Figs. 3.9–3.15 were also determined on the basis of this method.

Estimates of quantitative characteristics of Upper and Lower Weddell Sea Deep Waters, as well as Circumpolar Bottom Water and Warm Deep Water in the Drake Passage and Scotia Sea, are given in Tables 3.4 and 3.5. Figures 3.17 and 3.21 show boundaries of the spreading of these waters and the sector boundaries for calculating Tables 3.4 and 3.5. Figure 3.18 shows climatic salinity distribution at the ocean bottom. Figure 3.19 shows topography of the upper boundary of Upper Weddell Sea

Table 3.4 Volume, range of variations, and mean values of UWSDW and LWSDW characteristics in the Scotia Sea and Drake Passage

Ocean region	Volume, 10^4 km^3	Square, 10^5 km^2	——	θ, °C		S, psu		γ^n, kg/m^3		[O_2], ml/l	
LWSDW	13.8	2.5	Mean value	−0.30		34.660		28.324		5.32	
(2)			Range of variations	−0.70	0.00	34.650	34.675	28.260	28.410	5.05	5.85
LWSDW	24.8	4.1	Mean value	−0.29		34.659		28.320		5.39	
(3)			Range of variations	−0.75	0.00	34.645	34.675	28.265	28.420	5.15	5.90
LWSDW	0.1	0.1	Mean value	−0.28		34.663		28.319		5.23	
(4)			Range of variations	−0.45	−0.15	34.655	34.670	28.300	28.345	5.20	5.25
LWSDW	38.7	6.7	Mean value	−0.30		34.659		28.322		5.36	
(total)			Range of variations	−0.75	0.00	34.645	34.675	28.260	28.420	5.05	5.90
UWSDW	2.4	0.5	Mean value	0.12		34.686		28.265		4.99	
(1)			Range of variations	−0.15	0.45	34.665	34.705	28.240	28.295	4.80	5.25
UWSDW	31.5	5.7	Mean value	0.01		34.671		28.267		5.11	
(2)			Range of variations	−0.50	0.50	34.650	34.695	28.205	28.330	4.85	5.40
UWSDW	26.3	5.1	Mean value	−0.08		34.666		28.276		5.18	
(3)			Range of variations	−0.35	0.20	34.655	34.685	28.230	28.320	4.95	5.40
UWSDW	1.0	0.3	Mean value	−0.07		34.671		28.281		5.14	
(4)			Range of variations	−0.30	0.15	34.660	34.680	28.245	28.325	5.00	5.25
UWSDW	61.2	11.6	Mean value	−0.02		34.670		28.271		5.14	
(total)			Range of variations	−0.50	0.50	34.650	34.705	28.205	28.330	4.80	5.40

Table 3.5 Volume, range of variations, and mean values of Circumpolar Bottom Water and Warm Deep Water characteristics in the Scotia Sea and Drake Passage

Ocean region	Volume, 10^4 km^3	Square, 10^5 km^2	——	θ, °C		S, psu		γ^n, kg/m^3		[O_2], ml/l	
CBW	28.3	2.3	Mean value	0.54		34.706		28.208		4.80	
(1)			Range of variations	0.05	1.25	34.655	34.740	28.060	28.275	4.75	5.10
CBW	75.1	7.4	Mean value	0.57		34.698		28.194		4.84	
(2)			Range of variations	0.00	1.25	34.650	34.725	28.080	28.265	4.45	5.20
CBW	32.5	3.3	Mean value	0.37		34.682		28.205		4.94	
(3)			Range of variations	−0.10	0.95	34.665	34.710	28.125	28.275	4.60	5.20
CBW	9.6	1.7	Mean value	0.55		34.697		28.197		4.81	
(4)			Range of variations	−0.10	1.20	34.670	34.720	28.110	28.265	4.50	5.15
CBW	145.5	14.7	Mean value	0.52		34.696		28.199		4.85	
(total)			Range of variations	−0.10	1.25	34.650	34.740	28.060	28.275	4.45	5.20
WDW	7.4	0.6	Mean value	0.28		34.654		28.167		5.02	
(5)			Range of variations	−0.05	0.75	34.600	34.685	28.060	28.255	4.75	5.30
WDW	20.6	1.3	Mean value	0.29		34.661		28.180		4.95	
(6)			Range of variations	−0.10	0.85	34.585	34.680	28.030	28.270	4.70	5.20
WDW	15.5	1.1	Mean value	0.36		34.667		28.178		4.88	
(7)			Range of variations	−0.05	1.10	34.590	34.690	28.030	28.270	4.65	5.10
WDW	43.5	3.0	Mean value	0.31		34.662		28.177		4.94	
(total)			Range of variations	−0.10	1.10	34.585	34.690	28.030	28.270	4.75	5.30

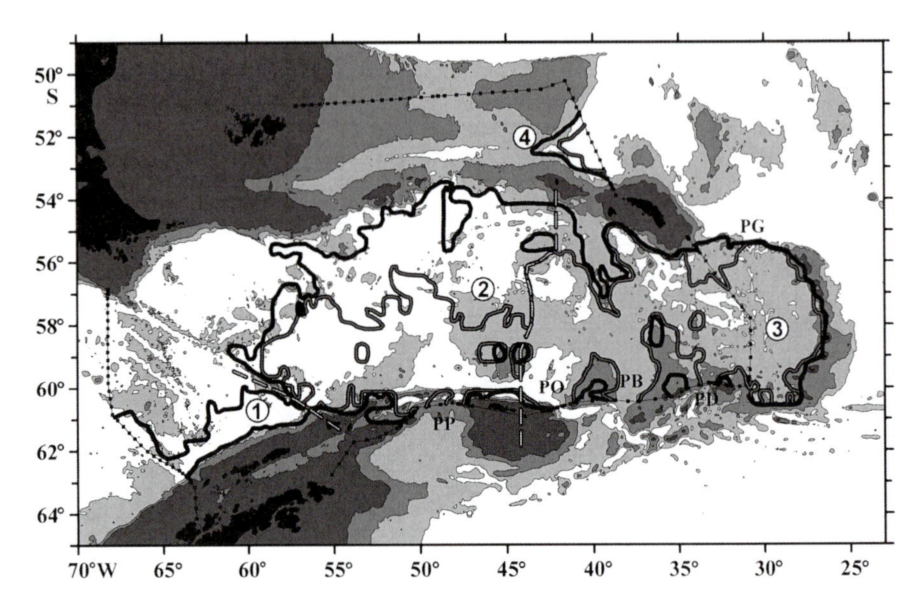

Fig. 3.17 Limits of UWSDW and LWSDW spreading in the Drake Passage and Scotia Sea (*black and dark gray solid lines*, respectively). *Dashed lines* and numerals in *circles* denote boundaries and numbers of sectors (Table 3.3). Isobaths 1,000, 2,500, and 3,500 m are shown. See Fig. 3.7 for other notations. (Modified and redrawn from Tarakanov 2009)

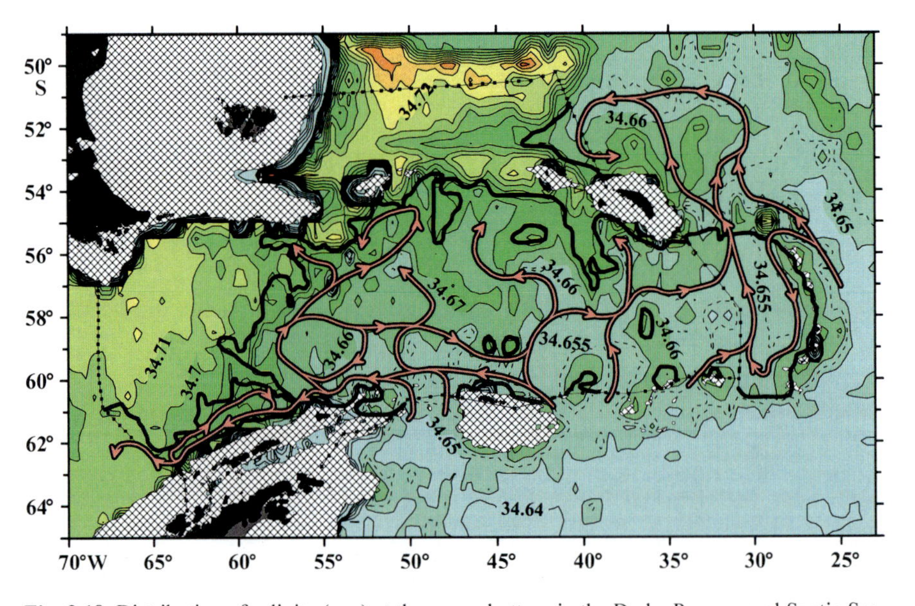

Fig. 3.18 Distribution of salinity (psu) at the ocean bottom in the Drake Passage and Scotia Sea. Regions with depths shallower than 1,000 m are hashed. *Arrows* show the authors' scheme of WSDW spreading in the bottom layer. See Fig. 3.7 for other notations

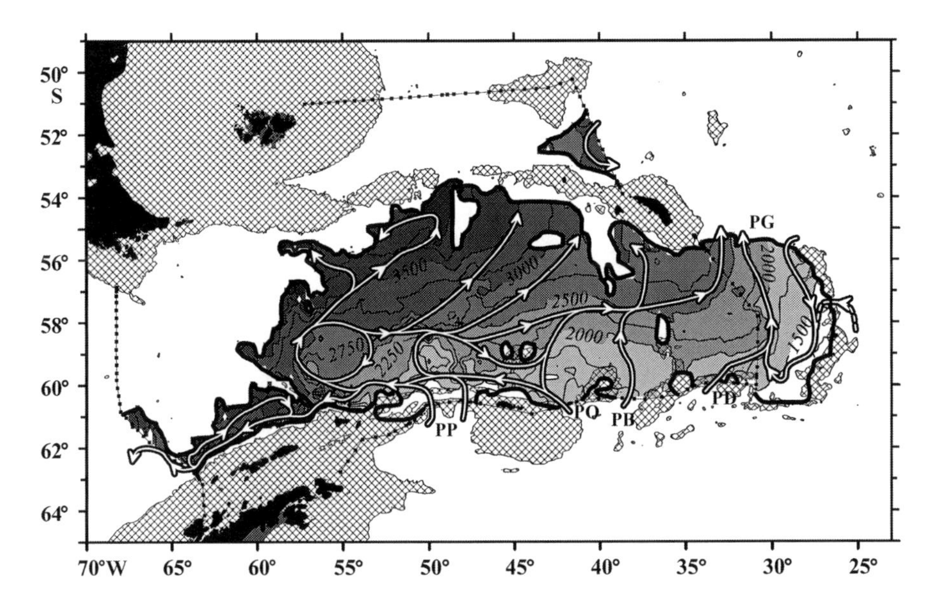

Fig. 3.19 Topography of the upper boundary surface of Upper Weddell Sea Deep Water (m) in the Drake Passage and Scotia Sea. Regions with depths shallower than 2,000 m are hashed. *Arrows* show the authors' scheme of UWSDW spreading near the upper boundary surface. See Figs. 3.7 and 3.9 for other notations

Deep Water. Figure 3.22 shows topography of the upper boundary of Circumpolar Bottom Water and Warm Deep Water. The schemes of spreading of Weddell Sea Deep Water (Figs. 3.18 and 3.19), Circumpolar Bottom Water, and Warm Deep Water (Fig. 3.22) are based on the detailed analysis of data in this region and published data (Fig. 3.7) (Arhan et al. 1999, 2002b; Nowlin and Zenk 1988; Orsi et al. 1993, 1999; Naveira Garabato et al. 2002a; Locarnini et al. 1993); distributions shown in Figs. 3.18, 3.19, and 3.22; peculiarities of bottom topography; and direct LADCP measurements of currents. Differences in transport schemes of lower and upper layers of Weddell Sea Deep Water in Figs. 3.18 and 3.19, respectively, are related to the following fact: the motion of bottom layers (Fig. 3.18) is governed by the forms of bottom topography, whereas the overlying waters are strongly influenced by jets of the Antarctic Circumpolar Current (Fig. 3.19). It is noteworthy that the horizontal scale of mesoscale (synoptic) variability of the ocean is comparable with the horizontal sizes of individual basins in the study region. Therefore, schemes shown in Figs. 3.18 and 3.19 should be considered as schemes of spreading of Weddell Sea Deep Water rather than mean current of this water.

3.4.3.2 Upper and Lower Weddell Sea Deep Water

As shown in Fig. 3.17, Lower Weddell Sea Deep Water in the Scotia Sea does not spread west of the Shackleton Ridge (the elevated part of the Shackleton Fracture

Zone). In the northern direction, it spreads much farther than the Southern Front of the Antarctic Circumpolar Current. The Lower Weddell Sea Deep Water region is divided by the Pirie Bank (BP in Fig. 3.7) into two virtually isolated sectors, which are connected only by the South Orkney Trough. Weddell Sea Deep Water is found practically in the entire Scotia Sea excluding the Yaghan Basin (Figs. 3.7 and 3.17). West of the Shackleton Ridge, Weddell Sea Deep Water is found only in the South Shetland Trench and a deep depression in the Hero Fracture Zone (Figs. 3.7 and 3.17).

Since Weddell Sea Deep Water is fresher than the overlying waters, the tongues of low salinity at the bottom should indicate directions of Weddell Sea Deep Water spreading from the sources of its propagation to the Scotia Sea. Together with the chart of bottom topography (Fig. 3.7), topography of the upper boundary surface (Fig. 3.19) gives information about the thickness of Weddell Sea Deep Water layer in the Scotia Sea. Minimum depths (1,250–1,750 m) of the upper boundary surface and bottom salinity (<34.655 psu) in the southern part of the Scotia Sea (Figs. 3.18 and 3.19) correspond to Weddell Sea Deep Water propagating through the passages in the South Scotia Ridge.

Tongues of low salinity at the bottom (<34.66 psu) extending from this region to the east and west (Fig. 3.18) show the main directions of Weddell Sea Deep Water propagation in the Scotia Sea. The tongues of higher salinity (but less than 34.670 psu) in the northern part of the sea mark the propagation of Weddell Sea Deep Water to the southeast and north in the zone between the South Polar Front and Southern Front of the Antarctic Circumpolar Current. Deflection to the west from this direction in the northwestern part of the Scotia Sea (Figs. 3.18 and 3.19) fits the tongue of low salinity related to deep depressions in the Endurance and Quest fracture zones. Salinity distribution at the bottom in the Georgia Basin is associated with anticyclonic bottom circulation in this region (Fig. 3.18). It is worth noting that this conclusion agrees with the results of LADCP measurements over the quasi-meridional part of section Falk_99 (Arhan et al. 2002b). Distributions of salinity and potential temperature over this section are shown in Fig. 3.15.

The AJAX section passes along the entire part of the Scotia Sea from west to east from Elephant Island to the South Sandwich Islands and follows practically along the pathways of Weddell Sea Deep Water to the west and east (Figs. 3.8 and 3.14). It is seen from Fig. 3.14 that thermohaline properties of two Weddell Sea Deep Water flows propagating to the west from the Protector Basin (BaPr in Fig. 3.7) and to the east from the Orkney Passage are almost similar. Upwelling of Weddell Sea Deep Water up to 1,000 m is observed in the eastern part of the AJAX section near the South Sandwich Islands (Fig. 3.14) corresponding to an isolated elevation (to less than 1,500 m) at the upper boundary of Weddell Sea Deep Water (Fig. 3.19). This fact evidences a possibility of Weddell Sea Deep Water inflow to the Scotia Sea, in addition to the aforementioned pathways through the Hoyer Passage with depths exceeding 2,000 m southwest of these islands (Fig. 3.7).

The westward flow of Weddell Sea Deep Water over the major part of sections in the Drake Passage and in the western part of the Scotia Sea (Dvtl97w, SR01a, Drk75b, western part of AJAX, SR01m, Drk05b, and A21_99) has a bimodal structure. Several stations are located near the Antarctic slope, where Weddell Sea Deep Water is characterized by low temperature and salinity compared to the offshore stations (Figs. 3.9, 3.11, 3.13, and 3.14). According to Naveira Garabato et al. (2002a), such structure of the western flow is related to the propagation of two types of Weddell Sea Deep Water with different scenarios of formation to the Scotia Sea. The young (cold and fresh) variety of Weddell Sea Deep Water formed in the western part of the Weddell Sea propagates to the Scotia Sea through the Philip Passage and partly through the Orkney Passage. The older (warm and saline) variety of Weddell Sea Deep Water, which propagates closer to the central part of the Weddell Gyre, propagates to the Scotia Sea through the Orkney, Bruce, and Discovery passages, and is later divided into the western and eastern flows (Naveira Garabato et al. 2002a). It was shown in Meredith et al. (2003) that convection was developing over the Antarctic slope north of Elephant Island (South Shetland Islands) as a result of ice formation during winter in the 1990s, resulting in the formation of water with a temperature of −0.9°C at a depth of 1,040 m in the years marked by extreme conditions. The density of descending water can reach the density range of Weddell Sea Deep Water. This water flowing to the west as plumes can also form a bimodal structure in the western Weddell Sea Deep Water flow in the Drake Passage.

As was mentioned above, the major part of the western flow of Weddell Sea Deep Water deflects to the north due to topographic obstacles in the western part of the Scotia Sea and then the flow turns to the east as a flow in the opposite direction. Such topographic obstacles for the westerly flow of Weddell Sea Deep Water are represented by the Shackleton Bank in the western part of the Protector Basin and the Shackleton Ridge (the depth over the ridge is generally less than 2,500 m). Indeed, the northwestern direction of Weddell Sea Deep Water motion along the northeastern slope of the Shackleton Ridge was found in 2005 on section Drk05b (Figs. 3.7 and 3.11) based on LADCP measurements.

Two main obstacles for the current exist along the pathway of the Antarctic Circumpolar Current through the Scotia Sea in the Southern Antarctic Current zone. Their depths only slightly exceed 3,000 m. The first is the Shackleton Ridge (Fig. 3.7). Two jets of the current overflow it. The southern jet passes through the Shackleton Passage (Barre et al. 2008). The second (quite large) region of the sea is located between the Bruce Bank and Georgia Passage (Fig. 3.7). Thus, owing to the fact that on the sections in the Scotia Sea the upper boundary of Weddell Sea Deep Water in the Southern ACC Front region is located at a depth of 2,500–3,000 m (Figs. 3.10–3.14), we can suppose that the upper layers of Weddell Sea Deep Water transported by the westerly Slope Antarctic Current are entrained by the southern branch of the Antarctic Circumpolar Current and transported from the western part of the Scotia Sea to its eastern part.

Naturally, Weddell Sea Deep Water could mix in the course of its motion with the overlying warm and saline circumpolar water. As a result, this flow of Weddell Sea Deep Water should be characterized by higher values of temperature and salinity by the moment of joining with Weddell Sea Deep Water flowing from the passages in the South Scotia Ridge directly to the east. Indeed, the warmer and more saline form of Weddell Sea Deep Water is observed in the eastern part of the Scotia Sea on sections AJAX and A23_99 at the northern periphery of the eastern flow of Weddell Sea Deep Water (Figs. 3.14 and 3.15). Based on this fact, the eastern direction of Weddell Sea Deep Water transport in the entire Scotia Sea in the Southern ACC Front zone is shown in Fig. 3.15. This pathway of Weddell Sea Deep Water motion is an additional (previously undescribed) transit pathway of transport in the Scotia Sea to the northern basins of the Atlantic Ocean.

Propagation of Weddell Sea Deep Water into the northern part of the Scotia Sea is likely governed both by the South Polar Front and bottom topography. Deep depressions of the bottom play important roles in this process: the rift valley of the West Scotia Ridge and Tehuelche and Endurance, and Quest fractures (Fig. 3.7). Quasi-meridional section SR01m located in the western part of the Scotia Sea crosses the northwesterly flow of Weddell Sea Deep Water in the Quest Fracture Zone (Fig. 3.13).

Weddell Sea Deep Water was found only at one station on section Drk07 occupied along the Shackleton Ridge crest (Figs. 3.8 and 3.12). The station was located at the southern side of the Shackleton Passage. The upper boundary of Weddell Sea Deep Water with a temperature of 0.00°C and salinity 34.666 psu was found at a depth of 2,500 m. The westerly transport of Weddell Sea Deep Water was estimated at 0.5 Sv, which agrees with the above-mentioned estimate 0.4–0.8 Sv. The reverse easterly transport in the northern part of the passage occurs in the circumpolar water layer. We also note that on all sections in the South Shetland Trench the depth of the upper boundary of Weddell Sea Deep Water location exceeds the threshold depth of the Shackleton Passage. This result agrees with the modern concepts that Weddell Sea Deep Water is transferred to the density range of circumpolar waters due to diapycnal mixing in the Drake Passage and only then transported to east by the Antarctic Circumpolar Current (Orsi et al. 1999).

It is likely that Weddell Sea Deep Water circulation in the South Shetland Trench is cyclonic quasi-isolated. This can be seen from a number of direct and indirect indications:

1. The South Shetland Trench with a threshold depth of about 4,000 m (Fig. 3.7) is a region of closed geostrophic contours f/H (f is the Coriolis parameter, H is the ocean depth). In these conditions, any upwelling over the trench would stimulate intense cyclonic circulation in the weakly stratified abyssal waters of the trench along the geostrophic contours (Johnson 1998). Upwelling is compensated by a water flow crossing the geostrophic contours in the bottom Ekman friction layer. We note that Weddell Sea Deep Water in the South Shetland Trench is distinguished for high density homogeneity, which can be characterized by quasi-conservative parameter fN^2 (N is the Brunt-Väisälä frequency) compared with the Ona Basin. The value of fN^2 is $1–2 \times 10^{-10}$ s^{-3} at stations in the trench and $3–6 \times 10^{-10}$ s^{-3} in the same isopycnal interval in the Ona Basin.
2. In addition to the westerly flow of Weddell Sea Deep Water, an easterly reverse current of Weddell Sea Deep Water was found on the northern slope of the trench using an LADCP instrument on sections Drk03 and Drk05a (Fig. 3.8). The currents in the easterly flow reached 10 cm s^{-1}.

3. If we assume that only the westerly flow of Weddell Sea Deep Water exists in the Southern Shetland Trench and take into account the available estimates of the transport volume (Table 3.4) and the Weddell Sea Deep Water transport through the Shackleton Passage, the renewal time of this water in the trench would be equal to 1.5 years. The length of the trench is approximately 400 km. Thus, the mean velocity of the westerly flow should be less than 1 cm s^{-1}. At the same time, velocities of the mean current along the Antarctic Slope in the Southern Shetland trench measured on moorings at depths of 2,700 and 3,590 m were equal to 10–20 cm s^{-1} (Nowlin and Zenk 1988).

The next obstacle on the pathway of the westerly flow of Weddell Sea Deep Water after the Shackleton Ridge is a 3,000 to 3,500-m-deep bank located west of the Hero Fracture Zone (Fig. 3.7). Since the depth of the upper boundary of Weddell Sea Deep Water in the South Shetland Trench exceeds 3,000 m, the slope flow of Weddell Sea Deep Water should be deflected to the north along the isobaths. This property of the Weddell Sea Deep Water slope current was found on section A21_99 (Figs. 3.8 and 3.9) that was occupied along the trough of the Hero Fracture Zone dividing this aforementioned bank and the ridge of the fracture zone, the depths over which are less than 3,000 m.

As was mentioned above, it is commonly considered that the westerly flow of Weddell Sea Deep Water does not reach the Pacific Ocean. However, traces of the cooler and fresher Weddell Sea Deep Water, as compared to the warmer and more saline Pacific type of Antarctic Bottom Water, were found, for example, in the eastern part of section S04P (see inset in Fig. 3.20) occupied in February and March of 1992 during the cruise of R/V *Akademik Ioffe* (Koshlyakov and Sazhina 1995). Water in the layer of $\gamma^n > 28.26$ with salinity lower by 0.001 psu and cooler by 0.005°C than the surrounding waters of this density range was found at station 690 close to Adelaide Island at a depth greater than 3,000 m (Fig. 3.20). The westerly direction of the flow at the bottom with a velocity of about 2 cm s^{-1} at this part of section S04P was reported in (Koshlyakov and Sazhina 1995).

Data in Fig. 3.9 indicate that characteristics of Lower Weddell Sea Deep Water in the western (sector 2, Fig. 3.17) and eastern (sector 3) parts of the Scotia Sea are very close to each other. The estimate of the total volume of Lower Weddell Sea Deep Water (Table 3.4) is slightly greater than the value $3.1 \pm 0.8 \times 10^4$ km^3 given in Meredith et al. (2008). Renewal time of Lower Weddell Sea Deep Water in the Scotia Sea at the known transport through the South Scotia Ridge (3.5 ± 1.2 Sv) (Meredith et al. 2008) is approximately 3.5 years, which takes into account the isolation of this water in the Scotia Sea from the Atlantic basins, and is the total time of Lower Weddell Sea Deep Water transition to the density range of Upper Weddell Sea Deep Water.

Characteristics of Upper Weddell Sea Deep Water are slightly different in the western and eastern parts of the Scotia Sea. Higher values of salinity and potential temperature in the western part of the sea are likely related to the fact that the region of spreading of Upper Weddell Sea Deep Water in this part of the sea is much greater than the spreading region of Lower Weddell Sea Deep Water. In addition, the region of spreading of the densest layers of Upper Weddell Sea Deep Water is

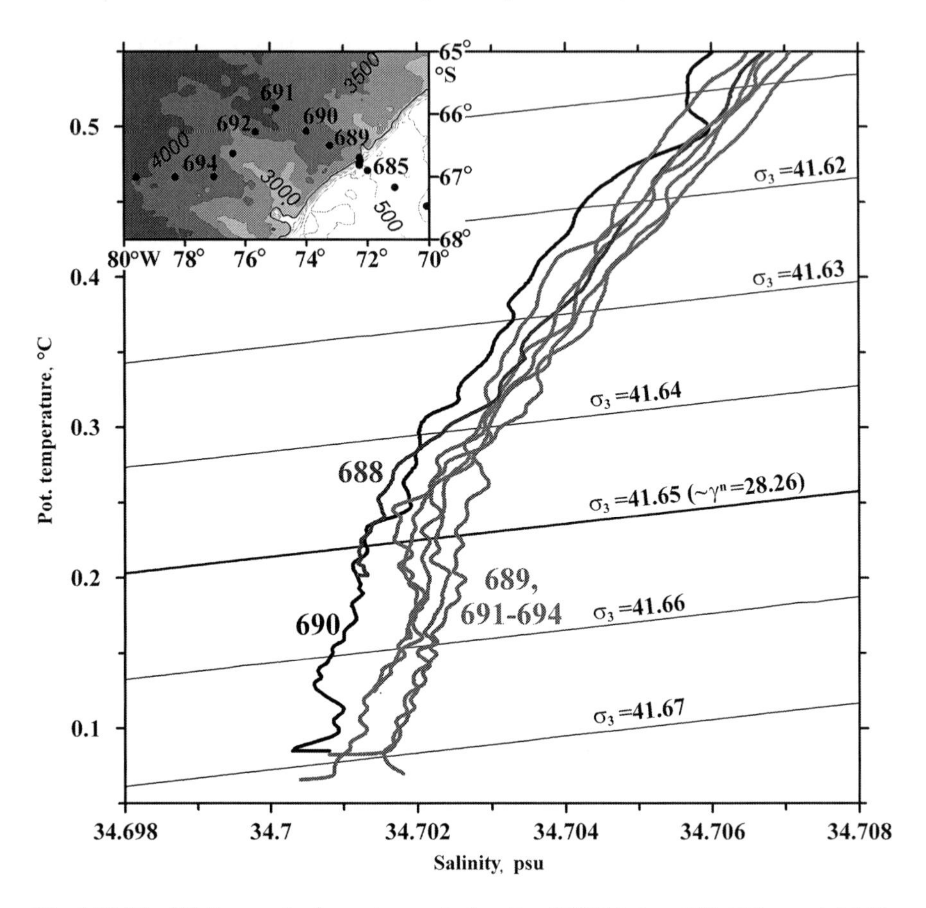

Fig. 3.20 The θ/S-diagram for the eastern part of section S04P (stations 682–696) near Adelaide Island in the Pacific Ocean. *Bold line* emphasizes isopycnal $\sigma_3=41.65$ ($\sim\gamma^n=28.26$) assumed as the upper boundary of Weddell Sea Deep Water. Scheme of stations along section S04P and bottom topography is shown in the *inset*. (Modified and redrawn from Tarakanov 2009)

strongly limited. This is seen from the fact that the neutral density is lower in the eastern part of the Scotia Sea.

Significantly higher values of potential temperature and salinity in the Southern Shetland Trench (sector 1, in Fig. 3.17) at almost the same values of neutral density as in the western part of the Scotia sea (sector 2) provide evidence of strong mixing of Upper Weddell Sea Deep Water in the trench with the overlying warmer and more saline circumpolar waters. Quasi-isolated cyclonic circulation in the trench facilitates the efficient mixing of these waters. Estimate of the total volume of Weddell Sea Deep Water (UWSDW + LWSDW) in the Scotia Sea and Drake Passage is 9.9×10^4 km^3 is also slightly greater than the value $8.7\pm1.9\times10^4$ km^3 given in Meredith et al. (2008). Total renewal time of Weddell Sea Deep Water in the Scotia Sea and Drake Passage is approximately 5 years.

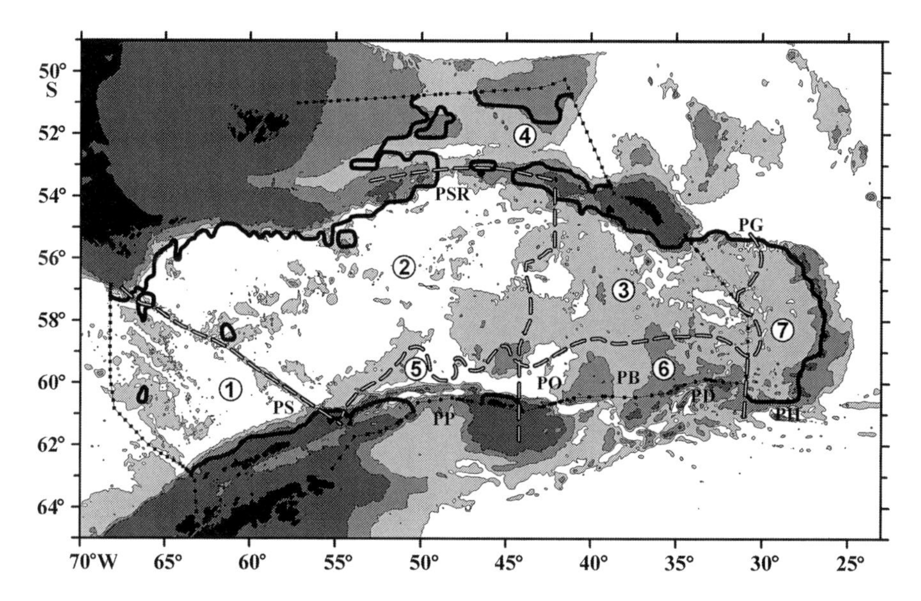

Fig. 3.21 Limits of Circumpolar Bottom Water and Warm Deep Water spreading in the Drake Passage and Scotia Sea are shown with *solid bold lines*. *Dashed lines* and numerals in *circles* denote boundaries and numbers of sectors (Table 3.5). Isobaths 1,000, 2,500, and 3,500 m are shown. See Figs. 3.7 and 3.8 for other notations. (Modified and redrawn from Tarakanov 2010)

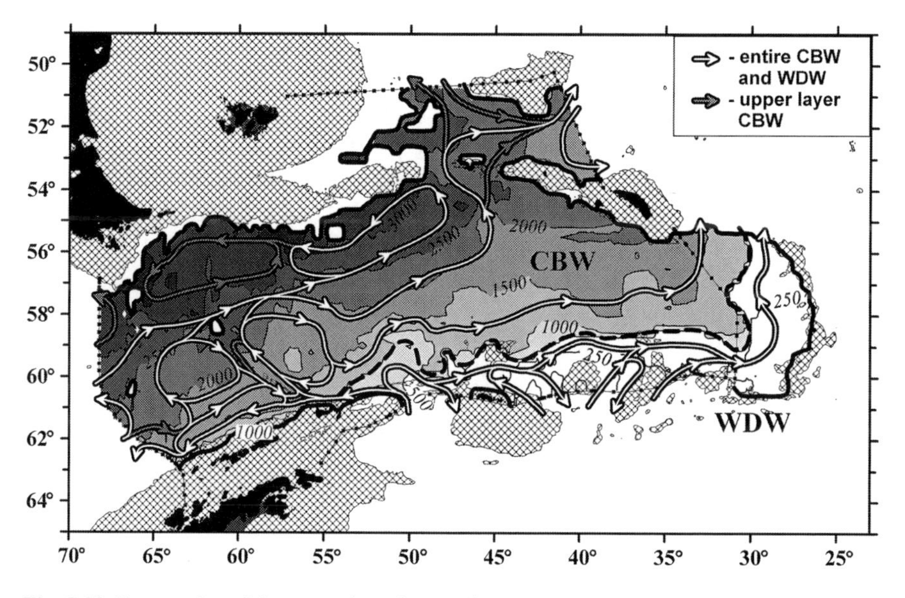

Fig. 3.22 Topography of the upper boundary surface of Circumpolar Bottom Water and Warm Deep Water (m) in the Drake Passage and Scotia Sea. Regions with depths shallower than 2,000 m are dashed. *Arrows* show the authors' scheme of Circumpolar Bottom Water and Warm Deep Water spreading in the region. See Figs. 3.7 and 3.9 for other notations. *Thick dashed line* divides the spreading regions of CBW and WDW. (Modified and redrawn from Tarakanov 2010)

3.4.3.3 Circumpolar Bottom Water

Figure 3.23 shows the θ/S-diagrams illustrating transformation of Circumpolar Bottom Water during its spreading from the southeastern Pacific to the central part of the Scotia Sea. Measurements were made with different instruments, and periods of research are divided by time periods exceeding 15 years. Therefore, we cannot formulate an exact concept about the transformation of waters in the course of their easterly propagation on the basis of these data. Variation of the form (slope) of the θ/S-curve during the motion of water with the Antarctic Circumpolar Current is significant in this case. On section P19 (R/V *Knorr*, (US), January and February, 1993) in the southeastern part of the Pacific Ocean in the Subantarctic Front zone and north of this front, θ/S-curves are straight lines in the Circumpolar Bottom Water layer and are bent to low salinities south of this front (Fig. 3.23a). Further to the east, the form of θ/S-curves in the northern part of section A21_99 does not almost change as compared with section P19, with the only difference that waters with $\theta < 0.55°C$ (Fig. 3.23b) do not appear on θ/S-curves. In the central part of the section, the slope of the curves slightly changes to lower salinities at temperatures $\theta < 0.35°C$, showing the influence of fresher cold waters, which flow to the west in the Slope Antarctic Current. Finally, in the southern part of section A21_99 located in the depression of the Hero Fracture Zone, we find significant freshening and cooling of waters due to the influence of waters of the Slope Antarctic Current that covers the lower layers of Circumpolar Bottom Water with $\theta < 0.5°C$. Further to the east, the same form of θ/S-curves is conserved in the part of the Drk03 section located south of the Shackleton Ridge (Fig. 3.23c). In the northern part of this section, we find not only strong variation in the slope of the θ/S-curves, but also lower salinities as compared with the southern part of the section (Fig. 3.23c), which was not found over sections P19 and A21_99 (Fig. 3.23a,b). This can only be explained by the transport of fresher waters from the Weddell Sea across the Antarctic Circumpolar Current. The same form of θ/S-curves is also repeated in the northern part of section Drk05b, located at a short distance to the east of the Shackleton Ridge (Fig. 3.23d). A more saline water type flowing from the zone south of the ridge is also found in the South Polar Front region and at one station in the southern part of this section (Fig. 3.23d). The salinity and temperature decrease, which is also observed in the southern part of this section along with the interstratified character of θ/S-curves (Fig. 3.23d), provides evidence of intense isopycnal mixing between Circumpolar Bottom Water and Warm Deep Water east of the ridge. The southern part of section Drk07 appeared slightly displaced to the west from the Shackleton Ridge crest. Here, we note numerous intrusions of fresher waters from the Weddell Sea, which cover the layer of Circumpolar Bottom Water and overlying Lower Circumpolar Deep Water (Fig. 3.23e). The absence of such intrusions on section Drk03, located further to the west, indicates that they do not propagate to the west much further than the Shackleton Ridge. On section SR01a (R/V *Polarstern*, Germany, September 1992), located further to the west from the Shackleton Ridge, θ/S-curves in the northern and southern parts of the section are virtually similar, excluding intrusions of fresher water in the south (Fig. 3.23f). In addition, their slope repeats the slope

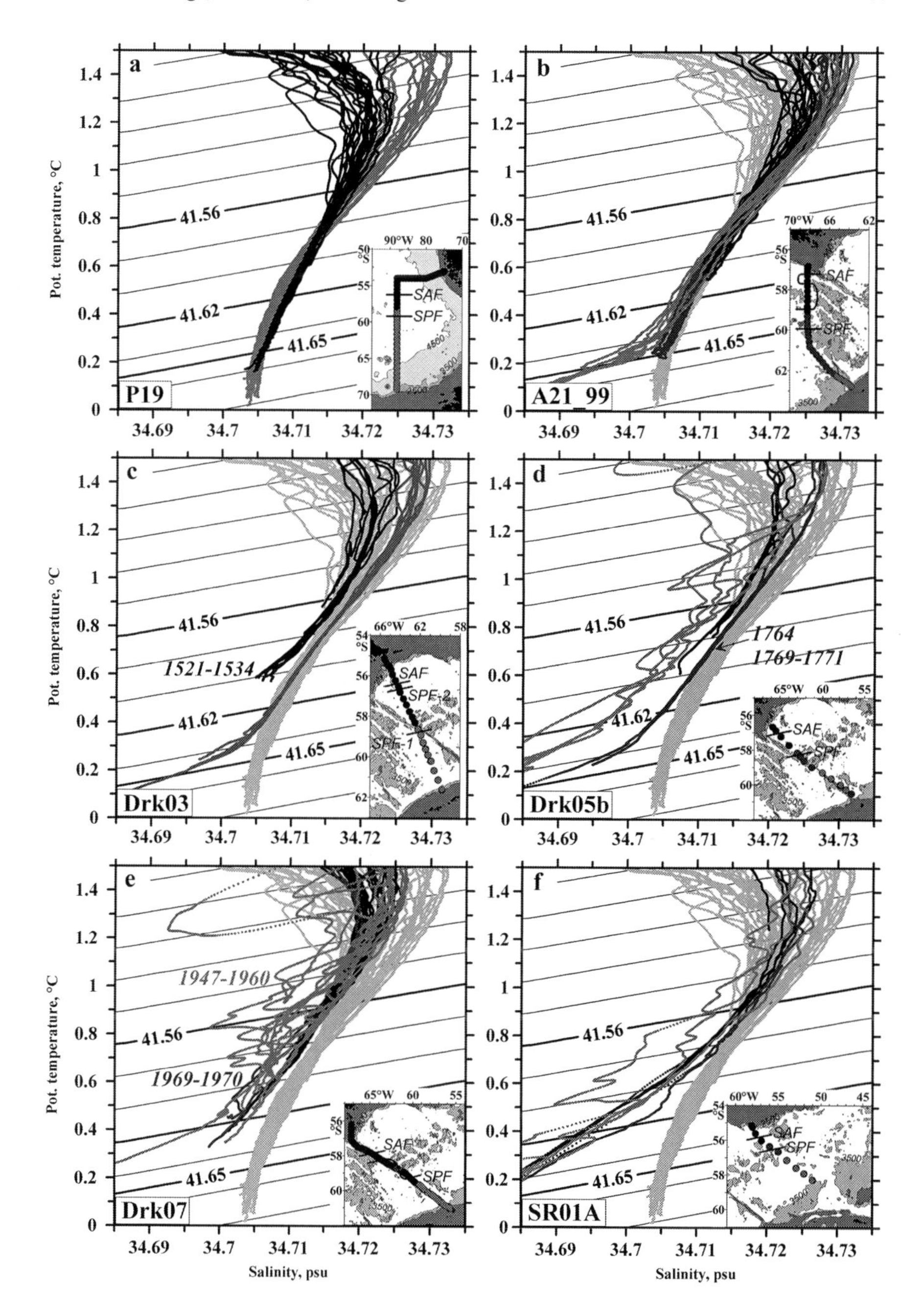

Fig. 3.23 The *θ*/*S*-diagrams for sections P19 (**a**), A21_99 (**b**), Drk03 (**c**), Drk05b (**d**), Drk07 (**e**), SR01a (**f**). *Bold lines* show isopycnals σ_3=41.56 ($\sim\gamma^n$=28.16), σ_3=41.62 ($\sim\gamma^n$=28.23), and σ_3=41.65 ($\sim\gamma^n$=28.26). Schemes of stations, locations of SPF and SAF, and bottom topography are shown in the *insets*. (Modified and redrawn from Tarakanov 2010)

of θ/S-curves in the northern parts of sections Drk03 and Drk05b (Fig. 3.23b, d, f), which points to the possible isopycnal nature of Circumpolar Bottom Water freshening in the northwestern part of the Scotia Sea. These intrusions provide evidence of continuing mixing between Circumpolar Bottom Water and Warm Deep Water at the southern periphery of the Antarctic Circumpolar Current. A slightly greater salinity in the South Polar Front zone is likely related to the remains of more saline water propagating from the west along the South Polar Front (Fig. 3.23f). Summarizing the aforesaid, we can state that the Shackleton Ridge creates conditions favorable for the transformation of the lower layer of circumpolar waters due to the presence of Weddell Sea waters that reach the northern boundary of the Antarctic Circumpolar Current.

A thick quasi-homogeneous layer is worth attention in the lower part of Circumpolar Bottom Water in the southern part of section Drk03 (Fig. 3.10). Its upper boundary, determined from the maximum of $|d^2\sigma_l/dz^2|$ (see relation for σ_l in Sect. 3.4.3.1), approximately corresponds to isopycnal $\gamma^n=28.23$ ($\theta \approx 0.4$°C). Such stratification of the lower layer of Circumpolar Bottom Water is almost missing on sections adjacent to the eastern ridge from the east. In contrast, the stratification is observed at all stations west of the ridge. This is illustrated by the distribution of fN^2 average over the lower layer of Circumpolar Bottom Water ($28.23 < \gamma^n < 28.26$) (Fig. 3.24b) that characterizes homogeneity of the vertical density stratification. This fact allows us to suppose that waters of the lower layer of Circumpolar Bottom Water at both sides of the Shackleton Ridge are strongly isolated by this ridge, in the sense that intensity of circulation at both sides of the ridge strongly exceeds water exchange over the ridge.

Comparison of water transport in this Circumpolar Bottom Water layer based on LADCP measurements on two sections (Drk07 along the crest of the Shackleton Ridge and Drk03 crossing the ridge) confirms this assumption. Water with density $28.23 < \gamma^n < 28.26$ was found on section Drk07 only at three stations 1957, 1958, and 1971 (Fig. 3.12). Estimate of the easterly transport of this water is 0.5 Sv and the westerly transport is less than 0.1 Sv, which is more than one order of magnitude smaller than the corresponding estimates on section Drk03 (8.9 Sv to the east and 9.6 Sv to the west). In the region of this section, water with such density is found only south of the ridge (Fig. 3.10).

We note that stations 1957 and 1958 on section Drk07 were located in the Shackleton Passage in the southern part of the section (Fig. 3.12). The easterly current was recorded only in the northern part of the passage at station 1957 (Fig. 3.16c). Transport in the lower layer of Circumpolar Bottom Water was 0.3 Sv. Station 1971 was occupied in the Southern Polar Current zone in the middle part of the section, where the rift valley of the West Scotia Ridge and Shackleton Ridge converge (Figs. 3.12 and 3.16c). The westerly water transport is estimated at 0.2 Sv. It is noteworthy that the traces of overflow of more saline Circumpolar Bottom Water with the quasi-homogeneous lower layer to the northeastern side of the ridge were found in the same places on section Drk05b (stations 1764, 1771, 1772) in 2005 (Figs. 3.11 and 3.23d), indicating stability of these deep currents. It is noteworthy that the current was directed to the northeast at station 1764 located opposite to the

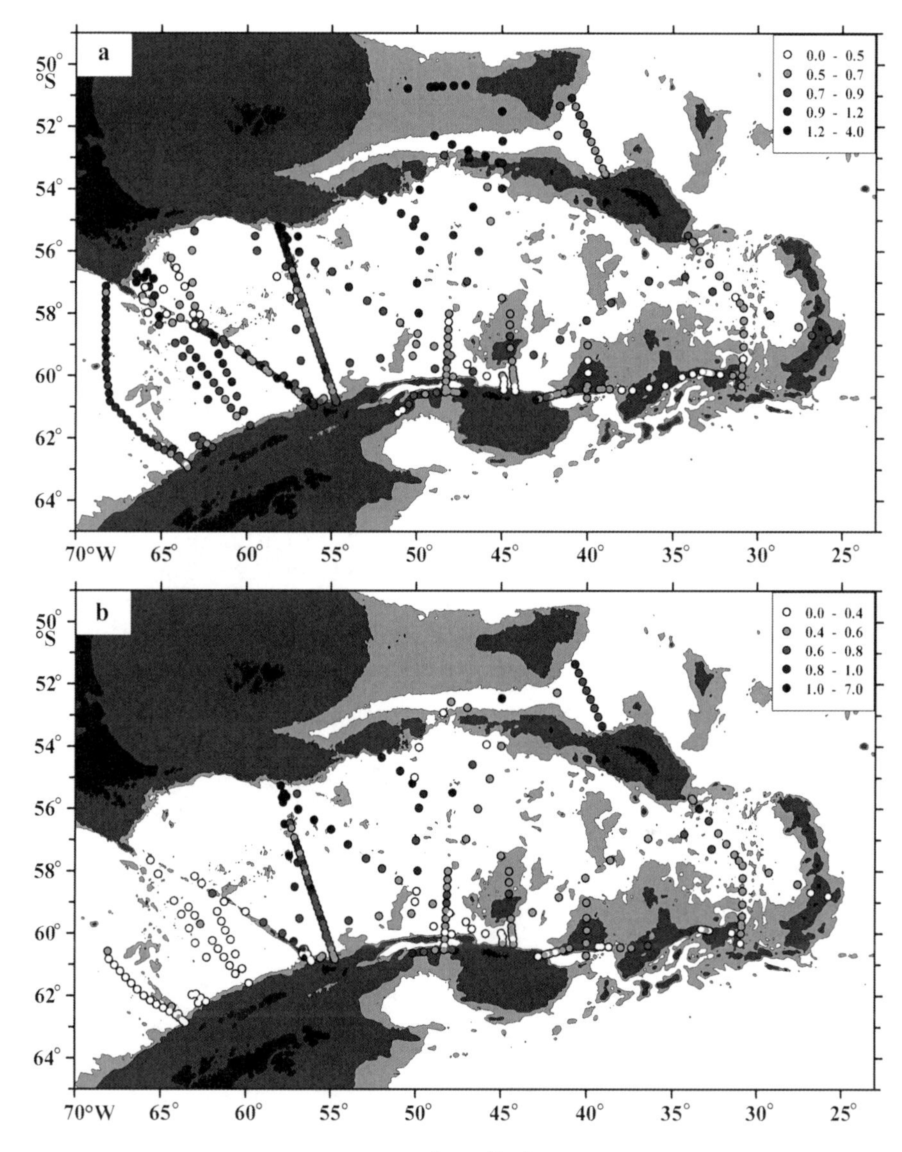

Fig. 3.24 Distribution of average values of fN^2 (10^{-11} s^{-3}) in the upper $28.16 < \gamma^n < 28.23$ (**a**) and lower $28.23 < \gamma^n < 28.26$ (**b**) layers of Circumpolar Bottom Water in the Scotia Sea and Drake Passage. (Modified and redrawn from Tarakanov 2010)

Shackleton Ridge (Fig. 3.16b) and directed to the northwest approximately along the slope of the ridge at stations 1771 and 1772 in the South Polar Current zone (Fig. 3.16b).

Data on transport across sections Drk03 and Drk07 (Table 3.6) show that water exchange intensity (transport in either directions) in the entire Circumpolar Bot-

Table 3.6 Transport of Circumpolar Bottom water (Sv) across sections in the region of the Shackleton Ridge based on LADCP data (see Fig. 3.25)

Section	Zone	Easterly transport	Westerly transport	Total
Drk07	Slope Antarctic Current	0.0	−0.8	−0.8
	South Antarctic Current	2.7	−4.5	−1.8
	South Polar Current	4.7	−2.0	2.7
	Subantarctic Current	1.1	−0.6	0.5
	Total	8.5	−7.9	0.6
Drk03	South of the Shackleton Ridge	27.8	−25.1	2.7
	North of the Shackleton Ridge	8.8	−10.3	−1.5
	Total	36.6	−35.4	1.2

tom Water layer is more than one order of magnitude higher than the total (unidirectional) transport of Circumpolar Bottom Water across both sections. Increased water exchange across section Drk03 in comparison with Drk07 indicates the existence of a more intense circulation in the basins adjacent to the Shackleton Ridge compared to water exchange over the ridge (Table 3.6, Fig. 3.25). However, it was already mentioned that water exchange over the ridge is small in the lower layer of Circumpolar Bottom Water relative to the water exchange across section Drk03. In the upper layer of Circumpolar Bottom Water, the transport values are comparable:

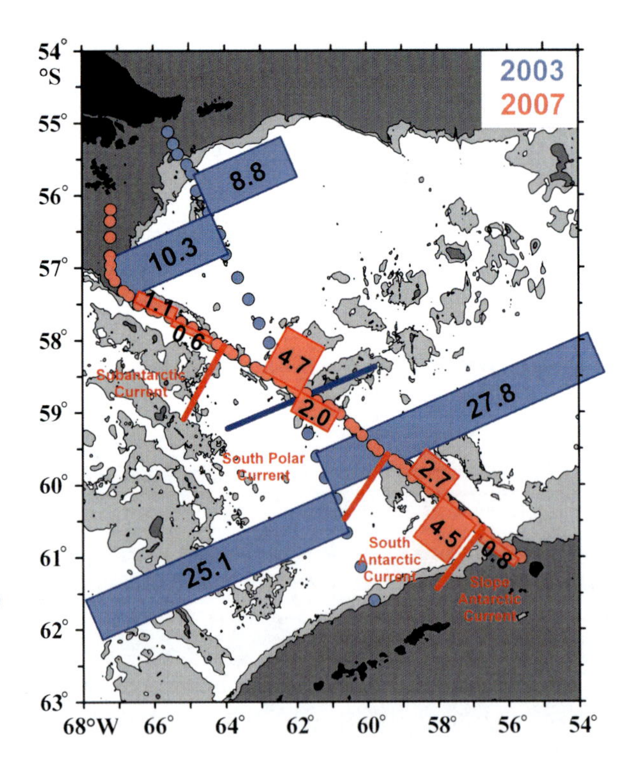

Fig. 3.25 Transport of Circumpolar Bottom water (Sv) across sections in the region of the Shackleton Ridge based on LADCP data in 2003 (*blue*) and 2007 (*pink*) (Table 3.6)

8.0 (7.9) Sv to the east (west) on section Drk07 and 27.7 (25.9) Sv on section Drk03. We note that these estimates do not agree with the published estimates of transport in this layer. According to the numerical analysis of circulation in the Scotia Sea (Naveira Garabato et al. 2003) based on isopycnal box-model, total transport in the Drake Passage in the layer $28.15 < \gamma^n < 28.26$ (approximately corresponding to the Circumpolar Bottom Water layer) is 9 Sv. The lower layer of Circumpolar Bottom Water was not resolved in the model.

It is likely that the Hero Ridge, like the Shackleton Ridge, significantly blocks the exchange of Circumpolar Bottom Water, in particular the exchange between its lower layers. This is evident from differences in the slopes of θ/S-curves, as well as values of potential temperature and salinity corresponding to the lower quasi-homogeneous layer in the southern part of section Drk03 and central part of section A21_99 that are located east and west of the Hero Ridge, respectively (Fig. 3.23b, c). All the facts mentioned above provide evidence that the Hero and Shackleton ridges facilitate the formation of a few quasi-stable and quasi-isolated circulations of abyssal and deep waters in the adjacent regions.

On the basis of LADCP measurements in 2003 and 2005 (Fig. 3.16a,b), we can state that at least two cyclonic gyres of Circumpolar Bottom Water exist in the southern part of the Drake Passage between the Hero and Shackleton ridges (Fig. 3.22). Circulation in the South Shetland Trough is the upper part of the Upper Weddell Sea Deep Water gyre. Its northern arc is related to the southern branch of the Antarctic Circumpolar Current. The southern arc is the Slope Antarctic Current, which is distinguished in Fig. 3.22 by shallow depths of the upper boundary surface of Circumpolar Bottom Water and closer location of isobaths. It is likely that the other circulation pattern located in the northern zone between the South Polar Front and Southern Front of the Antarctic Circumpolar Current does not correlate with the surface currents, which are mapped from the absolute dynamic topography of the ocean (Fig. 3.16).

According to the LADCP data in 2005, the northwesterly flow of Circumpolar Bottom Water, freshened due to mixing with Warm Deep Water, was observed along the northeastern slope of the Shackleton Ridge. This flow was confined to the zone between the Shackleton Passage and South Polar Front, which is the outflow sites of more homogeneous and saline Circumpolar Bottom Water from the region west of the ridge (Fig. 3.16b). This fact combined with the satellite altimetry data (Fig. 3.16) provides evidence that cyclonic circulation is formed in the western part of the Ona Basin in the water column from the surface to the bottom. This gyre is a meander of the southern branch of the South Antarctic Current, which crosses the ridge near the Shackleton Passage. Satellite altimetry data in the Ona Basin indicate that this meander is periodically separated from the main flow and split off into smaller eddies (Fig. 3.16). Existence of this gyre was shown in Barre et al. (2008). It is worth noting that LADCP data taken in 2007 and 2005 along the Shackleton Ridge, with the account of lack of traces of the fresher Circumpolar Bottom Water intrusions in the southern part of the Drake Passage over section Drk03, give grounds to state that anticyclonic circulation exists around the southern part of the Shackleton Ridge.

At least two quasi-isolated circulations of abyssal waters are formed in the western part of the Scotia Sea (Yaghan Basin) and in the northern part of the sea. These circulations are generally anticyclonic, but they are influenced by meanders and eddies of the Subantarctic Current and South Polar Current. Such conclusions are confirmed by a set of facts.

1. The depth of the upper boundary of Circumpolar Bottom Water in the northern part of the sea exceeds the threshold depth characteristic of the Shag Rocks Passage (2,900 m) and the eastern part of the Scotia Sea (3,200 m) (Fig. 3.22). Thus, part of Circumpolar Bottom Water volume and underlying Weddell Sea Deep Water appears to be blocked from eastern propagation with the Antarctic Circumpolar Current.
2. In the northern part of the Scotia Sea, the bottom layer is occupied by Antarctic Bottom Water, i.e., water with density $\gamma^n > 28.26$ ($\theta < 0.2$°C). In the Yaghan Basin, only the upper layer of Circumpolar Bottom Water is found with density $\gamma^n < 28.205$ ($\theta > 0.55$°C). Circumpolar Bottom Water of this density range is characterized in this basin by increased homogeneity of the vertical density stratification compared with the adjacent regions (Fig. 3.24a), excluding a few stations in the westernmost part of the basin, where the thickness of Circumpolar Bottom Water layer at the bottom is only 100–200 m. These differences in the water structure in the Yaghan Basin and northern part of the Scotia Sea indicate a relative isolation of the abyssal water circulations in these basins. The Subantarctic Front is an approximate boundary between them, which turns northward east of the Burdwood Bank and crosses the North Scotia Ridge.
3. Some measurements on the continental slope of South America in the northern part of the Scotia Sea, including moored measurements (Whitworth et al. 1982) and climatic distributions of some characteristics at the bottom, for example, salinity (Fig. 3.18) show that the slope current has a westerly direction, which forms the northern branch of anticyclonic circulations in the Yaghan Basin and northern part of the Scotia Sea. The core of this westerly Circumpolar Bottom Water flow (with lower salinity and temperature) located near the continental foothills of South America is clearly seen on section SR01m (Fig. 3.13). These characteristics of the layer can be related both to intense diapycnal mixing between Circumpolar Bottom Water and the underlying fresher Weddell Sea Deep Water over a very rough bottom topography in the northern part of the Scotia Sea and to isopycnal transport of fresher waters from the Weddell Sea across the Antarctic Circumpolar Current.
4. The LADCP measurements in the Yaghan Basin over sections Drk03 and Drk05a occupied along the same line showed that directions of Circumpolar Bottom Water flow were very different in the measurements periods (Fig. 3.16a,b). In 2005, the upper boundary of Circumpolar Bottom Water was displaced deeper approximately by 500 m, and temperature differences at the bottom were approximately 0.1°C: $\theta \approx 0.55$°C ($\gamma^n \approx 28.205$) in 2003 and $\theta \approx 0.65$°C ($\gamma^n \approx 28.195$) in 2005. (Koshlyakov et al. 2007). At the same time, the direction of currents in the Circumpolar Bottom Water layer at several stations correlates well with the

surface circulation in this basin. This is especially clearly seen on sections Drk03 and Drk05b (Fig. 3.16a, b).

Volumes of Circumpolar Bottom Water and Weddell Sea Deep Water, blocked for the easterly transport in the entire western part of the Scotia Sea (sector 2, Fig. 3.21), are estimated at the same value of 30×10^4 km^3, which is approximately two fifths of the total Circumpolar Bottom Water volume and two thirds of the total Weddell Sea Deep Water volume in this sector (Tables 3.4 and 3.5). Diapycnal mixing with the overlying waters, isopycnal transport to the southern periphery of the Antarctic Circumpolar Current due to the formation of anticyclonic eddies with their further dissipation, or formation of a westerly flow to the Pacific over the northern part of the Shackleton Ridge, can be considered as possible mechanisms of transport. According to the results of numerical analysis of circulation (Naveira Garabato et al. 2003), intense diapycnal mixing occurs in the Scotia Sea between Weddell Sea Deep Water and overlying layers of circumpolar waters. The authors of the paper cited above relate this fact to the interaction with rough bottom topography. A total of 1–3 Sv is transferred from Weddell Sea Deep Water to Circumpolar Bottom Water. Along with mixing, the above authors of the model note intense isopycnal ventilation of Lower Circumpolar Deep Water due to the influence of waters from the Weddell Sea and the formation of a local meridional circulation cell with an intensity of 8 Sv. Its lower limb, which includes practically the entire Circumpolar Deep Water layer, forms heat and salt flux directed from the northern periphery of the Antarctic Circumpolar Current to the polar region. The upper circulation limb directed to the north includes Antarctic Intermediate Water and the upper layers of Upper Circumpolar Deep Water. According to Naveira Garabato et al. (2003), eddy formation in the South Polar Frontal Zone is one of the possible mechanisms of the appearance of this cell. The existence of this cell plays an important role in the formation of heat and salt balance not only in the Scotia Sea, but also in the Weddell Gyre. We note that the latest estimates (for example, Klatt et al. (2005)) show that about two thirds of the entire heat flux from the ocean to the atmosphere in the Weddell Gyre should be provided by the transport of warm circumpolar waters into the gyre within the zone between the northern end of the Antarctic Peninsula and zero meridian.

According to the measurements in 2007 along the northern part of the Shackleton Ridge in the Subantarctic Current zone, all Circumpolar Bottom Water was concentrated near a deep narrow trough west of the ridge (Fig. 3.7). The upper boundary of Circumpolar Bottom Water was located only slightly above the threshold depth of the passage (approximately 3,600 m) separating the ridge from the continental slope of South America (Fig. 3.12). Hence, data of the Circumpolar Bottom Water transport (Table 3.6) based on the section in the Subantarctic Current zone give more information about the intensity of abyssal cyclonic circulation in this trough rather than water exchange over the ridge. Thus, the data of measurements in 2007 demonstrate that the Shackleton Ridge prevents the overflow of a significant part of Circumpolar Bottom Water flowing from the Pacific Ocean in the Subantarctic Current zone. Assuming this idea, anticyclonic recirculation in the Circumpolar

Bottom Water layer in this zone west of the Shackleton Ridge (north Drake Passage) is shown in Fig. 3.22. It is worth noting that this recirculation is anticyclonic as a whole, because it covers the entire Subantarctic Current zone (but not only the local abyssal circulation in the trough). This conclusion does not contradict the velocity field structure over section A21_99 shown in Naveira Garabato et al. (2003). This section is the western boundary of the Southern Ocean sector studied here. We note that, despite the absence of actual overflow of Circumpolar Bottom Water over the Shackleton Ridge in the Subantarctic Current zone, the data available in 2007 do not exclude water exchange, including the westerly overflow of Circumpolar Bottom Water that can be realized by meandering of the Antarctic Circumpolar Current jets.

It was discussed above that the Hero Ridge blocks water exchange of the lower layers of Circumpolar Bottom Water. Thus, part of this water flowing to the east with the Antarctic Circumpolar Current is not overflowing the ridge. Keeping this in mind, anticyclonic recirculation of Circumpolar Bottom Water west of the ridge between the South Polar Front and Southern Front of the Antarctic Circumpolar Current (central part of the Drake Passage) is shown in Fig. 3.22. We also note that this conclusion does not contradict the velocity field given in Naveira Garabato et al. (2003).

The structure of currents in Circumpolar Bottom Water layer near the Falkland Plateau, Georgia Basin, and North Scotia Ridge is based mainly on Arhan et al. (1999, 2002b); Naveira Garabato et al. (2002a); Nowlin and Zenk (1988) and Orsi et al. (1993, 1999) combined with the distributions of fN^2 shown in Fig. 3.24 and climatic distribution of salinity at the bottom (Figs. 3.7 and 3.18). According to the scheme of Circumpolar Bottom Water transformation in the Scotia Sea and Drake Passage, the data presented in Table 3.5 demonstrate gradual freshening and oxygenation in Circumpolar Bottom Water from the western part of the study region to its eastern part due to mixing with the waters of the Weddell Sea. Already in the western part of the region (sector 1, Fig. 3.22), Circumpolar Bottom Water appears fresher, cooler, and more enriched with oxygen than in the Antarctic Circumpolar Current zone in the eastern part of the Pacific Ocean. According to Koshlyakov and Tarakanov (2003b), mean values of potential temperature, salinity, neutral density γ^n, and dissolved oxygen were 0.57°C, 34.710 psu, 28.210 kg/m^3, and 4.70 ml/l, respectively. A slight temperature increase in sector 2 (Fig. 3.22) is related to the inclusion of a vast northern periphery of the Antarctic Circumpolar Current in the Scotia Sea into this sector.

3.4.3.4 Warm Deep Water

It was mentioned above that Warm Deep Water and Lower Circumpolar Deep Water are characterized by a salinity maximum. However, unlike Lower Circumpolar Deep Water, this maximum in the northwestern periphery of the Weddell Gyre appears displaced to the density range of Circumpolar Bottom Water, which is responsible for the main difference of Warm Deep Water from Lower Circumpolar

Deep Water and Circumpolar Bottom Water. In the Scotia Sea and at the northern tip of the Antarctic Peninsula, Warm Deep Water is located under the cold and fresher waters transported from the Bransfield Strait. This is why the upper boundary surface of Warm Deep Water is located more than 500 m deeper in this region (Figs. 3.14 and 3.22). As the waters spread eastward to the southern and southeastern parts of the Scotia Sea, the upper boundary surface of Warm Deep Water rises to a depth of approximately 250 m (Figs. 3.14 and 3.22), where Antarctic Surface Water is located above Warm Deep Water. As Warm Deep Water propagates further, it becomes more saline and warmer (Table 3.5) owing both to mixing with circumpolar waters and water exchange with the warmer and more saline Warm Deep Water from the central part of the Weddell Gyre. We note that, despite notable differences in temperature, salinity, and density between the waters from the Bransfield Strait and Antarctic Surface Water of the open ocean, these waters are included into Antarctic Surface Water in Fig. 3.14, because waters of the Antarctic shelf (with θ greater than $-1.7°C$) are modifications of Antarctic Surface Water.

3.5 Antarctic Bottom Water in the Argentine Basin

Bottom water structure in the Argentine Basin is likely the most complicated and less studied structure among all Atlantic regions considered here. This is caused by hydrological and dynamic fronts, branching of the pathways for the propagation of Antarctic waters from the south, and drastic change in the propagation direction of these waters (branching from the Antarctic Circumpolar Current, which covers the southern part of the Argentine Basin (Peterson and Whitworth 1989)) and waters of the North Atlantic origin propagating to the Argentine Basin from the north. The latter waters change the direction of their propagation in this region (approximately at 30°–40° S).

According to Reid et al. (1977), the relatively cold, fresh, and oxygen-rich Weddell Sea Deep Water is located in the Argentine Basin at the bottom. In compliance to the circumpolar classification by Gordon (1967) and Orsi et al. (1999) this water is equivalent to Antarctic Bottom Water. A thick layer above Weddell Sea Deep Water is occupied by the warmer oxygen-poor Circumpolar Deep Water that propagates from the Antarctic Circumpolar Current region. The more saline North Atlantic Deep Water wedges into this layer from the north (Reid et al. 1977), dividing Circumpolar Deep Water into Upper and Lower Circumpolar Water (according to terminology in Reid et al. (1977)). In general, lower oxygen concentration in Circumpolar Deep Water (with respect to the overlying and underlying waters in the Antarctic Circumpolar Current field and to the North Atlantic Deep Water) makes it possible to use the local oxygen minima as the main distinguishing indicators of Upper and Lower Circumpolar Water. We recall that Upper Circumpolar Water (according to classification by Reid et al. (1977)) and Upper Circumpolar Deep Water (according to the three-layer circumpolar classification by Gordon (1967) and Orsi et al. (1999)) is actually the same water, whereas Lower Circumpolar Water includes Circumpolar Bottom Water and partly Lower Circumpolar Deep Water. We also note that Lower Circumpolar Water and Weddell Sea Deep Water represent Antarctic Bottom Water according to classification by Wüst (1936).

Waters of the Antarctic origin propagate to the Argentine Basin along several pathways. It is considered that the Falkland Gap in the Falkland Ridge (Whitworth et al. 1991) is the main (and, probably, the only one) passage for Weddell Sea Deep Water to the Argentine Basin. However, trajectory of this water flow from the sources in the Weddell Sea to the Argentine Basin is divided into several branches (Fig. 3.18). It is also possible that Weddell Sea Deep Water propagates to the Argentine Basin east of the Islas Orkades Rise (Whitworth et al. 1991). However, the existence of this pathway has not been confirmed yet. Circumpolar waters propagate to the Argentine Basin together with the branches of the Antarctic Circumpolar Current related to the South Polar Front and Subantarctic Front mainly over the Falkland Plateau and east of the Maurice Ewing Bank over the Falkland Ridge (Fig. 3.26). The complex pattern and its branching indicate that the same waters passing along different pathways can acquire their specific properties. Thus, there is a possibility to distinguish the waters, which passed along different pathways, at the site where they merge again. The authors of Arhan et al. (1999) distinguished cores of five varieties of water masses at the northern side of the Falkland Plateau in the Argentine Basin: two cores each in Lower Circumpolar Deep Water (LCDW-1 and LCDW-2) and Weddell Sea Deep Water (Upper Weddell Sea Deep Water and Lower Weddell Sea Deep Water) and the Southeastern Pacific Deep Water layer

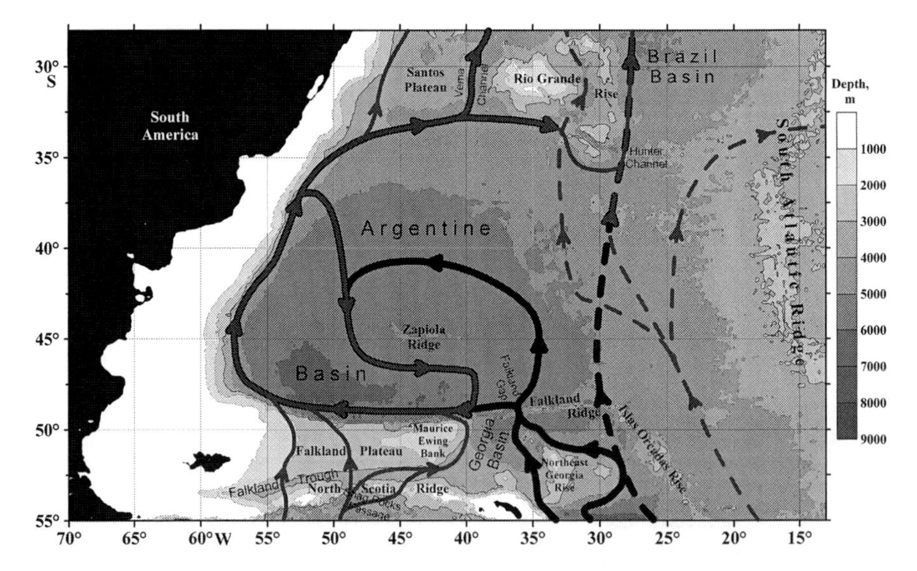

Fig. 3.26 Deep and bottom water circulation in the Argentine Basin. The *thick black solid line* shows Weddell Sea Deep Water pathways (Arhan et al. 1999; Smythe-Wright and Boswell 1998; Coles et al. 1996). The *thin gray line* shows Lower Circumpolar Water/Lower Circumpolar Deep Water pathways (Arhan et al. 1999; Coles et al. 1996). *Thick gray* shows WSDW, LCPW/LCDW pathways. *Dashed lines* represent circulation in the same layers based on Larque et al. (1997)

between them. All these water masses were separated by maxima of stability. We recall that Southeastern Pacific Deep Water is a special type of Circumpolar Bottom Water in the Drake Passage. Division of Weddell Sea Deep Water into two layers is related to the existence of a trajectory for the upper layer of Weddell Sea Deep Water that passes through the Scotia Sea (see the previous section). We also note that, precisely, Upper Weddell Sea Deep Water can be traced up to the equator (Reid 1989). Division of Lower Circumpolar Deep Water into the heavier LCDW-2 and the lighter LCDW-1 is related to the appearance of the upper layers of Lower Circumpolar Deep Water in the same density range with the more saline North Atlantic Deep Water (Arhan et al. 1999).

In general, the circulation of bottom waters in the Argentine Basin can be characterized as a cyclonic rather than closed (C-shaped) ring (Coles et al. 1996; Memery et al. 2000). After propagating to the Argentine Basin, the bottom waters turn to the west along the slope of the Falkland Plateau, flow to the north along the slope of South America, and then propagate to the Brazil Basin (Reid et al. 1977; Georgi 1981; Mantyla and Reid 1983; Whitworth et al. 1991). Along with this transport scheme in the central part of the Argentine Basin, anticyclonic circulation of bottom waters exists around the Zapiola Ridge with a center approximately at 45° S, 45° W. This fact was emphasized in Smythe-Wright and Boswell (1998). This scheme does not contradict the circulation pattern of bottom waters in the Argentine Basin. We also note that, according to Coles et al. (1996), the unclosed cyclonic "C"-shaped circulation in the northern part of the Argentine basin is accompanied by a water inflow that propagates from the Mid-Atlantic Ridge; part of this inflow propagates to the north through the Hunter Channel. However, the origin of this water was not analyzed by the authors of (Coles et al. 1996). Figure 3.26 also shows elements of the circulation scheme presented in Larque et al. (1997), which is based on the Optimum Multiparameter Analyses (Tomczak 1981) and seems the most simple one. In addition to the flow of Lower Circumpolar Deep Water and Weddell Sea Deep Water along the continental slope of South America, one can see quasi-meridional flows from the region of South Georgia to the Hunter Channel. Besides, Lower Circumpolar Deep Water branches off from this region and flows through the Mid-Atlantic Ridge and Walvis Ridge to the Cape Basin, which seems to be questionable. We think that conclusions in Larque et al. (1997) are caused by confusion of thermohaline indices corresponding to Weddell Sea Deep Water, Lower Circumpolar Deep water, and North Atlantic Deep water, resulting in overestimation of the role of Lower Circumpolar Deep Water in the circulation of bottom waters in the Southwest Atlantic.

It is difficult to compare Antarctic water transports in the Argentine Basin owing to the complex vertical structure of water masses, resulting in different locations of their upper boundaries. Available data from the high-precision zonal hydrographic sections in the Argentine Basin are scantier than, for example, in the Brazil Basin. That is why the estimates of transports are not numerous in literature. The western transport of Antarctic Bottom Water (Weddell Sea Deep Water) with potential temperature $\theta < 0.2$°C on the northern slope of the Falkland Ridge (41° W) was estimated at 8.2 Sv in Whitworth et al. (1991). The authors indicate that the major

Table 3.7 Transport of Antarctic Bottom Water in the Argentine Basin from literature data (rates of AABW formation)

Authors	AABW boundary	Transport, Sv	Comments
Georgi (1981)		5	Transport in the northern part of the South Sandwich Trench
Coles et al. (1996)		5–6.8	
Arhan et al. (1999)	$\sigma_4 = 45.78$	11	Integral transport of bottom waters from the south
	$\sigma_4 = 46.04$	4.0	Transport of WSDW with the deep current near the Falkland Plateau over A17 section. Combine geostrophic calculations and SADCP
	$45.98 < \sigma_4 < 46.04$	1.0	Transport of Southeast Pacific Deep Water with the deep current near the Falkland Plateau over A17 section
	$45.78 < \sigma_4 < 45.98$	6.0	Transport of LCDW with the deep current near the Falkland Plateau over A17 section
Sloyan and Rintoul, (2001a)	$\gamma^n = 28.30$	1.1 ± 0.5	Hydrographic sections and inverse model (47° S)
Hogg (2001)		8 ± 2	
Vanicek and Siedler (2002)	$\gamma^n = 28.12$	5 ± 2	Inverse model. Transport of LCDW and AABW across A11 section along 45° S
Ganachaud (2003)	$\gamma^n = 28.11$	5 ± 2	Inverse model (45° S)

part of this transport is related to Weddell Sea Deep Water recirculation in the Argentine Basin and also to the existence of a Weddell Sea Deep Water flow trapped by the Subantarctic front propagating from the Falkland Gap at 36° W to the east. The westerly transport of relatively new water in the basin with potential temperature θ less than −0.2°C is 2.5 Sv (Whitworth et al. 1991). Further to the west at the foot of the Falkland Rise at 55° W, the transport of Weddell Sea Deep Water with $\sigma_4 < 46.04$ is estimated at 4.0 Sv (Arhan et al. 1999). Transports of other water components of the bottom structure in this region are as follows: Southeastern Pacific Deep Water ($45.98 < \sigma_4 < 46.04$) 1.0 Sv; Lower Circumpolar Deep Water ($45.78 < \sigma_4 < 45.98$) 6.0 Sv (Arhan et al. 1999). Total transport of water with density $\gamma^n > 28.30$ (lower part of Weddell Sea Deep Water) across 47° S is 1.1 ± 0.5 Sv (Sloyan and Rintoul 2001b). At 45° S, the available estimates of Antarctic waters are close to each other: 5 ± 2 Sv (Ganachaud 2003) ($\gamma^n > 28.11$) and (Vanicek and Siedler 2002) ($\gamma^n > 28.12$) (Table 3.7).

Chapter 4
Exchange Between the Argentine and Brazil Basins; Abyssal Pathways and Bottom Flow Channels (for Waters of the Antarctic Origin)

4.1 General Description

The zonally aligned Rio Grande Rise separates the Argentine Basin in the south from the Brazil Basin in the north. It is a high topographic obstacle for bottom water propagation to the north. Two meridional gaps intersect the Rise at ~39° W and ~28° W (Vema and Hunter channels, respectively). The depth in the Vema Channel exceeds 4,600 m as compared to the background depths of 4,200 m. The Hunter Channel is much shallower and the greatest depth does not exceed 4,000 m.

Repeated observations from hydrographic stations and current-meter arrays deployed in the Vema and Hunter channels unambiguously confirm the prominent role of the Vema Channel for the transport of Antarctic Bottom Water with respect to the Hunter Channel in the east and the Santos Plateau in the west (see progressive vector diagrams in Fig. 9, curves 1, 2, 3, 4, 5, 6). In this section we use the definition of Antarctic Bottom Water by Wüst (1936) and divide this water into Weddell Sea Deep Water and Lower Circumpolar Water (Reid et al. 1977) Based on moored current-meter observations in combination with geostrophic velocity computations from hydrographic stations, the total Antarctic Bottom Water transport across the Rio Grande Rise and Santos Plateau is estimated at 6.9 Sv (Hogg et al. 1999). On the average, 58% of this volume passes through the Vema Channel. The remainder flows over the Santos Plateau and through the Hunter Channel. Earlier estimates in (Speer and Zenk 1993) inferred from hydrography alone yielded a northward net flow of 6.6 Sv distributed at a ratio of 30:59:11 on the western boundary current system, and the Vema and Hunter channels. Recent publications based on direct measurements of currents show that the Vema Channel plays an even more important role providing the northward transport of the dominating part of Antarctic Bottom Water rather than the flows over the Santos Plateau and in the Hunter Channel.

Some authors also distinguish the western slope of the Rio Grande Rise, as a structure similar to the Santos Plateau but less extended. It is considered as an independent pathway for bottom water penetration to the Brazil Basin, which is separated from the deep Vema Channel (McDonagh et al. 2002). The majority of publications consider this slope together with the Vema Channel.

E. G. Morozov et al., *Abyssal Channels in the Atlantic Ocean*,

DOI 10.1007/978-90-481-9358-5_4,

Figure 2.12 shows the forms of bottom topography and potential temperature at the seafloor, showing the northward warming of Antarctic Bottom Water as it flows toward the equator in the Brazil Basin. The climatology was compiled from all available high-quality CTD casts collected in WODB05 database. The dominating role of the Vema Channel for the bottom water exchange is apparent and it is clearly seen from Fig. 4.1.

Table 4.1 shows classic estimates of Antarctic Bottom Water transport over pathways in the region around the Rio Grande Rise based on different publications. As seen from the table, most researchers estimate the integral transport of Antarctic waters through the above mentioned channels at 6–7 Sv. The Antarctic Bottom Water transport through the Vema Channel is estimated at 1.3–5.3 Sv (McDonagh et al. 2002; Hogg et al. 1982) or 3.7 Sv (Cai and Greatbatch 1995). Calculations in Holfort and Siedler (2001) yielded a transport of 6.8 Sv below isopycnal surface $\sigma_4 = 45.93$.

Estimates for other channels and pathways are much smaller. Estimates of transport through the Hunter Channel are as follows: 0.7 Sv (Speer et al. 1992), 2.7 ± 1.3 (Lenz 1997), and 2.92 ± 1.24 (Zenk et al. 1999). Flow over the Santos Plateau ranges from − 1.5 to 2 Sv (McDonagh et al. 2002; Speer and Zenk 1993), indicating that water exchange over the Santos Plateau is possible in both directions. Flows of opposite directions were also found in the Hunter Channel (Zenk et al. 1999). According to McDonagh et al. (2002), the transport over the Santos Plateau is a branch

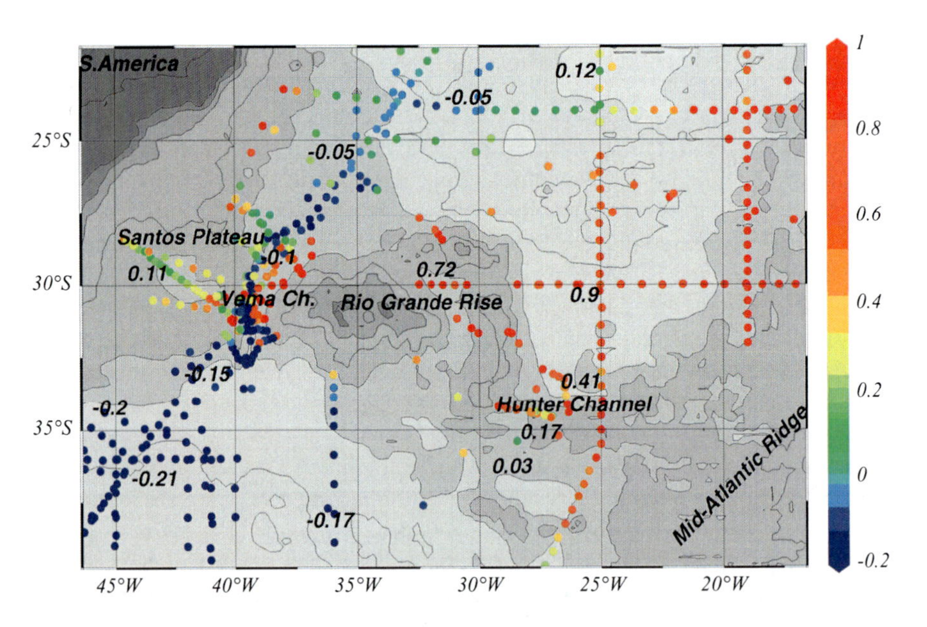

Fig. 4.1 Bottom topography (m) and distribution of potential temperature (°C) observations at the bottom in the Vema Channel region. Values of potential temperature below 3,500 m based on WODB-2005 are indicated at the points of measurements

Table 4.1 Transport of Antarctic Bottom Water over pathways in the region around the Rio Grande Rise: Vema Channel, Hunter Channel, and Santos Plateau

Authors	AABW boundary	Transport, Sv	Comments
Hogg et al. (1982)	$45.87 < \sigma_4 < 45.93$	4.15 ± 1.27 (Vema)	Mean geostrophic transport over six sections, reference surface: $\theta = 2°$
Reid (1989)		1.8 (Hunter)	
Zemba (1991)	$\sigma_4 = 45.85$	4.1 (Vema) 0.2 (Santos)	Geostrophic transport at 31° S, reference surface: $\sigma_4 = 45.85$
Speer and Zenk (1993)	$\theta = 2°$	3.9* (Vema) 0.7 (Hunter) 2 (Santos)	Geostrophic transport (Meteor 15, reference surface: $\theta = 2°$
Holfort and Siedler (2001)	$\theta = 2°$	5.1–5.3* (Vema)	Current meters data
Zenk and Hogg (1996)	$\theta = 2°$	2* (Vema) 1.8 (Santos)	Geostrophic transport over A10 section, reference surface: $\theta = 2°$
Lenz (1997)	$\theta = 2°$	2.7 ± 1.3 (Hunter)	Current meters data
Hogg and Zenk (1999)	$\theta = 2°$	4.5* (Vema)	Current meters data
		2* (Vema) −0.8 – 0 (Santos)	Geostrophic transport over A10 section. First estimate for Santos: Meteor 15
Zenk, Table 4.4	$\theta = 2°$	3.99 Vema	
Zenk, Table 4.4	Below 3,200 m	3.78 Vema	
Wienders et al. (2000)		3.5* (Vema)	Geostrophic transport over A17 section, reference surface is based on inverse method
Cai and Greatbatch (1995)		3.7	
Zenk et al. (1999)		2.92 Hunter	
McDonagh et al. (2002)		3.7 ± 0.7* (Vema)	A17 section
		0.1 ± 2.5 (Santos) 3.7 ± 0.7* (Vema)	A10 section
		0.1 ± 1.5 (Santos) 2.7 ± 0.7* (Vema) 1.3 (Vema)	A23 section
Zenk (2008)	<0.2°	2.07 ± 0.73	WSDW at Vema Sill only two current meters, direct observations

*Without inclusion or with partial inclusion of transport at the western slope of the Rio Grande Rise

of the upper part of a deep flow in the Vema Channel at 27° S, which turns to the north in the 31° S region and flows over the Santos Plateau.

According to McDonagh et al. (2002), a southerly flow exists over the western slope of the Rio Grande Rise that transports from 0.4 to 2.4 Sv. Other authors (Hogg et al. 1982; Zemba 1991; Speer and Zenk 1993) integrate this flow with the flow in the deep Vema Channel. Barnier et al. (1998) modeled the seasonal variability of bottom water transport at 30° S. According to their calculations, water transport changed only slightly from 1 to 2 Sv. A seasonal cycle was also found in the record of moored current meters in the Vema Channel deployed by IFM-GEOMAR (Kiel) in 2003–2007.

Analysis of the division of Antarctic waters into components shows that most publications distinguish Weddell Sea Deep Water and Lower Circumpolar Water. Weddell Sea Deep Water flows to the Brazil Basin only through the Vema Channel (Reid et al. 1977; Sandoval and Weatherly 2001). At the same time, Lower Circumpolar Water flows through both Vema and Hunter channels. A different scheme of circulation in the region is presented in Larque et al. (1997): Weddell Sea Deep Water propagates to the Brazil Basin through both channels. Speer and Zenk (1993) conclude that Weddell Sea Deep Water (bounded by $\sigma_4 > 46.05$) accounts for 40% of waters of the bottom layer flowing to the Brazil Basin. This agrees with the estimates in Reid (1989), according to which Circumpolar Deep Water accounts for 60%. A recent publication by Zenk and Morozov (2007) indicates that only Lower Circumpolar Water and Weddell Sea Deep Water with a boundary at isotherm $\theta =$ 0.2°C are found in the deep Vema Channel.

4.2 Vema Channel

4.2.1 Topography and General Description

For the first time this region was systematically investigated during the *Deutsche Atlantische Expedition* of R/V *Meteor* in 1925–1927 (Spiess 1932). The application of the newly invented echo sounding equipment allowed German researchers to reveal many objects of the bottom topography on a quasi-continuous track (Maurer and Stocks 1933). The *Meteor* found a narrow valley with a depth of 4,920 m at 29°47.1′ S, 39°27.5′ W (see Sect. 1.6). Decades later Cruise 22 of R/V *Vema* (for more historical remarks see Sect. 4.2.2) was especially dedicated to the study of bathymetry of this region (and the stratification of hydrological layers enabling water transport from the Argentine Basin to the Brazil Basin and vice versa).

The bottom topography around the Vema Channel is shown in Fig. 3. The Vema Channel is the deepest one among the passages existing for Antarctic Bottom Water (hereinafter we use the definition by Wüst (1936)); see the previous section). Therefore, the coldest water (Weddell Sea Deep Water) can exit the Argentine Basin in the equatorward direction only through this channel (Zenk et al. 1993). The Vema

Channel is a narrow conduit between two terraces located to the east and west of this canyon (Johnson et al. 1976). Its width slightly exceeds 18 km in the narrowest part of the channel.

A virtual "pipeline" for the bottom waters of the Antarctic origin with the highest density in the whole water column crosses the Rio Grande Rise. North of its entrance on the Argentine side, the Vema Channel depth contours follow an almost meridional direction. Depth in the Vema Channel exceeds 4,600 m against the background depth of 4,200 m. A small constraining sill (Vema Sill) located at 31°12′ S, 39°24′ W crosses the channel. At its shallowest spot, the depth of the Vema Sill is approximately 4,614 m and the width is roughly 15 km.

The presence of the Vema Channel explains the existing structure of deep waters in the Atlantic Ocean. Conditions of water formation in the Arctic and Antarctic regions do not differ significantly with respect to temperature at the surface. The Arctic waters flow into the Atlantic over the shallow thresholds between Greenland and Iceland and between Iceland and Scotland; the depths of the thresholds are 650 and 850 m, respectively. Therefore, only the intermediate waters of lower density from the Arctic Basin actually overflow the thresholds. On the other hand, depth of the sill in the Vema Channel allows the densest waters of the Antarctic origin to overflow this sill and occupy the bottom layers in the Atlantic Ocean.

The northern part of the Vema Channel is wider. Recently, the name Vema extension was suggested by Zenk and Morozov (2007) for this region at about 27° S, 34° W. In the region of 27–29° S, the channel turns northeast, filling a buffering reservoir for bottom water at 27° S and broadens slightly farther north of 29° S. The local topography facilitates the further easterly flow of the bottom waters after outflowing from the channel. In (Sandoval and Weatherly 2001), two branches of the bottom waters are distinguished in the Brazil Basin (depths greater than 5,000 m). The upper branch turns to the west immediately after outflowing from the channel.

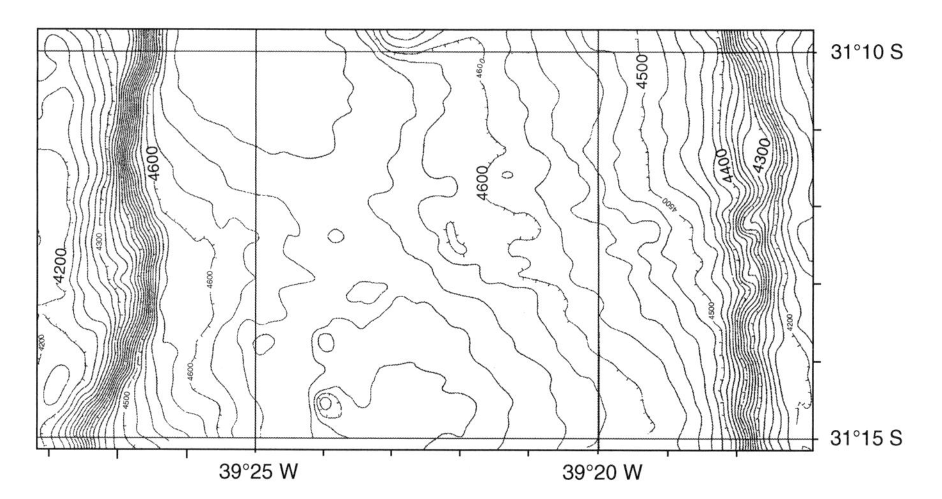

Fig. 4.2 Detailed bathymetry near the Vema Sill. (Based on Zenk et al. 1993 Depths are given in meters)

Possible cyclonic or anticyclonic rotation around the Rio Grande Rise is suggested in Hogg et al. (1982, 1996).

Seen from the bottom, the canyon-like walls on both sides of the saddle point at 31°12′ S are about 600 m high (Zenk et al. 1993). Relatively flat areas located on either side of the channel were called the eastern and western terraces (Johnson et al. 1976). Since 1991, the majority of revisits with CTD casts were concentrated on a narrow strip at that latitude. A detailed chart of the Vema Sill region is shown in Fig. 4.2. We have compiled a total of 79 CTD stations from 22 revisits to the Vema Sill. The first eleven successful visits to the sill region were already evaluated in the earlier work of Hogg and Zenk (1997).

4.2.2 History of Research and Datasets of Long-Term Observations

Initially, the Vema Channel was known as the Rio Grande *Rinne* or Rio Grande Gap (for more details see Zenk et al. (1993)). The width of this valley was estimated less than 30 km. In recognition of the intense work done by the former research vessel *Vema* the name Vema Channel was first suggested in Le Pichon et al. (1971) for this deep canyon of the Rio Grande Rise. The *Vema* (Fig. 4.3) made an outstanding contribution to physical oceanography and marine geology. Originally this 202-ft. pleasure yacht was built in Copenhagen in 1923 as a *luxury yacht Hussar*. Its name

Fig. 4.3 The *Vema* research ship that made an enormous contribution to the research of bottom topography

Vema was derived for the name ***Ma**ud **Ve**tlesen*, the first owner's wife. Displacement of the ship is 750 tons. In 1953–1981, it belonged to the former Lamont-Doherty Geological Observatory (now Earth Observatory) in Palisades, New York. The vessel covered 1,250,000 miles with research works in all parts of the world ocean. Information in the Internet reads that the ship served as the cruising *yacht Mandalay* in the Caribbean until the operator went out of business in 2008.

The first deep hydrological measurements in the Vema Channel region were made from the famous R/V *Meteor* in 1925. In the following years, this region was studied within the expedition of the International Geophysical Year, the "Geosecs" program, and a number of other expeditions. Twenty two (mainly German, Russian, and British) expeditions with deep water measurements were carried out in this region since 1990. The data used in this study were either collected by the authors of the present work or downloaded from the World Ocean Database (WODB 2005) dataset. During the World Ocean Circulation Experiment (WOCE), significant progress was achieved in observation of abyssal variability in the South Atlantic. The Deep Basin Experiment, a core project of WOCE, was focused on the key passages that control the equatorward interbasin exchange of bottom waters (Hogg 2001).

After the WOCE completion, further surveys were mainly concentrated on the Vema Channel. Investigators from many countries, including Germany and Russia, were involved in the research. Revisits of the site in this post-WOCE phase were inspired by a warming trend in bottom water at the Vema Sill, which was first noted in Zenk and Hogg (1996), and long-term variations in other properties of water and transport in the entire channel. To date, the slowly growing time series of the coldest water measurements in the Vema Sill (Zenk et al. 2003) are known as CLIVAR research topics of national programs in Germany ("Marin-2", terminated in 2005) and Russia ("Meridian") (Morozov et al. 2003). A compilation of international activities since 1972 is given in Table 4.2.

Recent studies in the channel continue the earlier measurements. Numerous expeditions to this region have confirmed its utmost importance for the entire water dynamics in the Atlantic Ocean. In our research, we divide datasets from the region into three parts: Vema Sill, Santos Plateau, and Vema Extension.

In the beginning of the time series (1972–1990), satellite-supported navigation with today's high accuracy was not permanently available on research vessels. The local topography was less well known. Hence, CTD casts were taken in a wider range south and north of the exact location of the sill. Since 1991, the lowest temperatures were encountered consistently on the eastern half of the sections across the Vema Sill on all repeated sections (Hogg and Zenk 1997) (see Sect. 3.5). This phenomenon reflects the frictionally induced thick bottom boundary layer on the right side of the throughflow jet in the Southern Hemisphere. The concurrent secondary or helical circulation sharpens the interface above the mixed bottom layer. It causes the well-known isotherm pinching in the vertical temperature distribution (Jungclaus and Vanicek 1999).

Data accuracy and stability are important issues for all long-term observations. A steady technical improvement of CTD-profilers was undertaken in the last decades. The data we analyzed had been obtained either with Neil Brown instruments or Sea-Bird instruments (in the last decade). In the search for systematic trends

Table 4.2 Measurements in the Vema Channel

Vema Sill visit no	Vema Ext. visit no	Cruise/Expedition	Country of ship	Date mmm/yyyy	Number of stations		
					Vema Sill	Santos Plateau	Vema Ext.
		Meteor, profile 2 Dt Atl Expedition	DE	Aug/1925	(1)		
		Meteor, profile 4 Dt Atl Expedition	DE	Dec/1925	(1)		
1		Cato, leg 6 GEOSECS	US	Dec/1979	1		
2		Atlantis II-107,leg 2	US	Dec/1979	2		
3		Atlantis II-107,leg 8	US	Jun/1980	4		
4		Thomas Washington	US	Dec/1984	4		
5		Meteor 15, leg 1	DE	Jan/1991	6		
6	E1	Meteor 22, leg 4	DE	Dec/1992	5		1
7		Polarstern ANT 12/1	DE	Nov/1994	1		
8	E2	Meteor 34, leg 3	DE	Mar/1996	3	1	1
9	E3	Meteor 41, leg 3	DE	Apr/1998	4		1
10		Meteor 46	DE	Mar/2000	2		
11		Ak. Ioffe 11	RU	Nov/2002	7	2	
12	E4	Ak. S. Vavilov 17	RU	Nov/2003	6	2	5
13	E5	RSS Discovery 276	UK/DE	Dec/2003	1	1	1
14	E6	Ak. Ioffe 16	RU	Nov/2004	5		1
15		Ak. Ioffe 17	RU	Mar/2005	5		
16	E7	Polarstern ANT 22/5	DE	May/2005	5	1	1
17		Ak. Ioffe 19	RU	Nov/2005	5		
18		Ak. Ioffe 22	RU	Oct/2006	5		
19		James Clark Ross 160	UK/DE	May/2007	1		
20		Polarstern ANT 24/4	DE	Apr/2008	1		
21	E8	Ak. Ioffe 27	RU	Apr/2009	5		2
22		Polarstern	DE	Apr/2009	1		
1–22	E1–E8	Sums			79	7	13

in abyssal bottom temperatures in the Weddell Sea, Robertson et al. (2002) also faced uncertainties in different cruises similar to our investigation. According to the WOCE standards, the accuracy of the high-quality CTD data is leveled nowadays at 0.002–0.003°C for potential temperature and approximately 0.003 psu for salinity. Absolute pressure uncertainties for deep-sea sensors are expected to lie below ±3 dbar. Most expeditions to the Vema area were carried out with pre- and post-cruise calibration checks of the operated CTD probes. Instrumental stability has improved significantly over the last three decades.

After the Deep Basin Experiment was completed as part of the World Ocean Circulation Experiment, the study of Antarctic Bottom Water transport was concentrated mainly in the Vema Channel. In the late 1990s and beyond, detailed study of the abyssal flow was carried out during several cruises of German research vessels *Meteor* and *Polarstern* to the Vema Channel. Besides long-term observations of the stratification and its conjectured changes, the studies were concentrated on the high-speed flow structure and the quantification of mixing between near-bottom water masses (Zenk 2008).

In 2002, Russian scientists from the *Shirshov Institute of Oceanology (Moscow)* joined the international research program in the region. In November 2002, the first Russian section across the Vema Channel was occupied within the Russian National Program "Meridian". These studies were planned together with the German colleagues from the *Institut für Meereskunde* in Kiel (now IFM-GEOMAR). The section crossed the Vema Channel along 31°14′ S. Seven stations were occupied using the Neil Brown CTD instrument. In November 2003, three sections across the channel were occupied in the northern, central, and southern parts of the channel. The sections were made across three narrow parts of the channel. In addition, during the cruise onboard R/V *Akademik Sergey Vavilov* in November 2003, a CTD section along the Vema Channel was occupied with a spacing of 30–40 miles. Stations along the 600-km-long sections were located at the eastern wall of the channel.

In November 2004, the section at the Vema Sill was repeated again. The section at the Vema Sill was displaced 2 miles northward to 31°12′ S to avoid collision with the German moorings at 31°15′ S. Two more stations were occupied in the northern (Vema extension) and central parts of the channel. In March, May, and October 2005, the section across the Vema Channel at 31°14′ S was repeated three times. In October 2006 and March 2009, the section was repeated again.

During the period from 2003 to 2009, the Vema Channel was alternatively revisited by the German and Russian scientists. In 2003, German scientists performed measurements in this region from R/V *Discovery* shortly after the Russian visit to the channel. This made it possible to intercalibrate the field measurements made by the German and Russian scientists. Differences in the measurements of temperature and salinity at one point in the coldest jet near the Vema Sill were equal to 0.001°C and 0.001 psu, respectively. In fact, similar insignificant observational differences were obtained in the spring of 2009. Potential temperatures and salinity values measured from R/V *Akademik Ioffe* and *Polarstern* only 3 weeks apart, differ by 0.001°C and 0.003 psu, respectively.

The deepest part of the southern entrance to the channel is located in its southern part at 33°33.6′ S. A seamount with its summit at 3,900 m forms the eastern slope of the channel here. A sedimentary ridge (Emery Drift) with the depth of the crest at 4,500 m against the background depth of 4,650 m is located at the western side of the channel. In October 2005, a section was occupied here from the deepest point of the seamount slope located at a depth of 4,900 m in the western direction to the slope of the Emery Drift.

In 2005, 2006, and 2009, the CTD measurements were supplemented by lowered ADCP profilers. All Russian and German measurements were conducted with the participation of the authors of the present work. Visits to the Vema Sill are schematically shown in Fig. 4.4 as sections of potential temperature across the channel.

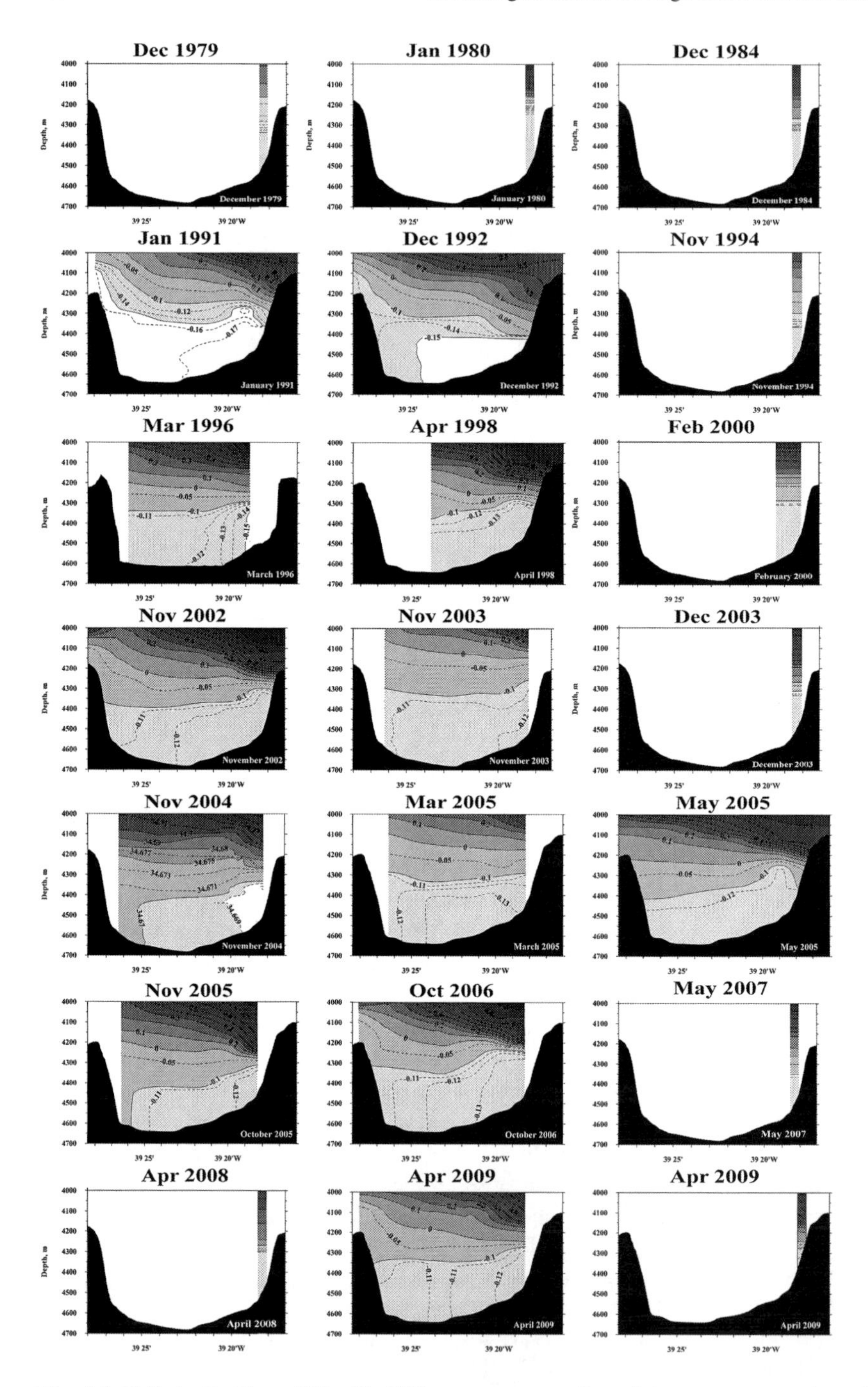

Fig. 4.4 Visits to the Vema Sill with CTD measurements. Potential temperature sections (°C) across the channel are shown if several stations were occupied in the channel. A narrow stripe shows individual stations in the Vema Sill region occupied when research vessels passed the region

4.2.3 Deep and Bottom Waters

The Vema Channel is interesting first of all as the natural pipeline for Antarctic Bottom Water propagation from the Argentine Basin to the north. Actually the channel in its topographic boundaries is a pipeline for Weddell Sea Deep Water, the densest and coldest component of Antarctic Bottom Water. Its flow through the channel is controlled by the depths of sills along the channel. This choke point of bottom water transport to the north is an excellent region for studying the mechanisms of bottom water propagation in the northern direction. The CTD and LADCP measurements reveal that the cold core of Weddell Sea Deep Water propagates as a well-mixed jet (occasionally split into two jets) displaced to the eastern wall of the channel. Thickness of the jet is 200–500 m. Within the sections of the jet, the variation range of its potential temperature and salinity is 0.1°C and 0.005 psu, respectively.

In this section, we show the thermohaline properties of bottom waters in the channel and waters above the channel. Figure 4.5 shows the θ/S-diagram for deep and abyssal waters. Names of water masses are indicated.

Researchers of Antarctic Bottom Water draw its upper boundary at different levels varying from $\sigma_4 = 45.87$ (Memery et al. 2000) to $\sigma_4 = 45.92$ (Fu 1981; Roemmich 1983). Thus, the difference between locations of the boundary isopycnal surface in the Vema Channel can be as large as 150 m.

Long-term measurements show that the Weddell Sea Deep Water core is characterized by minimum potential temperature θ equal to approximately −0.120°C (in situ temperature +0.225°C) and salinity 34.67 psu. The concentration of silicates is as high as 118 μmol/kg and phosphates is up to 2.3 μmol/kg. The local minimum of oxygen concentration (220 μmol/kg) is located at a distance of 500 m from the bottom, which allowed Vanicek and Siedler (2002) to divide the bottom water layer into two water masses.

4.2.4 Section Along the Channel

Let us consider the distribution of potential temperature in the bottom layer along the Vema Channel. In November 2003, the stations were set up along the line of maximum depths approximately at a distance of 700 m from the eastern slope of the channel. In order to find the desired location of the station at planned latitude, the ship moved westward along the slope with echo sounder measurements using three echo sounders: narrow beam ELAC LAZ 4700, multibeam ECHOS XDM, and Atlas Parasound. After the location of each future station determined at the bottom of the slope, the ship was slowed down, turned back, and stopped at the point to make a CTD cast. Figure 4.6 shows the latitudinal variation in the distribution of potential temperature. Figure 4.7 shows the distribution of salinity along the section in 2003.

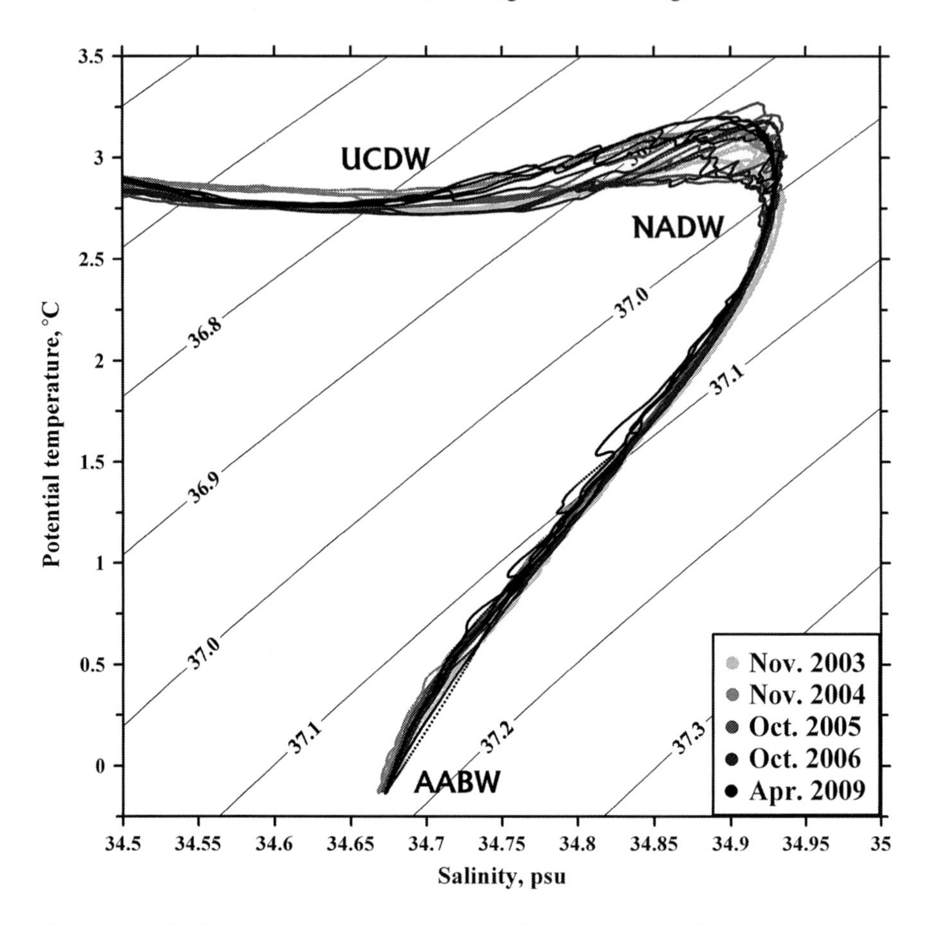

Fig. 4.5 The θ/S-diagram for deep and abyssal waters in the Vema Channel region. (Based on the results of CTD profiling in different years. Contour lines of potential density σ_2 are shown)

Measurements in the cold jet along the entire Vema Channel show that temperature in the jet increases due to mixing with overlying waters during the bottom water propagation from south to north. Latitudinal variation in temperature in the coldest part of the jet is shown in Fig. 4.8. Approximately the same linear trend of temperature was recorded in 2003, 2004, 2005, and 2009. However, the measurements in 2009 revealed a stronger warming of bottom waters in the northern part of the channel.

The channel widens in the Vema extension between 27° S and 28° S. It is noteworthy that the cold jet was not found here near the eastern slope of the channel. Therefore, the measured temperatures are not plotted on the graph. Due to widening of the channel here, the Ekman friction has less strength with which to displace the jet to the eastern slope.

The figures show a clear increase in the values of these characteristics. Data in 2003 show that potential temperature increases from south to north from −0.135

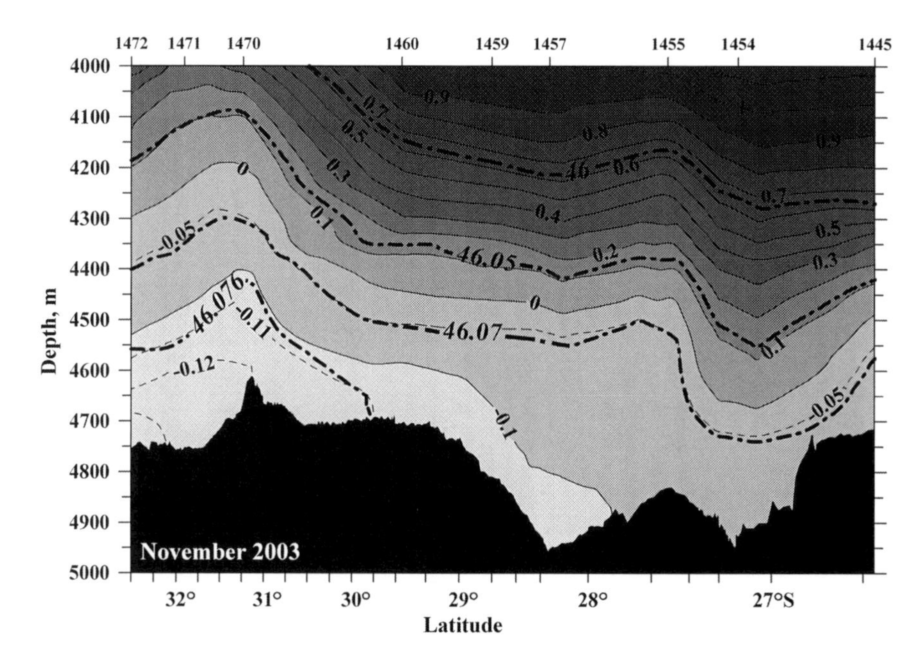

Fig. 4.6 Potential temperature (°C) along the Vema Channel in 2003 (measured at the foot of the eastern slope). *Contour lines* of density are shown with *thick dash-and-dot lines*

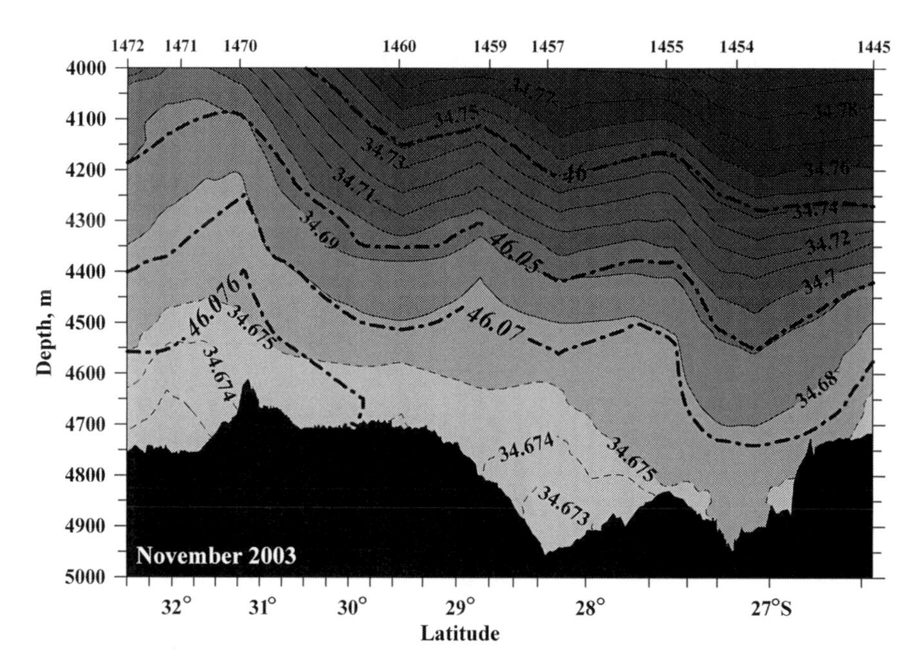

Fig. 4.7 Salinity (psu) along the Vema Channel in 2003. *Contour lines* of density are shown with *thick dash-and-dot lines*

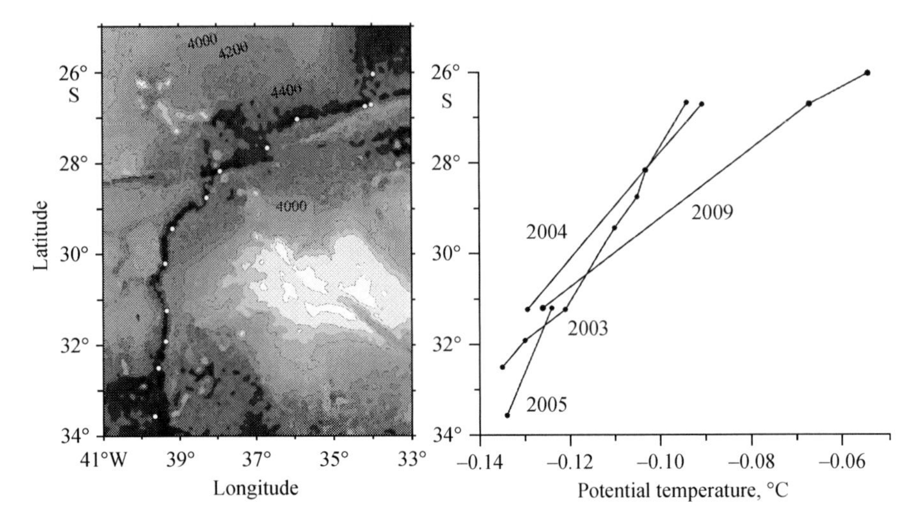

Fig. 4.8 Latitudinal variation in potential temperature (°C) along the Vema Channel in 2003, 2004, 2005, and 2009. *White dots* in the *left panel* show the stations at the bottom of the eastern slope. Latitude scales are the same in the *left* and *right panels* of the graph

to −0.094°C, while salinity increases from 34.67 to more than 34.68 psu at the extreme southern and northern points of the section. The maximum latitudinal gradients are confined to the narrowest part of the channel. The coolest and densest part of Weddell Sea Deep Water cannot overflow the sills in the channel. Moreover, mixing with the overlying waters increases the temperature and salinity of the flow.

Three temperature profiles along the channel below 4,000 dbar (in the northern, central, and southern parts) based on the measurements in 2003 are shown in Fig. 4.9. The greatest temperature increase is observed between 4,000 and 4,500 dbar at the boundary with North Atlantic Deep Water and the corresponding velocity shear due to oppositely directed flows. A temperature increase is not accompanied by strong changes in θ/S-characteristics. Graphs of θ/S-curves in the southern, central, and northern parts of the channel are shown in Fig. 4.10. The main variation is observed in the narrow channel.

Linear regression of the sequence of temperature minima of Weddell Sea Deep Water along the Vema Channel suggests a horizontal temperature gradient of $\sim 5.7 \times 10^{-8}$°C m^{-1}. Assuming pathway lengths of 220 and 760 km between the sill position and the downstream sites, one expects a spatially conditioned offset in potential temperature of ~0.013 and ~0.043°C, respectively. The difference between two salinity curves comprises the latitudinal increase of near-bottom salinity of $\sim 2 \times 10^{-9}$ psu m^{-1}. The salinity signal, however, is noisier than the temperature signal. Nevertheless, we note a density-compensating horizontal correlation between potential temperature and salinity along the flow path of Weddell Sea Deep Water.

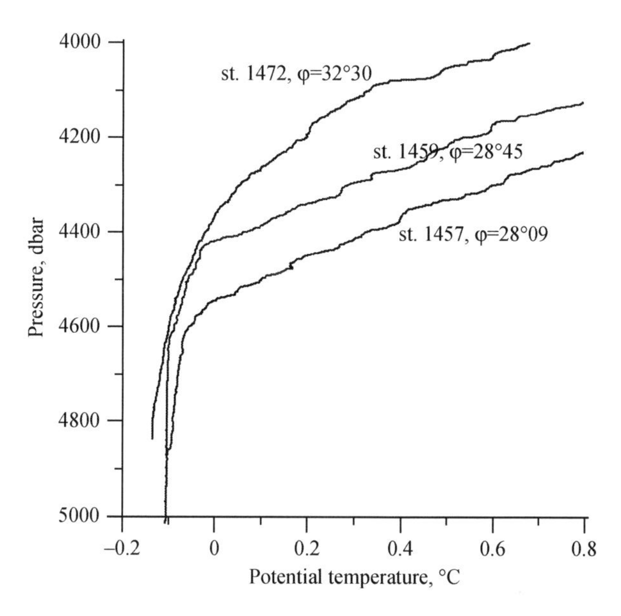

Fig. 4.9 Changes in the vertical profile of potential temperature (°C) along the Vema Channel in 2003

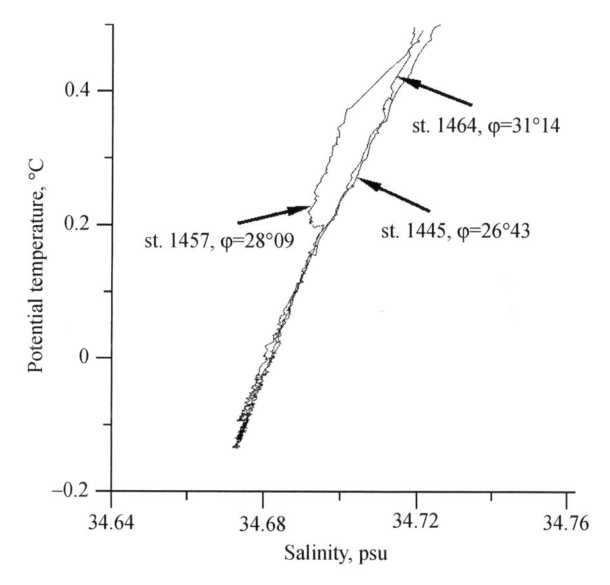

Fig. 4.10 Changes in θ/S-characteristics along the Vema Channel in 2003

4.2.5 Structure of the Flow. Sections Across the Channel

In November 2003, three CTD sections were made across the channel in the northern, central, and southern parts of the channel. Two sections of potential temperature and salinity across the channel occupied in 2003 in the northern and central parts of the channel are shown in Figs. 4.11 and 4.12, respectively. The third section

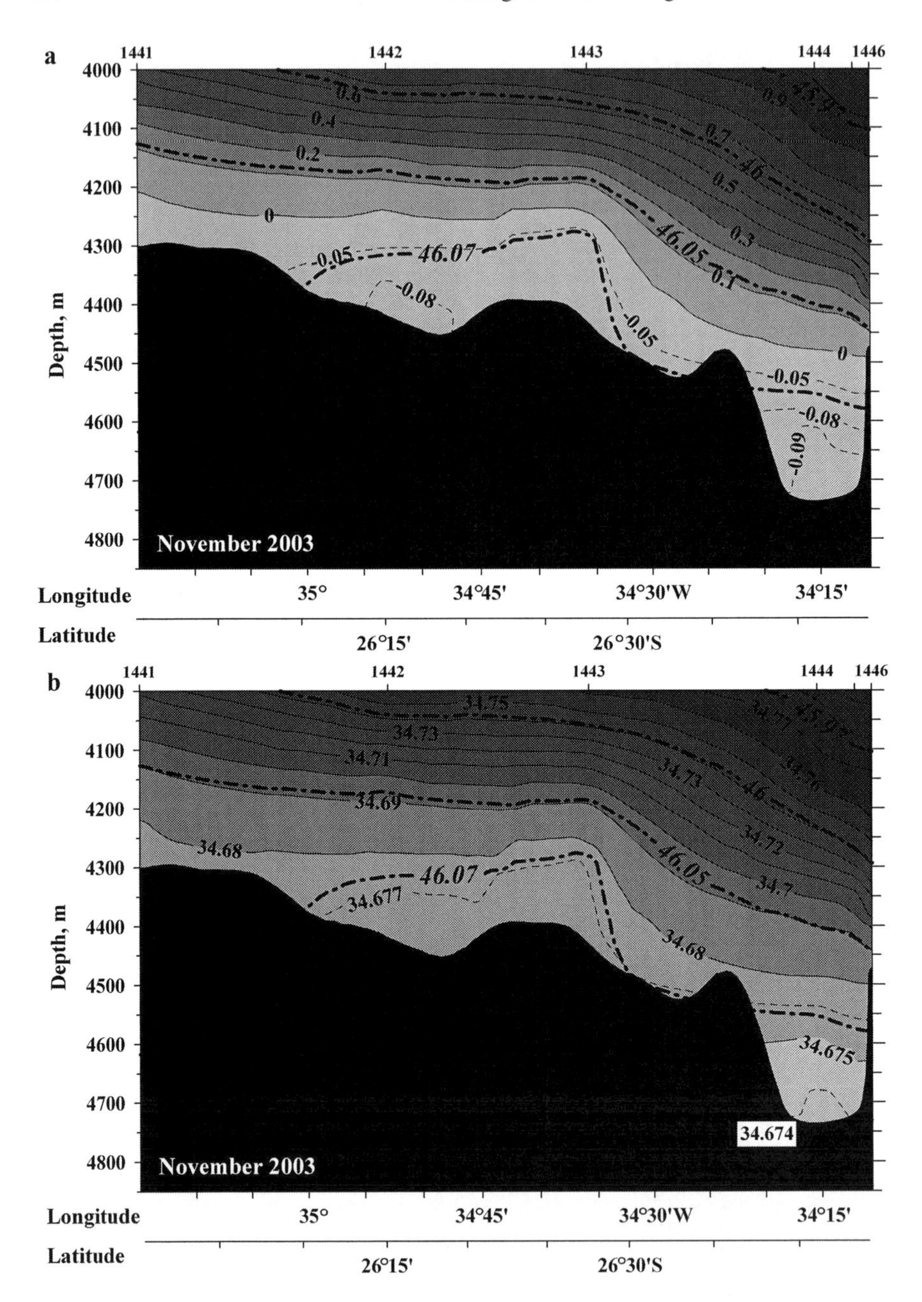

Fig. 4.11 Sections of potential temperature (°C) (**a**) and salinity (psu) (**b**) across the Vema Channel in 2003 at 28°09′ S (Vema extension). *Contour lines* of density are shown with *thick dash-and-dot lines*

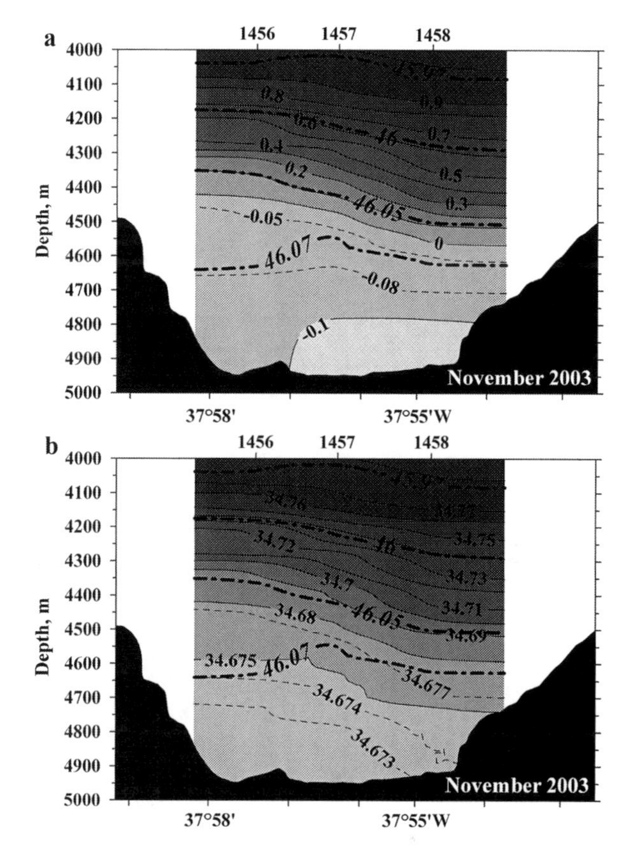

Fig. 4.12 Sections of potential temperature (°C) (**a**) and salinity (psu) (**b**) across the Vema Channel in 2003 at 26°43′ S. *Contour lines* of density are shown with *thick dash-and-dot lines*

at 31°14′ S is shown in Fig. 4.13f. All sections were made in the narrow parts of the channel. On all sections, the jet was displaced to the eastern slope of the channel owing to the Ekman bottom friction. Correspondingly, the temperature in the coldest parts of the jet was increasing from south to north.

Distributions of temperature and salinity over sections across the channel near the Vema Sill along 31°12′ S or 31°14′ S in 1991, 1992, 1996, 1998, 2002, 2003, 2004, 2005 (March, May, and November), 2006, and 2009 are shown in Figs. 4.13 and 4.14. Contour lines of density are also shown in the figures. Measurements made in many expeditions showed that the Antarctic Bottom Water core is usually displaced to the eastern slope of the channel. The core is a jet of well-mixed water. Thickness of the core is approximately 150–200 m. The flow is strongly ageostrophic.

Usually, the width of the cold core is approximately one fourth of the channel. The upper boundary of the core migrates between 4,300 and 4,400 m. High location of the boundary results in a sharp thermocline over the core as, for example, in 2009 (Fig. 4.13l). In 2003 and 2004, it was elevated over the bottom. Displacement

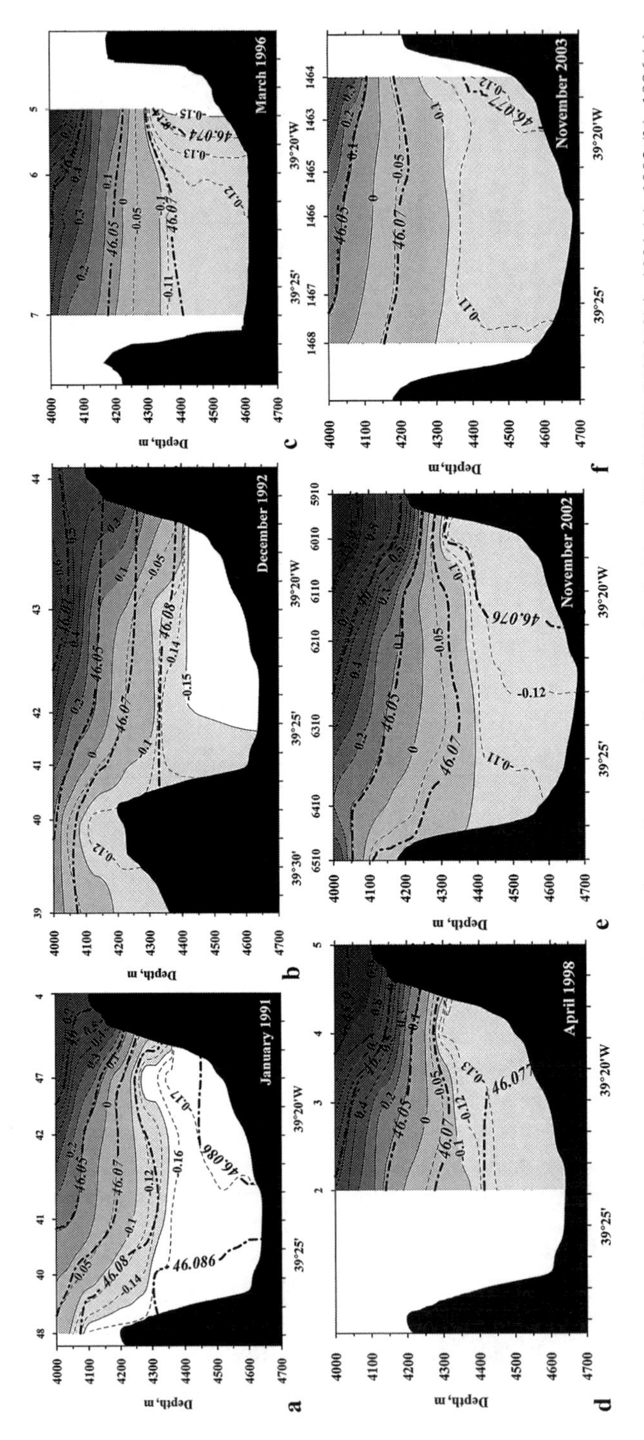

Fig. 4.13 Distributions of potential temperature (°C) over the sections across the channel at the Vema Sill (31°12′–31°14′ S) in 1991 (**a**), 1992 (**b**), 1996 (**c**), 1998 (**d**), 2002 (**e**), 2003 (**f**), 2004 (**g**), March 2005 (**h**), May 2005 (**i**), November 2005 (**j**), 2006 (**k**), and 2009 (**l**). Axes *y* and *x* show depth and longitude, respectively. The year and month of measurements are indicated at each section. *Contour lines* of density are shown with *thick dash-and-dot lines*

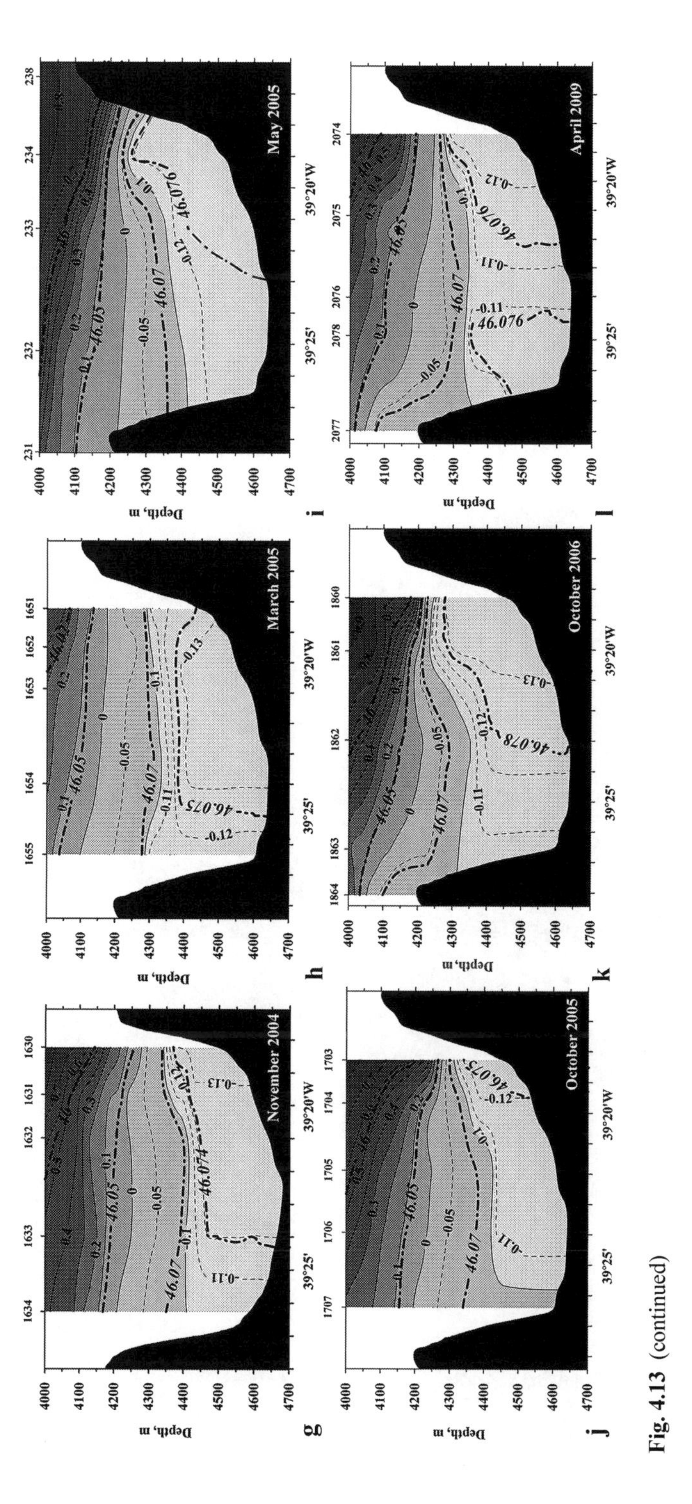

Fig. 4.13 (continued)

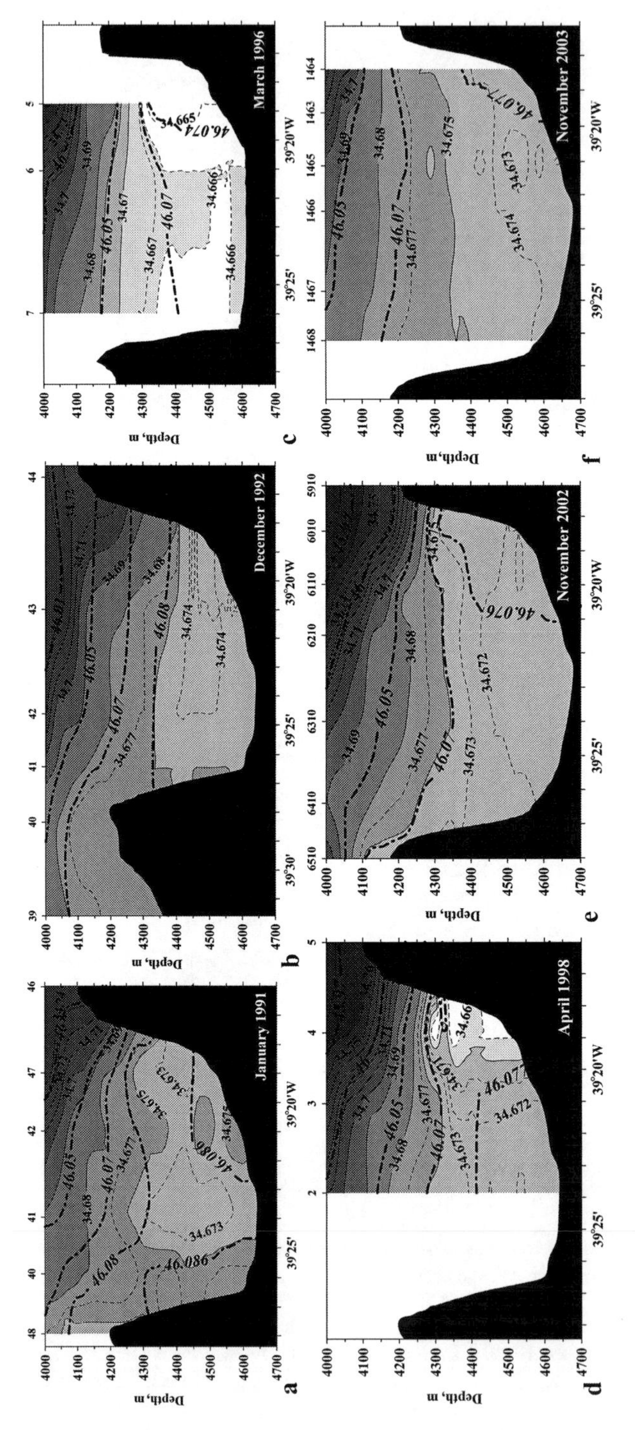

Fig. 4.14 Distribution of salinity (psu) over the sections across the channel at the Vema Sill (31°12′–31°14′ S) in 1991 (**a**), 1992 (**b**), 1996 (**c**), 1998 (**d**), 2002 (**e**), 2003 (**f**), 2004 (**g**), March 2005 (**h**), May 2005 (**i**), November 2005 (**j**), 2006 (**k**), and 2009 (**l**). Vertical axis: depth; bottom axis: longitude. The year and month of measurements are indicated at each section. *Contour lines* of density are shown with *thick dash-and-dot lines*

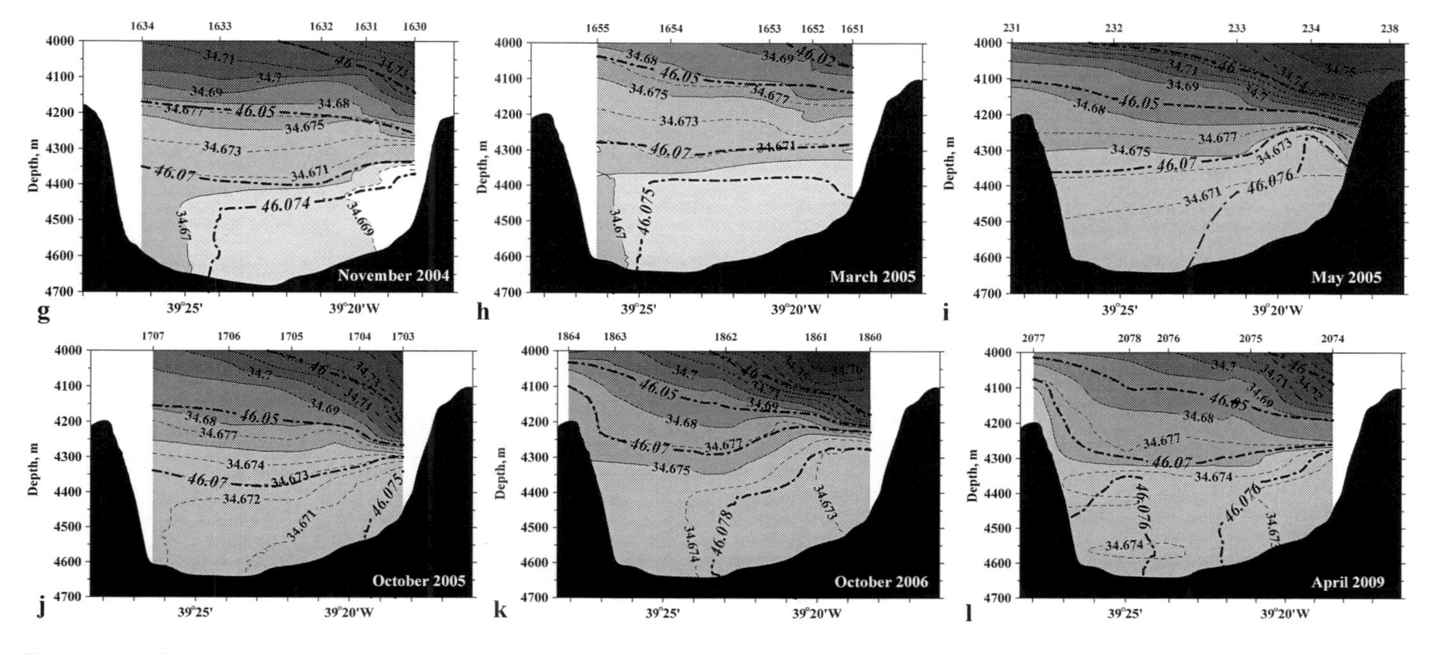

Fig. 4.14 (continued)

of the jet from the easternmost position, elevated above the bottom to a position closer to the middle of the channel, is likely caused by meandering of the jet. The eastern displacement of the coldest core is explained by the Ekman force due to bottom friction. In the Southern Hemisphere, the Ekman friction is directed to the right of the flow, which agrees with the modeling results of Jungclaus and Vanicek (1999). In March 2005, the core occupied the most distant position from the slope. This is the only observed case of such displacement. Possibility of meandering of the flow even in a very narrow channel is suggested by the fact that the jet can notably shift to the east and up the slope and simultaneously occupy a position closer to the middle of the channel. Mixing with the overlying waters is related to lateral and bottom friction.

In the upper part of the flow above the channel, the cold low saline core of the flow is displaced to the west. This core is found over the western wall of the channel. This phenomenon was observed in 1991, 1992, 2002, 2006, and 2009. In 2002, the low values of temperature and salinity observed over the western slope of the channel were as follows: potential temperature was less than −0.04°C and salinity was less than 34.676 psu. At the eastern slope, such values of temperature and salinity are observed at depths greater by 150–200 m. Only in these years, the measurements were made not only in the deepest part of the channel but also over the eastern and western slopes of the channel. We suppose that this core always exists. However, observations at the upper part of the western slope are not available for each section, because several times the western part of the section did not reach the western slope. This jet exists due to displacement by the Coriolis force.

Observations in different years show that the structure details of the flow changes every time. However, the general feature associated with the displacement of the jet to the eastern slope at the bottom is almost the same.

In 1991, we noted that the values of potential temperature and salinity were the lowest (θ less than −0.17°C, $S < 34.672$ psu). In 2003, the distribution of salinity was more homogeneous compared to the sections in other years, indicating the fact of stronger mixing in the bottom layer.

We also found variation of the position of isotherms and isohalines by depth, which can naturally be explained by different intensities of Antarctic Bottom Water transport. Variation in the position of 0°C isotherm can be as high as ~200 m between the uppermost and lowermost positions. The highest position of this isotherm was recorded in 2003 and March 2005.

The lowest values of temperature and salinity observed in November 2004 and March 2005 were related to greater amounts of Antarctic Bottom Water in the bottom layer of the channel. In 2004, the strongest horizontal stratification of the flow was observed compared to all other years of observation. Thermohaline properties of the flow in the eastern and western parts of the channel are slightly different. The western part of the flow is usually more saline and warmer.

Year to year observations show changes in the jet structure. The cold core is mixed well in the vertical direction, but horizontal stratification is always observed.

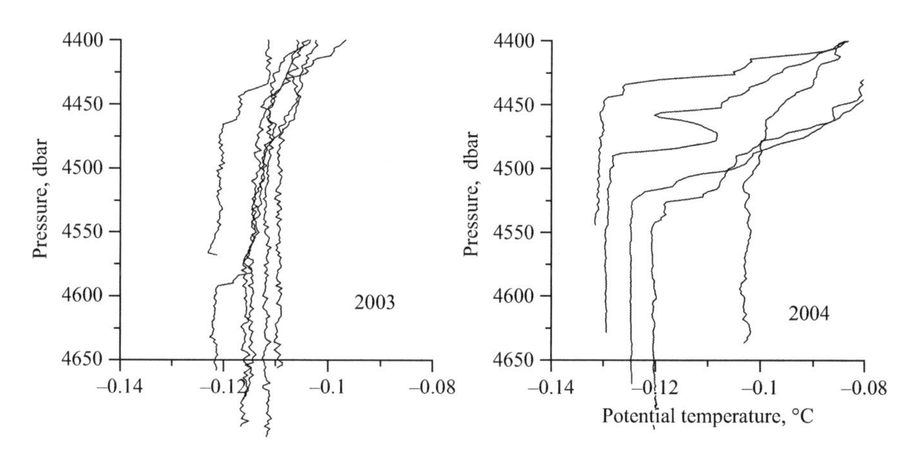

Fig. 4.15 Vertical profiles of potential temperature (°C) across the Vema Channel (section along 31°14′ S) in 2003 and 2004

In 2002, almost the whole jet was mixed, but the warmer water was found only in the western part of the channel (station 1964). In 2003, the cold part of the jet was located in the eastern part of the channel (stations 1463, 1464), while the remaining part of the jet was mixed. In 2004, stratification of the jet was greater than in the previous years. The entire jet was horizontally stratified. In 2005, the jet was mixed in the central part; cold water was observed in the eastern part of the jet; and the warmer water was located in the western part. Vertical profiles of potential temperature in 2003 and 2004 are shown in Fig. 4.15.

We determined the position of the Antarctic Bottom Water core using the classic principles: local minima of temperature and salinity (Wüst 1936). In many cases, they do not coincide on the sections across the channel. The most typical spatial structure of the flow is the location of waters with minimum temperature and salinity and maximum density in the lower part of the flow at the eastern slope with differences of potential temperature up to 0.02°C and salinity up to 0.003 psu across the deepest part of the channel. Sometimes, the distribution was even more homogeneous. In 1991 and 2009, density increased from the center to the eastern and western slopes. In 1991, the central part of the channel was occupied by anomalously saline waters. Thus, three jets could be distinguished on the basis of minimum salinity. In 1991, a strong temperature difference was observed between the eastern and western parts of the channel (>0.04°C).

When the cold jet is strongly displaced to the eastern slope and elevated, a strong temperature gradient is observed at the upper boundary of the slope, as was the case in 2006 (Fig. 4.16). This happens because the lower isotherms are elevated by the core of the flow, while the upper remains at their previous depths. The vertical gradient is 0.02°C/m. In previous years, such a strong gradient was not found. Strong layering is also characteristic of the gradient zone.

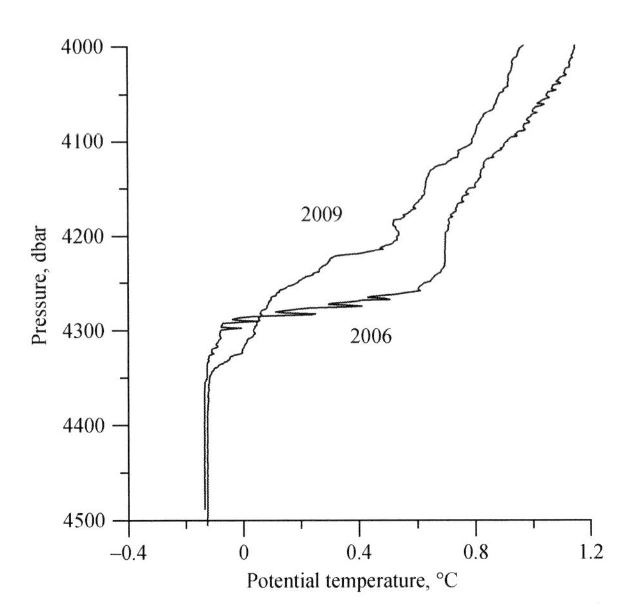

Fig. 4.16 Vertical profiles of potential temperature (°C) at the eastern slope of the Vema Channel. At the eastern station (station 1860 in 2006), a transition zone with sharp vertical temperature gradients was found above the mixed jet of Antarctic Bottom Water

4.2.6 Trends in Potential Temperature and Salinity of the Coldest Bottom Water Observed Since 1972

Antarctic Bottom Water represents the coldest and deepest layer of the western South Atlantic. Significance of abyssal pathways for global thermohaline circulation and ventilation, as well as the role of Antarctic Bottom Water in interhemispheric exchange, have turned the Vema Channel into an object for continued observations beyond the end of the WOCE decade. The ongoing motivation for numerous revisits (by now 22) is maintained by a general decadal warming trend in Antarctic Bottom Water that was first noted in Zenk and Hogg (1996). Due to the small magnitude of the expected signals, detection of significant changes in bottom water properties requires equipment with the highest accuracy and stability.

The length and accuracy of the CTD time series from the Vema Channel appear sufficient to allow an analysis of the associated salinity record as well. The latter record appears to be anti-correlated with the observed decadal temperature rise. In fact, this result implies a long-term trend of decrease in the density of abyssal waters in the Vema Channel and possibly the entire deep western South Atlantic.

Comparative analysis shows that it is difficult to isolate a trend in spatially and temporally aliased shorter series. As an example, Fukasawa et al. (2004) found a general warming of the deep Pacific along 47° N between 140° E and 120° W, which is negatively correlated with salinity changes in the depth range 500–3,000 m. However, below 5,000 m they also reported a basin-wide warming by 0.005°C in 14 years but with "no significant change in the salinity field".

Such long-term changes of deep water mass characteristics are needed as observational constraints in ocean circulation models. The objective of this section is to describe the evolution of the abyssal warming phenomenon in the Vema Channel in the light of the historic database and recent site visits one decade after the last progress report by Hogg and Zenk (1997). Annual visits of the Russian and German research vessels to this site, gave new data on the variability of Antarctic Bottom Water through the channel.

Time series of potential temperature from the two sites in the Vema Channel (standard section at 31°12′–31°14′ S and Vema extension in the northern part of the channel) and in the Santos Plateau are shown in Fig. 4.17. The bulk of the data collected in the Vema Channel originates from the sill region.

Furthermore, the potential temperature series from the Vema Extension with eight visits round off the more complete series from the sill. Two of the three potential temperature θ curves in Fig. 4.17 are shifted upward against the base curve (from the sill). Natural averaged offsets of 0.025 and 0.039°C were obtained for the period 1990–2009. Within the statistical error range, the implicit spatial increase in potential temperature θ coincides reasonably with the horizontal temperature structure from a quasi-meridional section in November 2003.

The section used to compare temperature measurements in different years is located at 31°12′–31°14′ S. Long-term observations reveal a trend of increase in potential temperature in the core of Antarctic Bottom Water with time. Before 1993, potential temperature at 31°14′ S was −0.18°C, while the deviations did not exceed 0.01°C. From 1972, potential temperature increases (Fig. 4.17). During the period

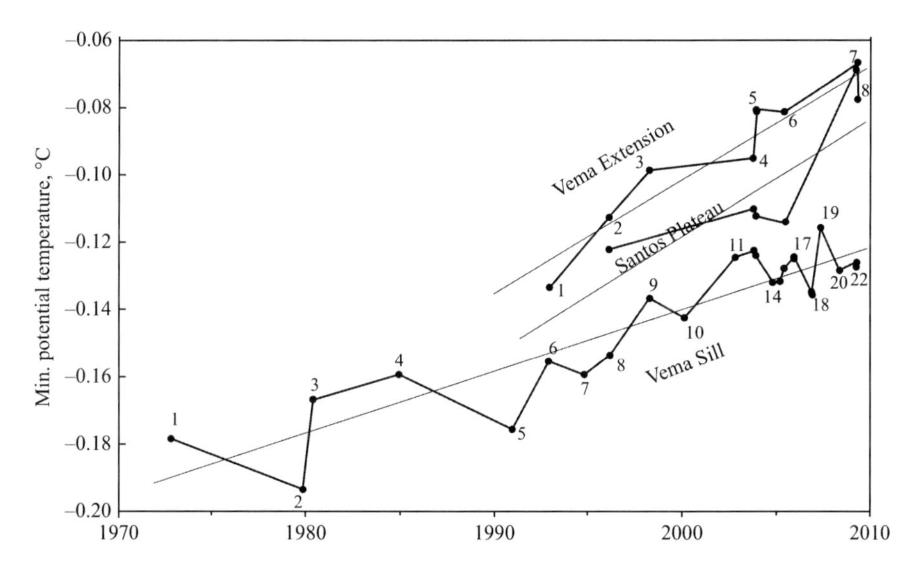

Fig. 4.17 Evolution of potential temperature (°C) of the coldest Antarctic Bottom Water in the Vema Channel and Santos Plateau. Numbers 1–22 (Vema Sill region, see Table 4.2; the reference point of observations is located on the eastern slope of the Vema Channel within 31°12′–31°14′ N, 39°17′–39°19′ W) and *1–8* (extension) designate revisits. The linear regression lines for Vema Extension and Santos Plateau start in 1990. (Modified and updated from Zenk and Morozov 2007)

from 1993 to 1998, the temperature increased to −0.135°C. This was the period of significant temperature rise reported in Zenk and Hogg (1996) and Hogg and Zenk (1997). Actually, we can speak about the temperature rise that started after 1991. From this period, observations were started at the standard section located within 31°12′–31°14′ N, 39°17′–39°19′ W. In November 2002, the measurements recorded a temperature of −0.123°C. In 2003, the temperature increased to −0.120°C. Unexpectedly, potential temperature in the coldest part of the jet decreased to −0.131°C in 2004. At the same time, a strong horizontal stratification of the jet was observed in 2004. In 2005, the temperature remained at the level of 2004. During the following years, temperature fluctuations of the cold jet were observed with an amplitude of approximately 0.02°C.

The graph shown in Fig. 4.18a illustrates the time evolution of the vertical distribution of potential temperature on the standard section across the channel at 31°12′–31°14′ S, while Fig. 4.18b shows the time evolution of the vertical distribution of salinity. It seems that the warming trend observed before 2003 was related to the propagation of a large mass of cold fresh water, which changed later to fluctuations of smaller water masses propagation.

Next, we investigate temporal evolution of the coldest Antarctic Bottom Water (Weddell Sea Deep Water) along the Vema Channel in more detail. The first part of the record has already been analyzed in Hogg and Zenk (1997). They discussed in

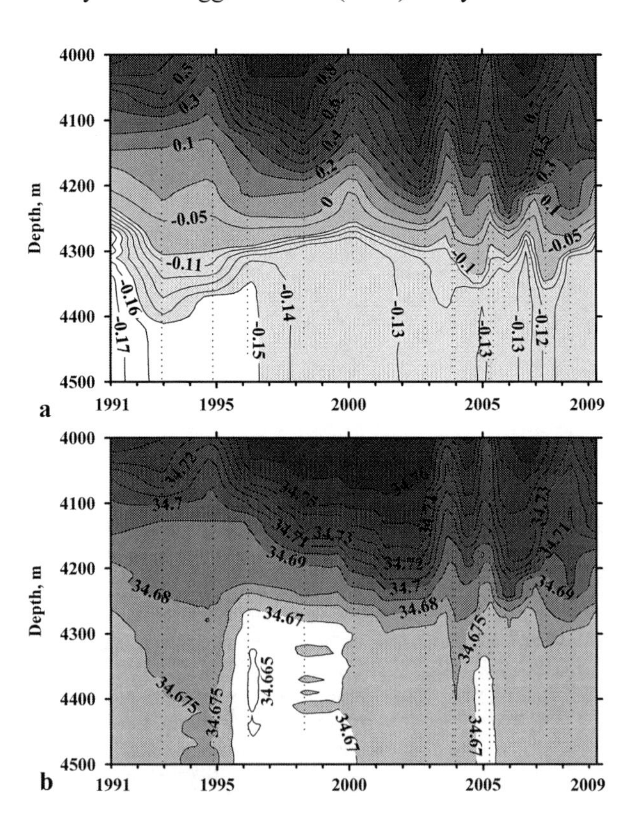

Fig. 4.18 (a) Vertical section variations in potential temperature (°C) versus time at the Vema Sill at 31°12′–31°14′ S. (b) Vertical section variations in salinity (psu) versus time at the Vema Sill at 31°12′–31°14′ S

detail at least 14 visits to the Vema Channel with high-quality hydrographic measurements until 1996. Eleven datasets of the total 14 measurements were judged successful. Zenk and Hogg (1996) found little variation of the coldest Antarctic Bottom Water passing through different pathways across the Rio Grande Rise before 1990. In the Vema Channel, fluctuations were smaller than 0.005°C. After such stagnant temperatures over 20 years, a change of +0.030°C in the interval 1991/1992 was rated as "quite significant."

Our measurements show that the annual temperature increase near the Vema Sill is a few mK. Usually, time gradient of temperature is in the range of 0.002–0.003°C year^{-1} (though with occasional fluctuations of O(15) 10^{-3}°C year^{-1} on shorter scales (Zenk and Hogg 1996)).

A most relevant observation was reported in Robertson et al. (2002). Based on data from 1912 to 2000, they found indications for a slight temperature rise at the depth of 1,500–3,500 m in the eastern Weddell Sea, i.e., at the Weddell Sea Deep Water source. The weak temperature signal seemed to be uncompensated or insignificantly compensated by salinity. The issue of how far the warming is related to changes in the seasonal cooling or to advection modulated by the Weddell Front remains open in this study.

Let us consider time variations in temperature in the bottom layer over the entire length of the Vema Channel. Figure 4.19 shows temperature measurements at different points of the Vema Channel as a function of latitude and time. Different symbols indicate measurements in different years. One can conclude from Fig. 4.19 that the positive temperature trend is not at all related to temperature sampling in different regions of the channel. The character of time variability over the entire length of the channel is the same as was previously noted on the standard section across the channel at 31°12′ S. The line in the lower part of the graph shows the bottom depth at each station. The measurement in 1979 made at a depth of 4,200 m over the upper part of the slope demonstrates anomalously low temperature. One can suppose that the temperatures are even lower down the slope. Minimum values

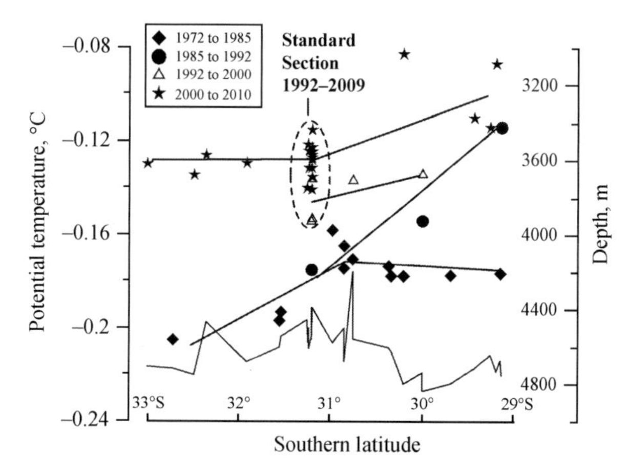

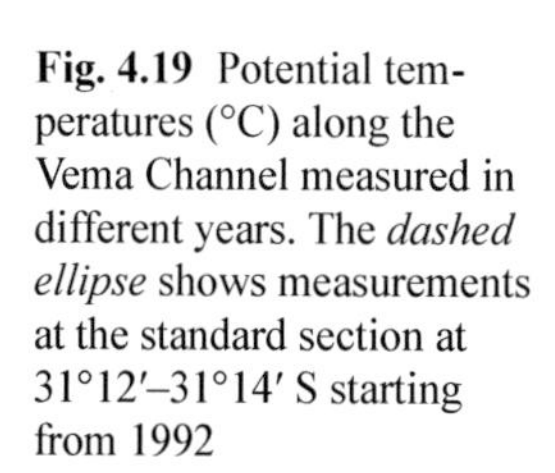
Fig. 4.19 Potential temperatures (°C) along the Vema Channel measured in different years. The *dashed ellipse* shows measurements at the standard section at 31°12′–31°14′ S starting from 1992

of potential temperature are recorded at the eastern slope of the channel rather than at the maximum depths, similarly to the measurements at the standard section.

We distinguish three periods of potential temperature change in the channel (see also Figs. 4.17 and 4.18a). The minimum temperatures over the entire length of the channel were recorded in 1972–1982. A temperature rise was recorded in 1991–2003. After 2000, potential temperatures along the entire channel remained high, but small time fluctuations were found. From 1972 to 1991, potential temperatures were low and no significant fluctuations were found. No doubt, this lack of fluctuations can be explained by the fact that the measurements were rarely taken.

We note a specific character of potential temperature variations along the channel. During the period from 1979 to 1980, potential temperatures increased from the south to the Vema Sill and did not almost change in the northern part of the channel. After 2000, potential temperatures only slightly increased from the south to the Vema Sill and then increased rapidly.

Bottom water properties are subjected to interannual and long-term variations (Fahrbach et al. 2004; Rintoul 2007) in the Southern Ocean, which are recorded even in the northern parts of the Pacific (Fukasawa et al. 2004) and Atlantic oceans. Increase in potential temperature near the Weddell Sea bottom is one of the possible explanations for the temperature trend observed in the Vema Channel. It is likely that this trend is related to the general warming in the Southern Hemisphere in the 20th century; mean temperature of air, sea surface, and land in the Southern Hemisphere increased approximately by half a degree (New et al. 2000). Temperature variation over the globe from 1850 up to the present day (2009) is shown in Fig. 4.20. Analysis of satellite data from 1979 to 1998 demonstrates that this warming was accompanied with an increase in Weddell Sea surface temperature by 0.01–0.02°C/year (Comiso 2000).

A correlation between the variability of conditions at the Southern Ocean surface and flux of bottom water was reported in a series of publications (Comiso and Gordon 1998; Fahrbach et al. 1995; Jacobs and Guilivi 1998). As to the deep and bottom waters in the Weddell Sea, Weddell Sea Deep Water was becoming warmer near the source of its formation at a rate of +0.003°C/year (Fahrbach et al. 2004). From 1970 to 1990, the warming trend of Weddell Sea Deep Water was not statistically significant (Robertson et al. 2002). According to Robertson et al.

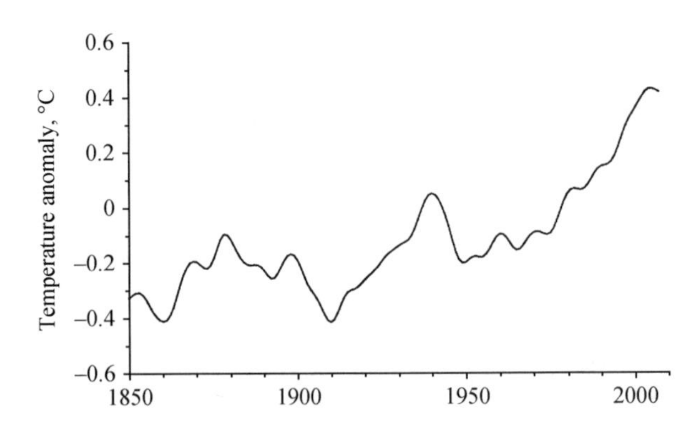

Fig. 4.20 Evolution of the anomaly of mean temperature (°C) at the Earth's surface from 1850 to 2009

(2002), Warm Deep Water in the Weddell Sea has a significant statistical warming trend of +0.012 (±0.007)°C/year from 1970 to 1990 and a negative trend from 1998 to 2002 (Fahrbach et al. 2004). The temperature increase trend in the Weddell Sea is not monotonic: this fact suggests that the explanation of temperature increase in the Vema Channel and its correlation with the bottom temperature in the Weddell Sea is not straightforward.

The increase in water temperature in the Vema Channel can be considered a remote response (with a time shift of tens of years) to the aforementioned temperature increase in the Antarctic region. Anomalies formed in the Antarctic region spread along the slope of South America to the Southwest Atlantic (Whitworth et al. 1991; Locarnini et al. 1993; Coles et al. 1996; Orsi et al. 1999). According to Coles et al. (1996), these anomalies were recorded near 18° S approximately 5–11 years after their appearance in the Weddell Sea. Analysis of hydrographic data in the Argentine Basin in 1988–1989 and their comparison with the results of measurements 8–10 years before this period showed a decrease in the amount of coldest waters with potential temperature θ less than –0.2°C (Coles et al. 1996). On the other hand, Smythe-Wright and Boswell (1998) reported that the age of bottom waters in the Argentine Basin is approximately 30 years. Thus, the mean velocity of Weddell Sea Deep Water spreading in the basin can be estimated at ~0.7 cm s^{-1}, which seems reasonable in order of magnitude. A warming trend in the bottom waters of approximately +0.04°C was found in 1989–1995 and 2005–2003 in the Brazil, Argentine, and South Georgia basins (Johnson and Doney 2006).

The temperature of Weddell Sea Deep Water near the Argentine Basin bottom increased significantly from 1994 to 2003. It is especially well manifested in the southern part of section A17 WOCE. In 1994, the mean potential temperature in the zone between 37° S and 48° S in the layer deeper than 4,800 dbar was −0.130°C, which is 0.024°C lower than the mean potential temperature of water (−0.106°C) in the same layer, calculated from the measurements in 2003. Figure 4.21 presents the difference between section A17 in 1994 and 2003 in the Argentine Basin. A temperature increase in the bottom layer is clearly seen. Coles et al. (1996) reported a temperature increase in the bottom waters in the 1980s and concluded that this climatic signal has already reached the Brazil Basin.

One can also see in this figure that potential temperature variations in the Argentine Basin at the level of Vema Channel depths are ±0.02°C. The long-term record of temperature fluctuations in the cold jet of the channel also demonstrates an alternating temperature with fluctuations within ±0.02°C. This fact is interpreted as a consequence of temperature increase in the bottom layer of the Weddell Sea (Fahrbach et al. 1998) that was responsible for temperature increase of the Weddell Sea Deep Water flow in the Vema Channel.

Here, we describe another interpretation of temperature variations in the Argentine Basin that can correlate with the temperature fluctuations in the Vema Channel, because bottom waters in the Argentine Basin are sources of the flow through the Vema Channel. The idea put forward in Meredith et al. (2008) based on the idea in Coles et al. (1996) seems promising. Changes in atmospheric cyclonic forcing over the Weddell Gyre influence the Weddell Sea Deep Water export across the South Scotia Ridge. Isopycnal surfaces in the Weddell Sea become either steeper or flatter

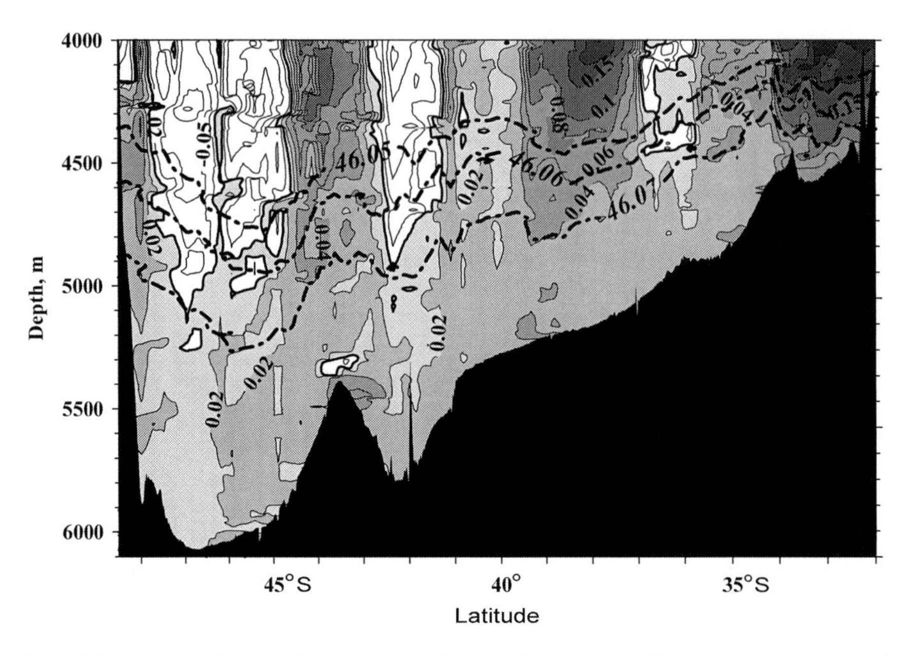

Fig. 4.21 Variation of potential temperature (°C) in 2003 compared to 1994 in the southern part of section A17. *Contour lines* of σ_4 are also shown

in response to spinup or spindown of the gyre induced by winds, resulting in changes in the density of the deepest Weddell Sea Deep Water exported across the South Scotia Ridge to the Scotia Sea. The stronger Weddell Gyre entraps the coldest water south of the South Scotia Ridge and produces the less dense (warmer) Weddell Sea Deep Water that crosses the South Scotia Ridge and enters the Scotia Sea through the Georgia. During the period of not very strong winds, the weaker Weddell Gyre leads to flattening of isopycnals and denser (colder) Weddell Sea Deep Water crossing the South Scotia Ridge. Two pathways for Weddell Sea Deep Water from the Weddell Sea exist: one through the South Sandwich Trench, the other through the Scotia Sea. The Pathway through the Scotia Sea is longer (see Sect. 3.4) and water is subjected to mixing with saline circumpolar water. Later both pathways merge in the Georgia Basin and later propagate to the Argentine Basin. Waters that passed different pathways attain different properties and this process produces an additional modulation of bottom water properties. Then, volumes of colder or warmer water that passed through the Scotia Sea and South Sandwich Trench circulate in the Argentine Basin and change temperature of the flow in the Vema Channel, depending on the temperature of water that appears at the southern entrance to the channel. A scheme of isopycnals within the Weddell Sea adapted from Meredith et al. (2008) and Coles et al. (1996), together with a section of potential temperature and density across the Weddell Sea, is shown in Fig. 4.22. Two types of isopycnals are shown: during "normal" deep water formation under moderate wind forcing (dashed lines) and during string forcing leading to gyral spinup and midgyre convection (solid lines).

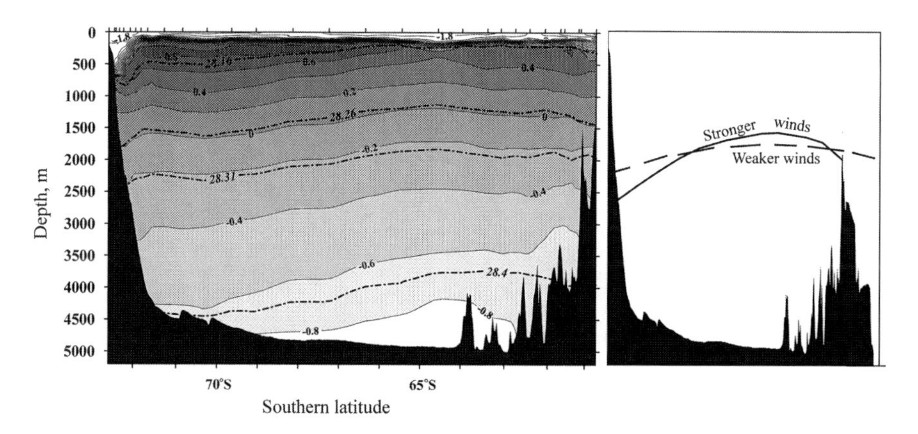

Fig. 4.22 Section of potential temperature (°C) and neutral density across the Weddell Sea and scheme of isopycnals within the Weddell Sea. (Adapted from Meredith et al. 2008; Coles et al. 1996). The section is plotted from the A23 section data (southern part) obtained in March and April 1995. Isopycnals on the graph approximately correspond to the boundaries of water masses (from *bottom* to *top*): WSBW/LWSDW, LWSDW/UWSDW, UWSDW/CBW, and CBW/LCDW

4.2.7 Salinity Variations

Coles et al. (1996) reported a freshening of bottom water upstream in the northwestern corner of the Argentine Basin; Hogg and Zenk (1997) were unable to find a significant change in salinity within the initial part of the available time series. After 1997, the number of stations in the Vema Channel increased and allows us to perform an analysis of salinity variations in time.

We consider salinity variations based on the measurements on the standard section at 31°12′–31°14′ S since 1980. Salinity variations in the bottom layer of the Vema Channel and instrumental errors are shown in Fig. 4.23. Salinity variations are very close to the instrumental accuracy, especially those measured in the 1980s–1990s. There is no significant salinity trend, although salinity variations by 0.002–

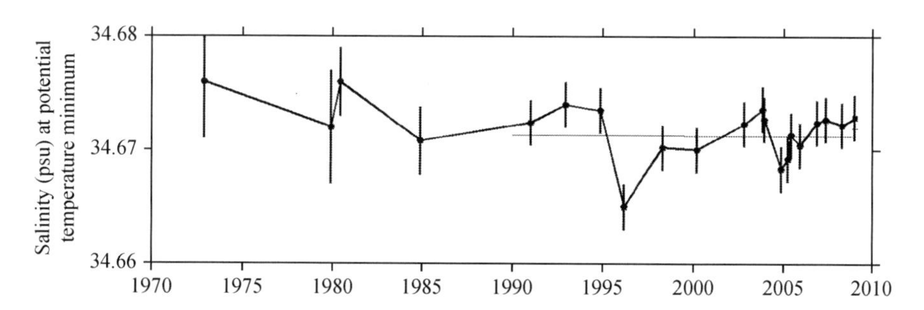

Fig. 4.23 Salinity variations (psu) near the Vema Sill in the bottom layer of the Vema Channel. Potential instrumental errors are indicated with *vertical bars*

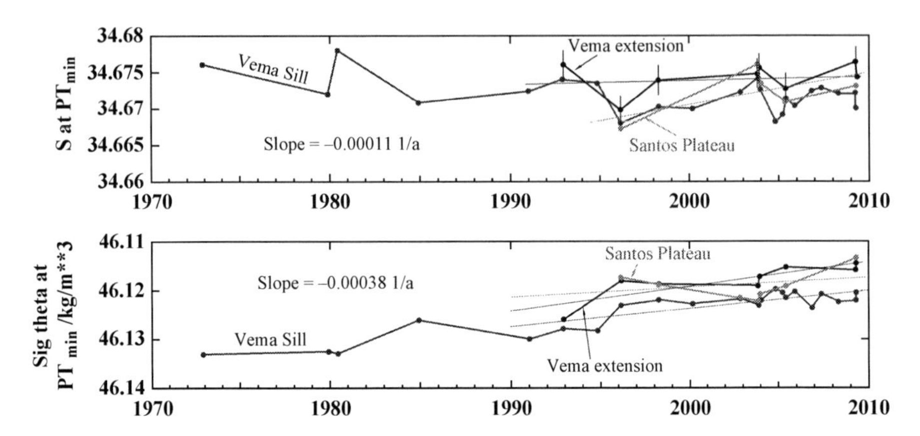

Fig. 4.24 Evolution of salinity and density at the reference points of potential temperature minimum in the Vema Channel (Vema Sill and Vema extension) and Santos Plateau

0.005 psu were recorded in 1994–1998 and 2003–2006. We note that minimum salinities were measured in 1996. It is possible that many variations can be caused by small deviations of the instrumental calibration from the standards. However, no salinity variations were found in the layer above the depths of the Vema Channel.

The graphs in Fig. 4.24 show time evolution of salinity and density at the reference points of potential temperature minimum in three major regions of observations from the 1970s to 2009. Potential temperature θ rises systematically over three decades, but the overall series in salinity begin to show a negative tendency ranging from -0.3 to -1.2×10^{-4} psu year^{-1}. The associated density (σ_4) decrease measures about -0.4 g m^{-3} year^{-1}.

We can distinguish a few characteristic states of the bottom layer based on variations in the θ/S space (Fig. 4.25) over the section along 31°12′–31°14′ S. The aforementioned potential temperature increase started in 1991. This allows us to separate the data in 1991–1994 that differ from the previous ones. The stations in 1992 and

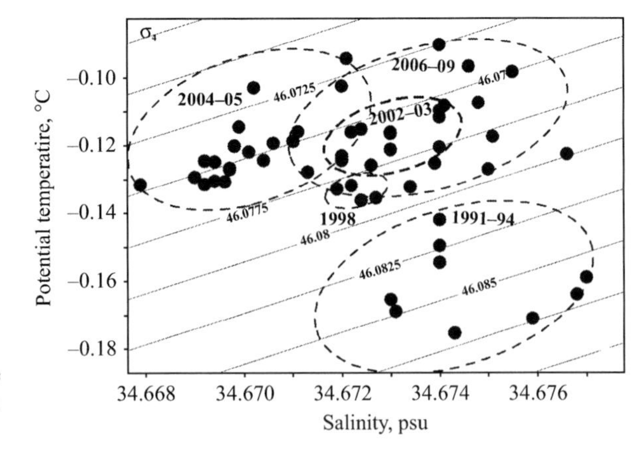

Fig. 4.25 Time variations in the temperature-salinity relations: θ/S-points are based on measurements in the Antarctic Bottom Water (Weddell Sea Deep Water) core in different years. Stabilization periods of temperature salinity relations are marked in the θ/S space with *dashed ellipses*

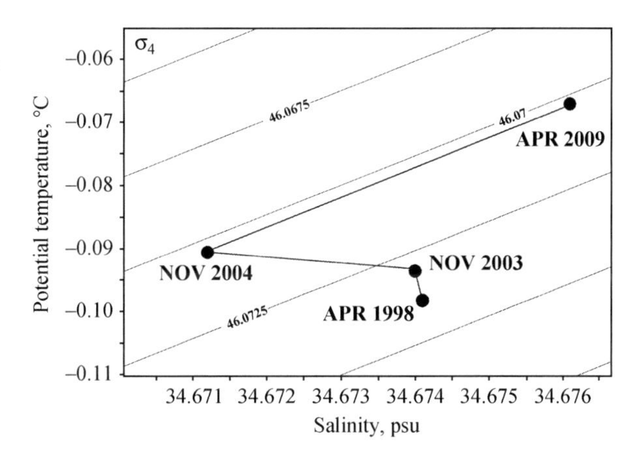

Fig. 4.26 Time variations in the temperature-salinity relation in the Antarctic Bottom Water (Weddell Sea Deep Water) core in the northern part of the Vema Channel

1994 did not likely reach the coldest part of the flow, because no measurements were made at the eastern slope of the channel. CTD-casts in 1991–1994 were made without specific plans for investigating the cold water jet. They correspond to the cold state of water in the Vema Channel flow. The group of stations in 2004–2005 stands alone owing to a strong freshening of water. It is noteworthy that these measurements were made by different scientific teams from various countries. Experiments were thoroughly planned and the close results evidence their high accuracy. Data in 1998, 2002–2003, and 2006–2009 can be considered as the background state of the flow that characterize the usual state of the mixed jet. The period of 2004–2005 is related to the fresh state of the flow. Measurements in 1996 with minimum salinities are not included in the graph.

The time series in the northern part of the Vema Channel is much shorter than in the Vema Sill region. If we consider variations of θ/S-relations in the Vema extension, we can also conclude that freshening was observed in 2004 and salinity increased in 2009 (Fig. 4.26).

Our new data on increase in minimal temperatures in the Vema Channel are substantially more robust and they confirm the previously reported rising trend. At the present state and in view of intrinsic instrumental errors, we have to admit that the observed salinity changes are not compelling. However, with more planned visits to the Vema Channel, we hope in the future to confirm or disprove the recognizable anti-correlated trends in salinity and density. They demonstrate a step-by-step transition from a colder and saltier state to a warmer and slightly fresher one. The pertinent decrease in density is obvious from the overlaid isopycnals.

4.2.8 Flow in the Southern Part of the Channel at the Boundary with the Argentine Basin

The southern region (approximately 33°10′–33°40′ S) of the Vema Channel is the deepest part of this channel. The eastern slope of the channel here is a seamount

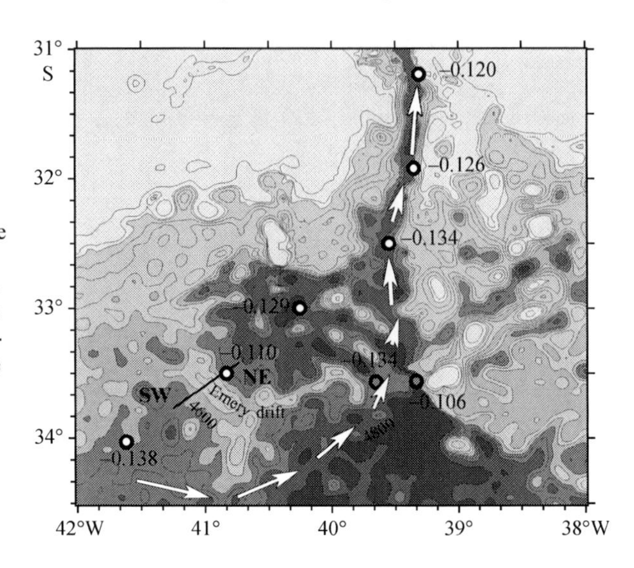

Fig. 4.27 Potential temperatures at the bottom (°C) at characteristic points of measurements in the Vema Channel. Two stations in 2005 are highlighted with *gray color* in the center of circles. Stations in 2003 have white centers. *Arrows* show the most likely AABW flow. Potential temperatures at the bottom at key points of measurements are indicated. The *black line* shows the section in 2003 across the Emery Drift (see Fig. 4.28)

with a summit at 3,900 m under the ocean surface. The western slope is a sedimentary ridge (Emery Drift) with a crest at 4,480 m over the background depths of 4,650 m (Fig. 4.27).

In 2003 and 2005, several stations were occupied in the southern part of the Vema Channel. Potential temperature at the bottom measured at these stations and shown in Fig. 4.29 indicates that the flow of the coldest water is concentrated between the Emery Drift and aforementioned seamount. The data of two stations in 2005 at 33°34′ S, (Fig. 4.29), show that the flow of cold and less saline bottom water is displaced from the maximum depth of the channel to the western slope. Temperature at the deepest point was −0.106°C, which is warmer than at the slope (−0.134°C). The densest water is also located not in the deepest part of the trench. Such structure of the flow indicates the dominating role of the Coriolis force in displacing the core of the flow across the channel here, unlike Ekman friction near the Vema Sill in a narrow and long channel. Most likely, the coldest water does not overflow the Emery Drift, because the temperature north of the ridge is higher by almost 0.03°C than south of the ridge (−0.138 and −0.110°C, respectively).

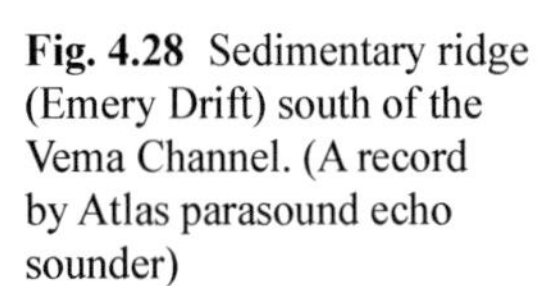
Fig. 4.28 Sedimentary ridge (Emery Drift) south of the Vema Channel. (A record by Atlas parasound echo sounder)

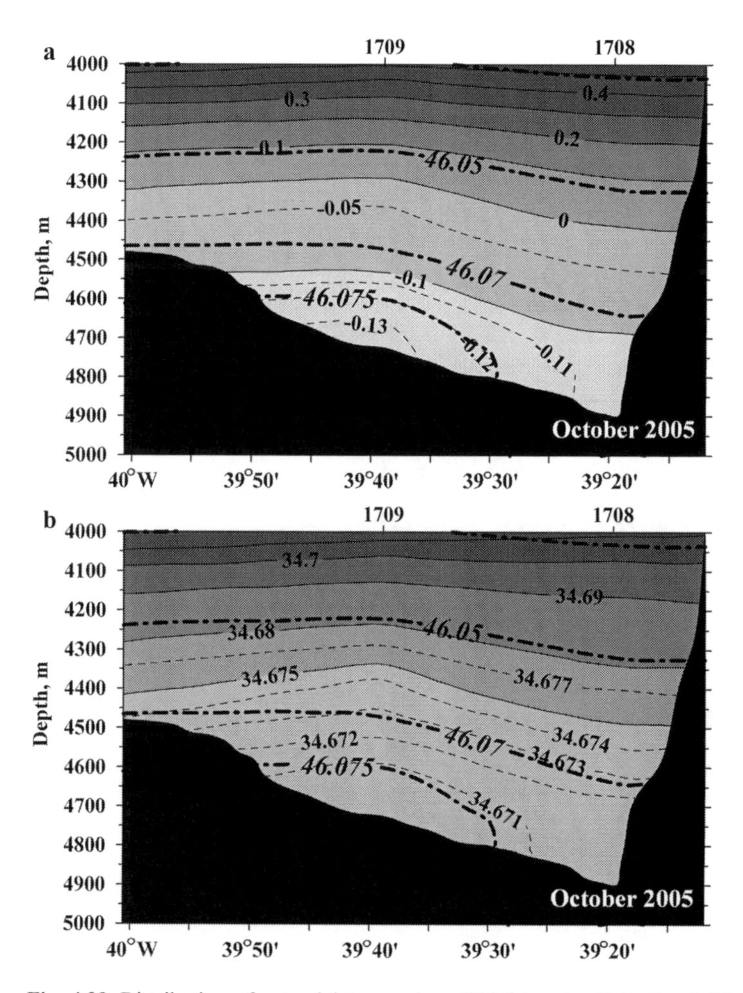

Fig. 4.29 Distribution of potential temperature (°C) (**a**) and salinity (psu) (**b**) over the section across the deep passage (33°34′ S) in 2005. Isopycnals referenced to 4,000 m (σ_4 in kg/m^3) are shown

4.2.9 Moored Observations of Velocities in the Channel

In the early 1990s, the horizontal flow of Antarctic Bottom Water across the Rio Grande Rise of the South Atlantic was intensely studied during the Deep Basin Experiment (DBE) of WOCE (Hogg et al. 1996). A core objective of this campaign was to describe and quantify the circulation and mixing of the sub-thermocline water masses while entering and exiting the Brazil Basin through its major abyssal passages. Starting in January 1991, an array of 13 moorings was set in an easterly direction at its southern end, roughly along 29° S between the continental rise off Brazil and the eastern terrace of the narrow Vema Channel at about 39° W. After recovery in December 1992, a subgroup of instruments was shifted farther eastward, where they were redeployed primarily in five moorings across the wider Hunter Channel. This second installation lasted until early 1996 (Zenk et al. 1998). In most cases, individual records covered a period of at least 18 months. Comprehensive results from these moored arrays were published in Hogg et al. (1999).

In fact, the basin-scale nature of the Deep Basin Experiment allowed a reasonable resolved quantification of the total horizontal transport, its variability, and temperature fluxes across the Rio Grande Rise. However, with only one available current-meter set from the Vema Channel, the database was too sparse to investigate the variable vertical and horizontal structures of the abyssal flow across this choke point passage between the Argentine and Brazil basins.

Twelve years later, a new effort was started by IFM-GEOMAR to observe the Vema Sill overflow of Antarctic Bottom Water in a 3-D approach (longitudinal, vertical, time). Part of the rationale for the new array was to establish an ocean site for long-term current observations based on the representative numbers of earlier records since 1979. After a significant change in the abyssal temperature field of the northern Argentine Basin (Coles et al. 1996) and analysis of the flow at the entrance to the Vema Channel (Hogg and Zenk 1999; Zenk and Morozov 2007) the problem of how a change in mass and temperature of the flow could manifest themselves remained open.

During three visits between December 2003 and May 2007, a twin array of current meters and temperature recorders was installed and completely redeployed on opposite flanks of the Vema Sill, the narrowest section of the whole Vema Channel (Zenk et al. 1993). The mooring sites were zonally separated by 14 km at a sill width of about 20 km. Figure 4.30 displays the potential temperature stratification during the middle cruise of R/V *Polarstern* in May 2005. Positions of the initial twin moorings are overlaid. Bottom contours are inferred from nocturnal Hydrosweep surveys on the site during mooring preparations. Instruments were concentrated on the deepest 1,500-m-thick segment of the water column.

After the turn-around of equipment, we collected seven current records distributed on the western (three records) and eastern (four records) posts of the gateway. Time series on the west side were continued as planned. For a better resolution in the vertical, the east side array was even enhanced by two additional instruments. During the second deployment, we collected at least 16 simultaneous records, most of them covering by then in total a period of 3.5 years. Table 4.3 shows the data inventory.

Alternatively to a beam diagram, we display the available data set in stick plot form in Fig. 4.31. Only each fifth current vector out of the daily mean is displayed. The vector series are further ordered by location (west on top) and instrument depths shown at both ends of the curves. Averaged water depths of the two successive deployments are 4,451 and 4,539 m on the western and eastern site, respectively. The maximal sill depth in the middle of both locations amounts to 4,630 m.

A qualitative inspection of the low-frequency variability in Fig. 4.31 confirms several aspects of earlier observations. Down to 3,500 m, abyssal currents are primarily aligned in an alternating meridional direction, though southern currents seem to prevail. Below about 4,000 m, the topographic influence starts to dominate

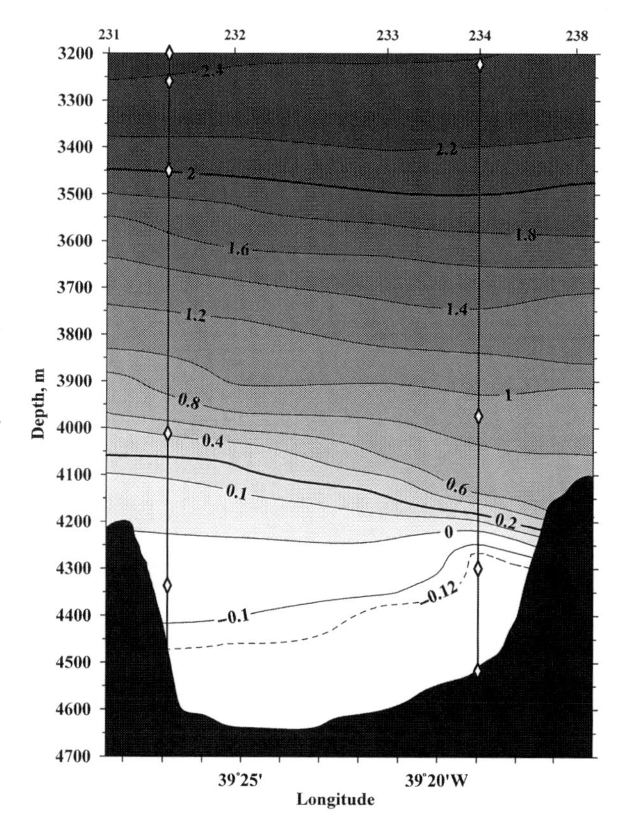

Fig. 4.30 Zonal deep distribution of potential temperature (°C) across the Vema Sill. The data were collected in May 2005 from R/V *Polarstern*. Two isotherms are labeled as boundary layers of Antarctic Bottom Water (Lower Circumpolar Water) (2°C) and Weddell Sea Deep Water (0.2°C). Station locations are depicted at the *top* axis. Positions of the moored self-recording current meters are indicated by *diamonds* on both sides of the channel

Table 4.3 Mooring inventory Vema Sill 2003–2007, IFM-GEOMAR Kiel

Latitude south	Longitude west	Date start	Date end	Instrument depths, m	Location
31°15′ S	39°19.2′ W	Dec 18, 2003	May 27, 2005	3,222 3,974 4,299 4,517	Vema East I
31°15′ S	39°19.2′ W	May 31, 2005	May 17, 2007	3,077 3,679 3,922 4,255 4,479 4,513	Vema East II
31°15.6′ S	39°27′ W	Dec 18, 2003	May 27, 2005	3,259 4,012 4,336	Vema West I
31°15.6′ S	39°19.2′ W	May 30, 2005	May 18, 2007	3,255 4,033 4,367	Vema West II

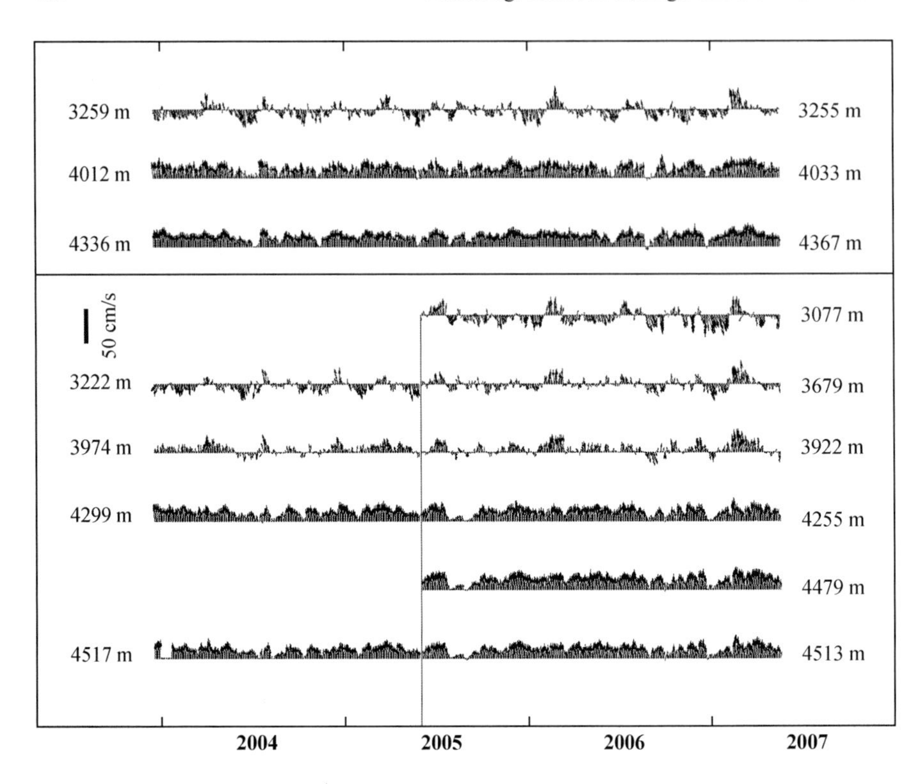

Fig. 4.31 Current vectors (cm s^{-1}) of abyssal flows over the period 2003–2007 on positions shown in Fig. 4.30. Records from the western and eastern sides of the channel are on the *top* and *bottom*, respectively. Numbers on the *left* and *right* rims indicate the depth of instruments. The intermediate visit of R/V *Polarstern* for instrument maintenance is given by the *vertical line*. A speed reference line (50 cm s^{-1}) is given on the *left* side

the flow direction. With a few exceptions, all southern components appear rectified and enhanced by the near-bottom inter-basin exchange. The unidirectional flow speed increases systematically toward the bottom. A modest reduction of the flow speed was recorded only on the deepest level (26 m above the ground) on the east side. Perhaps the most striking aspect of Fig. 4.31 is the period of several weeks long with highly correlated features in the vertical and the horizontal. Particularly evident are horizontal correlations among both post sides, seen in the long-lasting coherent phases in the beginning and end of the second deployment period.

Figure 4.32 displays estimates of horizontal flow in more detail. The data were objectively mapped (Hiller and Käse 1983). After subtraction of the large-scale field, the remaining field was assumed to be Gaussian with a horizontal correlation length of 9 km i.e., approximately the half-width of the channel. The 120-m correlation distance in the vertical represents an acceptable compromise between resolution and instrument separations. The root-mean-square flow error was set to 2 cm s^{-1}. Due to the applied objective interpolation scheme and the pertinent parameter choice, absolute current components appear smaller then in local averages of current components as observed by moored current meters.

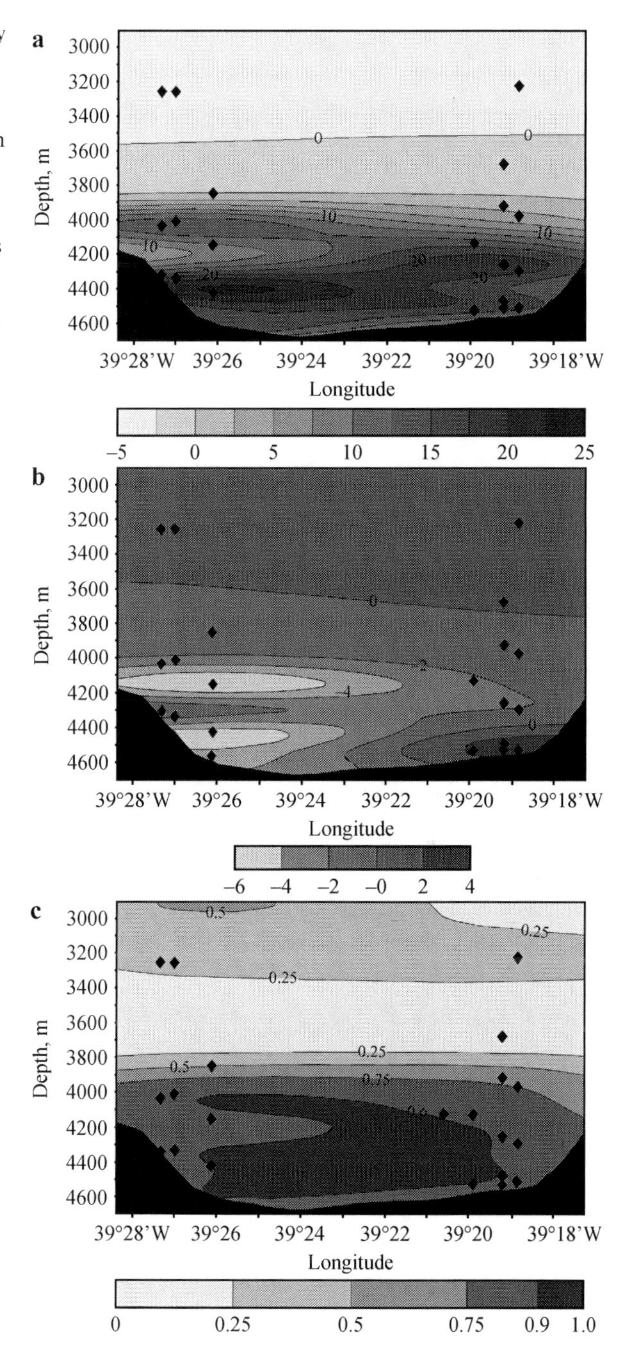

Fig. 4.32 Map of objectively interpolated meridional (**a**) and zonal (**b**) velocity components (cm s^{-1}) across the Vema Sill from long-term current observations. Positive numbers indicate equatorward or eastward flows, respectively. Panel (**c**) shows the associated vector steadiness distribution. Current meter levels are indicated by *diamonds*. Note the level of almost no mean flow above the Antarctic Bottom Water (depths above ~3,200 m)

The structure of the along-passage (meridional) current component shows an almost horizontal course of the zero line at ~3,550 m separating the averaged poleward drift of North Atlantic Deep Water from the equatorward flow of Antarctic Bottom Water. The corresponding cross-channel component has its insubstantial zero line at ~3,650 m. While searching for a reference level of no motion for geostrophic analyses, we find the 3,600-m level to be a good choice for the whole Vema Sill based on the 3.5 years average of direct current observations.

The collected time series reveal an uneven current structure on both sides of the sill area. In the west, the flow has two cores of about equal intensities. They are centered at 4,000 and 4,400 m. Only the lower core extends as a robust flow band across the whole channel. On the eastern side one can clearly see a concentration close to the bottom. The strongest intensity was encountered just beneath the eastern side convergence of isotherms shown, for example, in Fig. 4.30 on the right side. In the eastern corner of the section, one finds the coldest and swiftest vein of water originating directly from polar latitudes, i.e., Weddell Sea Deep Water.

As expected, zonal currents in the deepest part of the channel are much weaker under strong topographic guidance. The current shear with changing signs on the eastern side expresses the helical or secondary circulation due to bottom friction (Hogg 1983; Jungclaus and Vanicek 1999). The westward component extends in the 4,000-m core on the energetic western side of the channel. The distribution of directional steadiness, i.e., the ratio of the magnitude of the vector mean current to the averaged speed, reflects the impact of topographic constraints. Both sides show high steadiness paralleled by a robust band of high values across the channel. With the loss of topographic influence farther up in the water column, the steadiness decreases systematically. It disappears totally at the identified averaged level of no motion. Here, weak alternating currents prevent any steadiness of this direction parameter.

Transports from the first deployment (2003–2005) of current meters were studied by Denker (2007). She optimized the results of testing different geometric choices of partial cross sections centered around the instrument levels (without applying any objective analysis algorithm) (see Table 4.4).

Table 4.4 Partition of transports inferred from moored current meter records on the western and eastern sides of the Vema Sill during 2003–2005 according to Denker (2007). For details see period Vema I in the mooring inventory in Table 4.3

Layer	Volume transport, Sv (10^6 m^3/s) ± Standard deviation			Comments
	Sum	West	East	
0.2° – Bottom	2.51 ±0.82	1.35	1.16	WSDW fraction of AABW with high directional steadiness
				Total
2.0° – Bottom	3.60 ±1.52	2.19	1.41	Classic Antarctic Bottom Water (AABW)
				Total
3,200 m – Bottom	3.52 ±1.64	2.15	1.37	Classic Antarctic Bottom Water (AABW)

Due to the strong vertical shear, the upper integration boundary with sluggish currents appears rather insensitive to depth variations. Here, transports of Antarctic Bottom Water beneath a fixed upper level of 3,200 m (3.52 ± 1.64 Sv) differ insignificantly from those with a changing level at the 2°C contour line of potential temperature (3.60 ± 1.52 Sv).

With a record length of 3.5 years, we can build on Denker's (2007) calculations and extract an annual transport signal by a harmonic approximation of the filtered original time series. The cut-off period for this analysis was chosen at 4 months. The filtered series in Fig. 4.34 resembles distinctly a seasonal signal. Its optimal fit is reached at a 6 weeks delay between the beginning of the calendar year and the calculated transport series. Basically, the maximum Antarctic Bottom Water transport rates occur in summer (mid-February). Both curves are shown as overlays in Fig. 4.34 on the recorded data. Annual transport fluctuations with an amplitude of ~1.2 Sv are evident. The averaged transport from the second deployment (Vema II) increased by ~10% in comparison to the first record-long series. However, this difference appears insignificant with respect to the standard deviations of both records. The whole series is still too short to allow a trusted perception about multiyear speed and transport fluctuations.

In search of additional periodicity, as seen in the upper curves in Fig. 4.31, we calculate spectra from the transport time series. In the sub-inertial range, the power spectrum in Fig. 4.36 shows an eminent peak at a period of about 112 days. It reproduces those oscillations already seen in the level of sluggish currents at 3,200 m and above. Outstanding coherence in current vector series from both sides of the Vema Channel is obvious among those levels that are not yet superimposed by the steady near-bottom exchange flow, i.e., above 3,200 m.

Comparable oscillations were also reported from the Samoan Passage of the South Pacific. In this region, bottom water of Antarctic origin enters low-latitudes (10° S, 170° W) through a channel: a situation strongly resembling the Vema Channel outflow across the Rio Grande Ridge. Direct velocity measurements in the Pacific (Rudnick 1997) revealed transport fluctuations near 30 days. The 30-day variability might be in an expression of resonance due to the channel geometry. Multi-week variability is also reported from the abyssal transport of the western boundary current system in (Whitworth et al. 1999). They found spectral peaks at 50, 20, and 10 days, which presumably show little or no concordance with the proposed Samoan dynamics.

More random flow perturbations at the Vema Sill may also be caused by sporadically advected sub-mesoscale abyssal eddies entering the Brazil Basin on their equatorward journey originating from the southern rim of the Argentine Basin (Arhan et al. 2002a). Such solitary passages of circumpolar water eddies were observed one or two times during 2 years of WOCE observations in the near-bottom layers of the western South Atlantic. Indications as event-like current reversals with a duration of about 1 week are again reflected at both sides of the Vema Sill. Two paramount examples (Fig. 4.31) about 1 year apart were recorded coherently on both channel sides (at depths of 4,513 and 4,033 m) during the second deployment period (2005–2007).

Regardless of strong mean currents (with seasonal modulation and perhaps aperiodic advected perturbations), substantial fluctuations are observed especially close

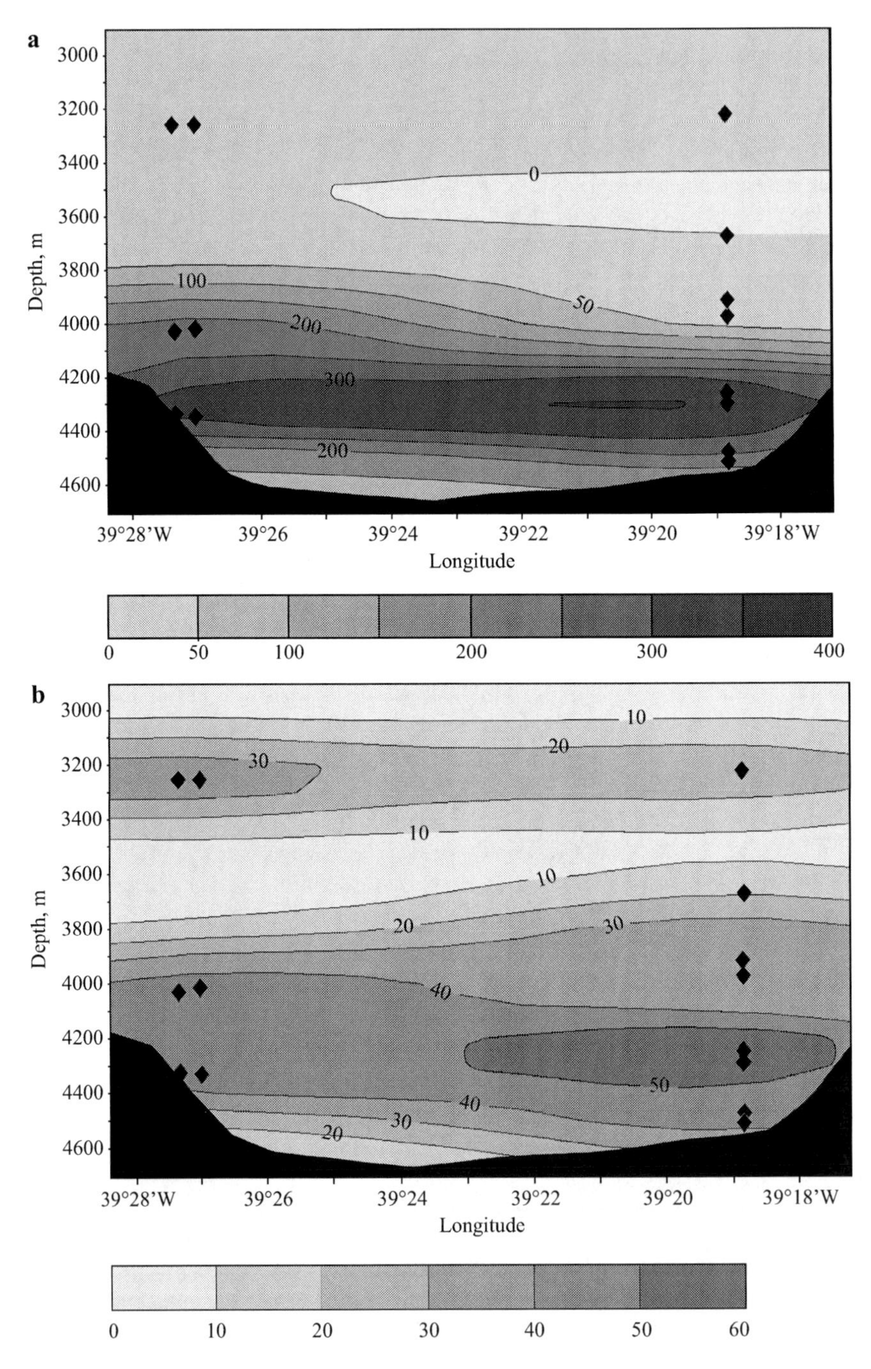

Fig. 4.33 Maps of objectively interpolated distributions of mean (**a**) and eddy (**b**) kinetic energies across the Vema Sill. Current-meter levels are indicated by *diamonds*

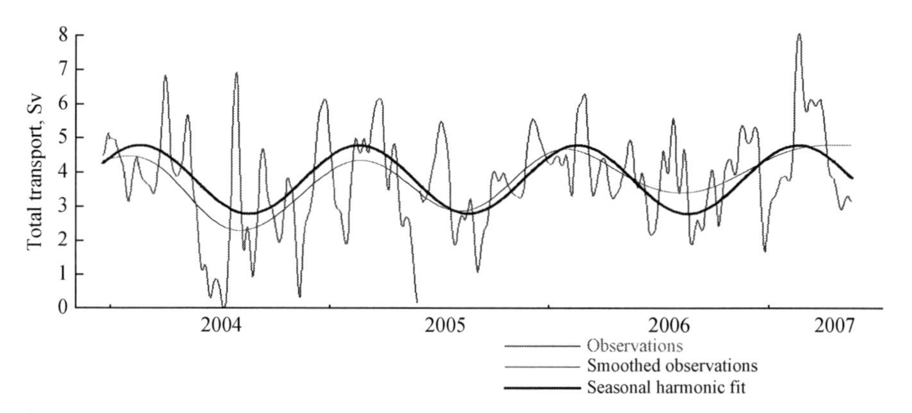

Fig. 4.34 Seasonal fluctuations of transport rates of Antarctic Bottom Water based on 3.5 years of direct observations (*thin line*). The low-pass filtered measurement series is approximated by a seasonal harmonics (*thick line*)

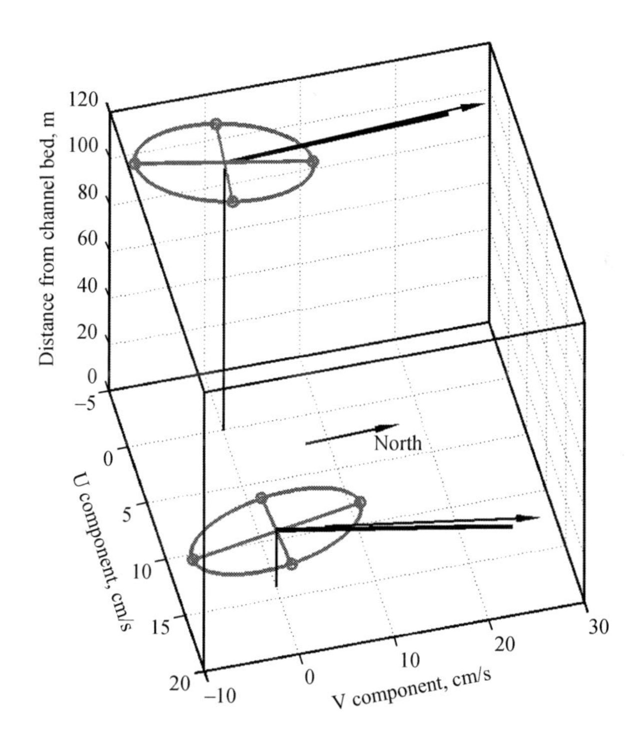

Fig. 4.35 Three-dimensional presentation of mean and fluctuative current components from the seafloor of the Vema Channel. Flow direction toward the equator is indicated by a North *arrow* to the right. Topographic constraints are reflected in the orientation of the variance ellipses

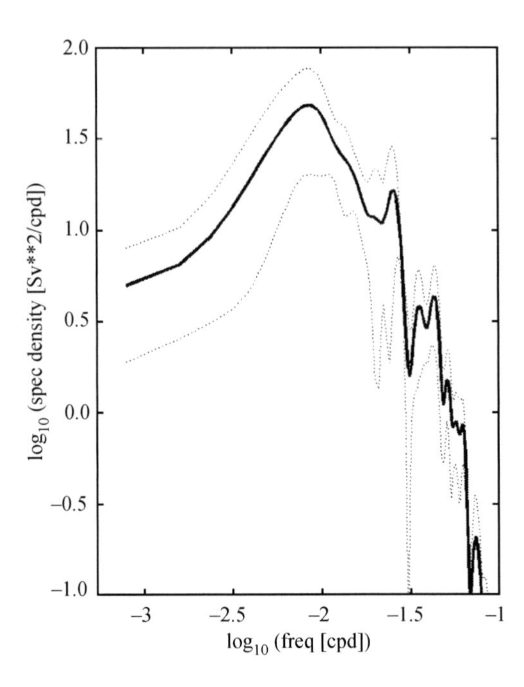

Fig. 4.36 Power spectrum of AABW transport through the Vema Channel. Note the peak at 112 days likely caused by resonance due to the channel geometry

to the seafloor. Figure 4.35 shows averaged deep velocity vectors at the western and eastern sides in a 3-D view. The third dimension contains the instrument distance from the bottom. Also included are the two variance ellipses of the observed time series. They have been rotated such that the semi-major axes grew to their maximal lengths, while the semi-minor axes shrunk to minima. The ellipses are plotted at the origin of the mean vectors to the same scale. Both near-bottom instrument depths differ by 90 m, mitigating the measured topographic impact on the western side. Generally, all fluctuations in the deeper time series are massively masked by the steady Antarctic Bottom Water overflow across the Vema Sill.

The averaged partition between the mean and fluctuating energies per unit mass from the whole cross section below 2,900 m is depicted in Fig. 4.33. Both energy forms increase toward the bottom. The zone of highest energy emerges at the flank on the eastern side concurrent with the pinching isotherms (Fig. 4.30). Comparison with other deep channel flows (Hogg et al. 1999; Whitworth et al. 1999) emphasizes the extremely high mean energy conditions (>300 cm^2 s^{-2}) superimposed on eddy fluctuations of magnitudes that are more typical for abyssal passages (40–70 cm^2 s^{-2}) at other locations.

4.2.10 Measurements with the Lowered ADCP

In October 2005, October 2006, and April 2009, the CTD measurements in the Vema Channel were accompanied with velocity measurements using a Lowered

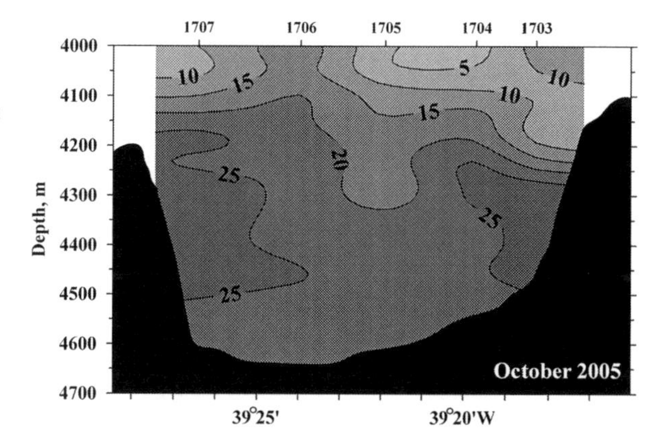

Fig. 4.37 Distribution of velocities (cm s^{-1}) over the section across the Vema Channel at 31°12′ S based on the LADCP data in October 2005

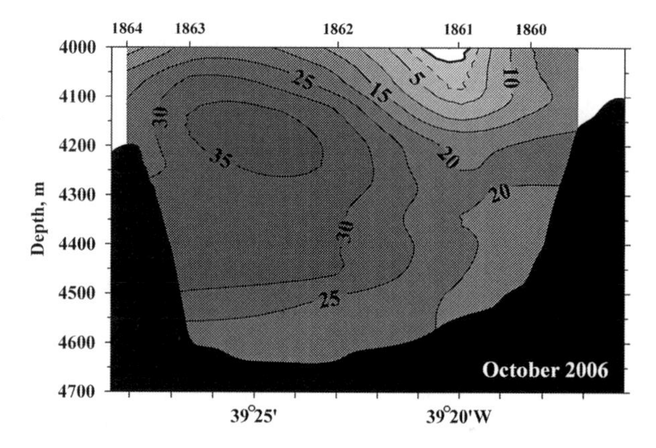

Fig. 4.38 Distribution of velocities (cm s^{-1}) over the section across the Vema Channel at 31°12′ S based on the LADCP data in October 2006

Acoustic Doppler Current Profiler (LADCP). These measurements allowed us to distinguish the spatial structure of the flow in the channel. The structure of the current field changes in time. The maximum measured velocities were 29 cm s^{-1} in 2005, while in 2009 they were as high as 35 cm s^{-1}. Figure 4.37 shows the velocity profile as measured by the LADCP instrument in 2005. Distribution of velocities was almost uniform with a slight increase at the eastern (up to 29 cm s^{-1}) and western (up to 26 cm s^{-1}) slopes of the channel. In 2006, the maximum velocities were observed in the upper western part of the channel (Fig. 4.38). In 2009, two regions of maximum velocities were detected at the eastern and western slopes of the channel. In 2009, a strong current (up to 45 cm s^{-1}) was recorded over the upper part of the western slope, while a southerly countercurrent was found over the eastern slope of the channel (Fig. 4.39).

In 2006, the northerly flow was intensified in the central part of the channel. Normal velocities to the section across the channel (northern component) reached 40 cm s^{-1}, which is almost 1.5 times greater than those measured in October 2005. During measurements in 2006, the current at the western slope appeared stronger than at the eastern slope. These figures correlate with the section of velocity based on moored measurements (Fig. 4.32a).

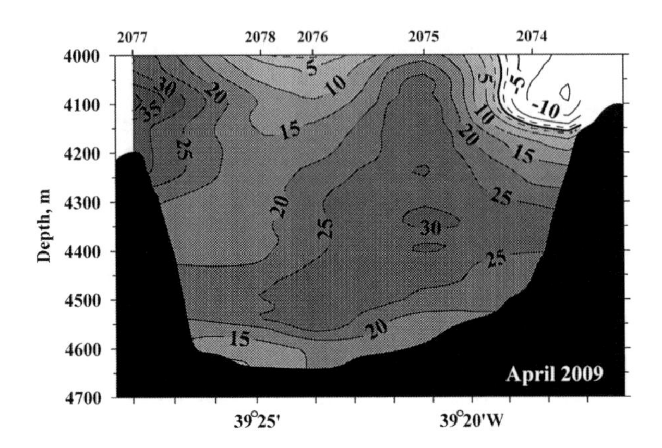

Fig. 4.39 Distribution of velocities (cm s^{-1}) over the section across the Vema Channel at 31°12′ S based on the LADCP data in April 2009

Transport of Antarctic Bottom Water in the Vema Channel was estimated from the LADCP data. Table 4.5 gives the values of northerly transport in 3 years of measurements for the waters bounded by the 0.2 and 2.0°C isotherms of potential temperature. The transports are calculated for two versions of lateral boundaries. One version is limited by the stations of LADCP profiling. The second version is limited by the walls of the channel.

Quite a wide range of Antarctic Bottom Water in the Vema Channel transport is found in the literature. Sometimes, the publications include the region between the channel and Rio Grande Rise, thus including the southerly transport. According to McDonagh et al. (2002), this transport is estimated at 1.4 ± 1.0 Sv. All estimates range between 2 and 4.5 Sv (Hogg et al. 1999; McDonagh et al. 2002; Wienders et al. 2000; Zenk and Hogg 1996; Speer and Zenk 1993). Our estimates based on LADCP data show a range of 1.6–2.7 Sv, which is closer to the lower boundary of this interval and close to the range in Hogg et al. (1999) and McDonagh et al. (2002).

Vertical distributions of velocity on the section at each station in 2005, 2006, and 2009 are shown in Fig. 4.40. The maximum velocities are observed in the middle of the channel at a depth of 4,250 m.

The data of LADCP measurements reveal a strong southerly countercurrent in the depth interval 3,000–4,000 m. The velocities of this current reach 20 cm s^{-1}.

Table 4.5 Transport of Antarctic Bottom Water (Sv) in the Vema Channel at the standard section measured by LADCP in different years

Years of measurements	Limits of LADCP profiles		Western limit 39°28′ W; Eastern limit 39°16′ W	
	Water transport below $\theta < 0.2$°C	Water transport below $\theta < 2.0$°C	Pure northward transport $\theta < 2.0$°C	Integrated transport (northward + southward) $\theta < 2.0$°C
2005	1.54	2.41	2.92	2.36
2006	1.97	2.71	3.69	3.18
2009	1.82	1.62	2.44	1.44

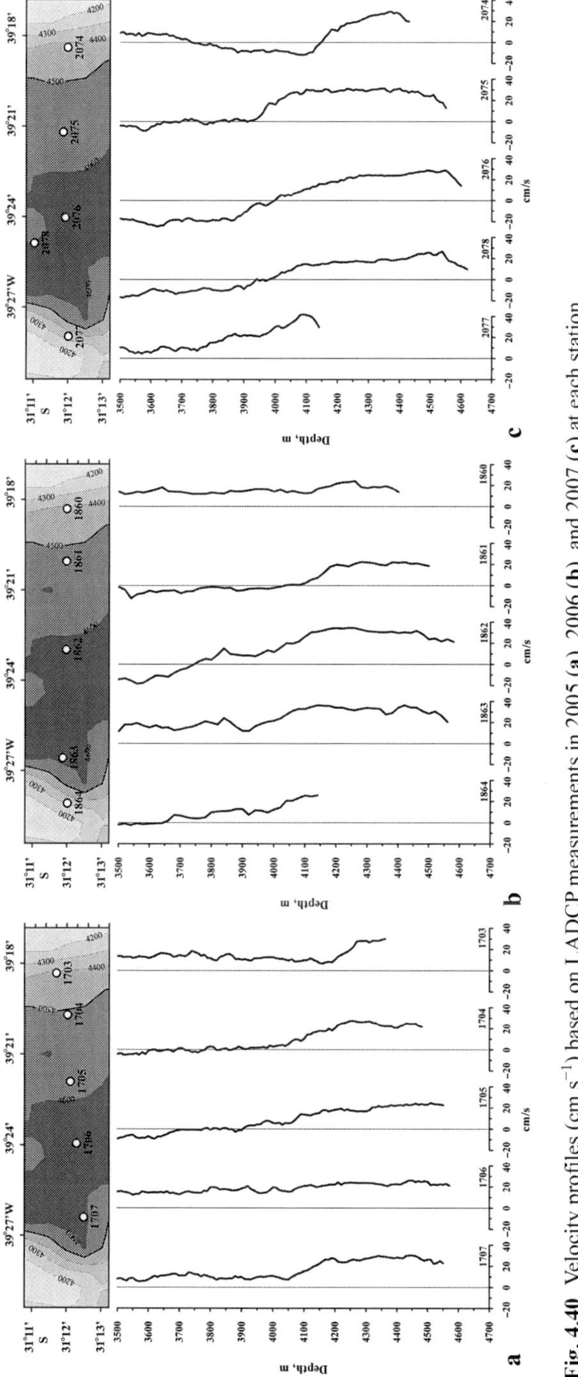

Fig. 4.40 Velocity profiles (cm s^{-1}) based on LADCP measurements in 2005 (**a**), 2006 (**b**), and 2007 (**c**) at each station

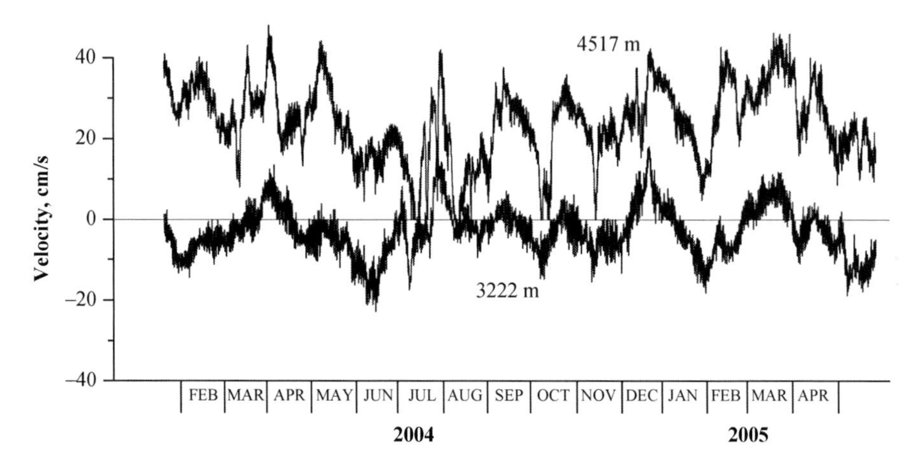

Fig. 4.41 Meridional component of current (cm s^{-1}) at the mooring in the Vema Channel operating in 2004–2005. Measurements at levels of 3,222 and 4,517 m are shown

A similar countercurrent was recorded by the moored instruments (Fig. 4.41). The mooring records indicate that this countercurrent appears when the velocities of the bottom flow in the channel are not very high. This southerly current can be generated by the inflow of water through the Vema Channel and elevation of water due to this inflow. The countercurrent can feed the circulation over the part of the Santos Plateau adjacent to the Vema Channel.

Current measurements using an LADCP instrument in the northern part of the Vema Channel revealed low velocities close to 10 cm s^{-1}. This fact can be interpreted as splitting of the flow to the branch that flows down to the abyssal depths in the Brazil Basin and retroflection of the flow in the layer above the channel, which was recorded at the standard section.

4.2.11 Moored Measurements and Satellite Topography

In the beginning of the WOCE period, i.e. 1991–1992, mooring #338 was deployed in the Vema Channel near the Vema Sill with current meters set at 750, 1,100, 2,900, 3,850, 4,150, 4,425, and 4,625 m. The mooring deployment depth was 4,675 m. The buoy was deployed at 31°08′ S, 39°26′ W and operated from January 12, 1991, to December 5, 1992. Deployment of moorings was repeated in 2003. Two moorings were deployed: at 31°15.3′ S, 39°19.0′ W (#438) and 31°15.8′ S, 39°26.8′ W (#439). The buoys operated from December 18, 2003, to May 28, 2005. The moorings were deployed for the third time at the same sites and they operated from May 30, 2005, to May 18, 2007.

An attempt was made to correlate the velocities in the Vema Channel with the surface satellite observations and calculations of geostrophic velocities due to eddy activity in the region. Charts and data of dynamic heights of geostrophic currents and the calculations of geostrophic velocities at the surface based on the TOPEX/POSEIDON satellite altimetry data of altimetry are available on the Internet (http://las.aviso.oceanobs.com/las/servlets/dataset). We wanted to find a correlation between mesoscale eddies in the region that could either increase or decrease bottom velocities, assuming that the eddies penetrate to the bottom of the ocean.

The other idea was to correlate intensification of velocities in the channel with the location of strong eddies at the southern entrance to the channel, assuming that mesoscale eddies either elevate or depress isopycnals in the entire water column, keeping in mind that the flow in the channel is forced by the level of Antarctic Bottom Water at the entrance to the channel.

The graphs of meridional velocity variations at three levels are shown in Fig. 4.42. Mean values of velocity at 750 m were close to zero, since no long-term transport is known in the region and the velocities are determined by the passing mesoscale eddies. In the bottom layer, the mean velocity is 29 cm s^{-1}, which is determined by the flow of Antarctic Bottom Water. The velocity of the flow ranges from 56 cm s^{-1} to nearly zero. A short period of reverse flow was observed with southern velocities reaching 3 cm s^{-1}. The spectra of velocity fluctuations show that a maximum of periodicity is observed at a period of 40 days.

Analysis of Fig. 4.42 shows that a correlation exists between the flow at the bottom and flow at 3,850 m. In order to find the correlation between the flow at the bottom and at 750 m, we carried out cross-spectral analysis between the time series of the meridional component. The coherence spectrum is shown in Fig. 4.43. At frequencies lower than three cycles per month (10-day period and greater), the coherence is significant in the entire range of low frequencies, while the phase shift is close to zero. At higher frequencies, coherence decreases to an insignificant level. Therefore, the phase shift is unreliable and not shown here. Thus, powerful eddies and currents induced by eddies, which do not decay to the bottom, may influence the bottom flow in the Vema Channel.

However, analysis of the long-term velocity measurements on moorings and satellite observations did not show any correlation between the intensification of currents in the channel and the presence of mesoscale eddies with northerly currents in the region above the Vema Channel. A high correlation between time series of currents at 750 m and near-bottom time series at periods greater than 10 days provides evidence that the possibility of a correlation between intensification of currents at the surface due to mesoscale eddies and currents in the deep channel. However, we did not find such a correlation.

The second supposition that location of eddies near the southern entrance to the channel can influence the flow in the channel was fruitful. The record of currents at the bottom demonstrated a long period intensification of currents in February and March 2005. The meridional component of velocity measured on mooring 438 from April 2004 to May 2005 is shown in Fig. 4.44. One can see intensification of currents in February–April, 2005.

We analyzed the charts of dynamic topography based on satellite altimetry from TOPEX/POSEIDON data (Fig. 4.45). A mesoscale cyclonic eddy approached the southern end of the Vema Channel in February 2005. Presumably, this caused eleva-

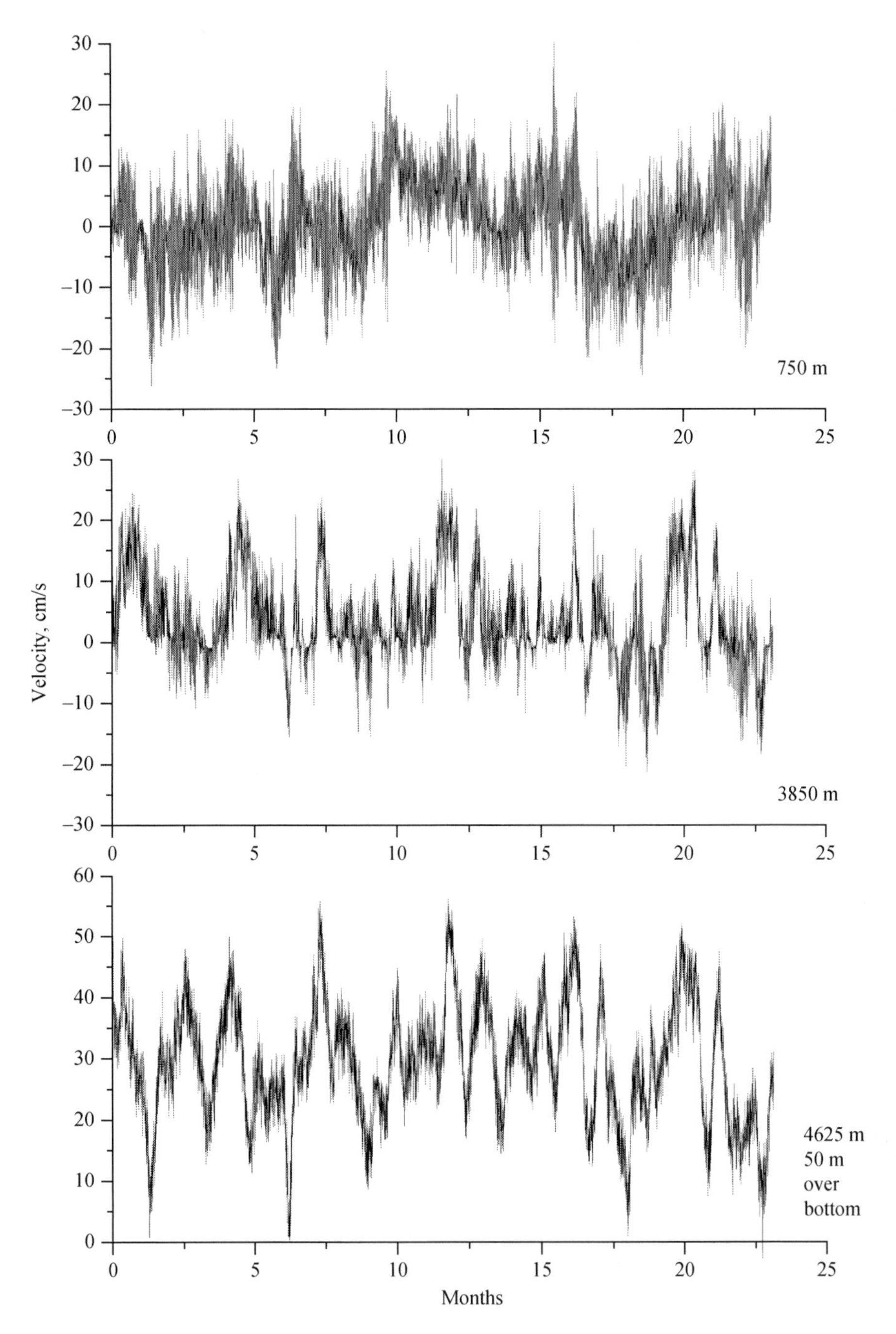

Fig. 4.42 Variations in northern velocity (cm s^{-1}) at three levels on a mooring deployed at 31.137° S, 39.433° W in the Vema Channel from January 12, 1991, to December 5, 1992

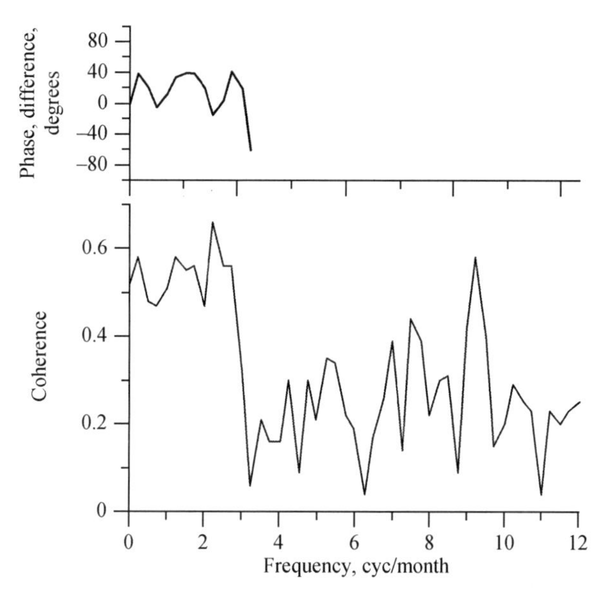

Fig. 4.43 Phase shift and coherence between fluctuations of the meridional velocity component at 750 and 4,625 m (50 m *above* the bottom) on a mooring in the Vema Channel from January 12, 1991, to December 5, 1992. Only reliable phase shift is shown for high coherence levels at low frequencies of the spectrum

tion of isopycnals in the entire water column, including the upper level of Antarctic Bottom Water. Location of the upper boundary of Antarctic Bottom Water can influence the velocity and transport in the channel. This correlation is seen in Figs. 4.44 and 4.45. Later in May 2005, the cyclone displaced to the south and velocities decreased from 40 to 20 cm s^{-1}.

In addition, Hogg and Zenk (1997) suggest that intrathermocline lenses observed over the Santos Plateau can influence the dynamics of the flow in the channel. It is likely that the flow is subjected to the influence of many factors: elevation of Antarctic Bottom Water in the Argentine Basin at the beginning of the Vema Channel, strong surface eddies, and intrathermocline lenses.

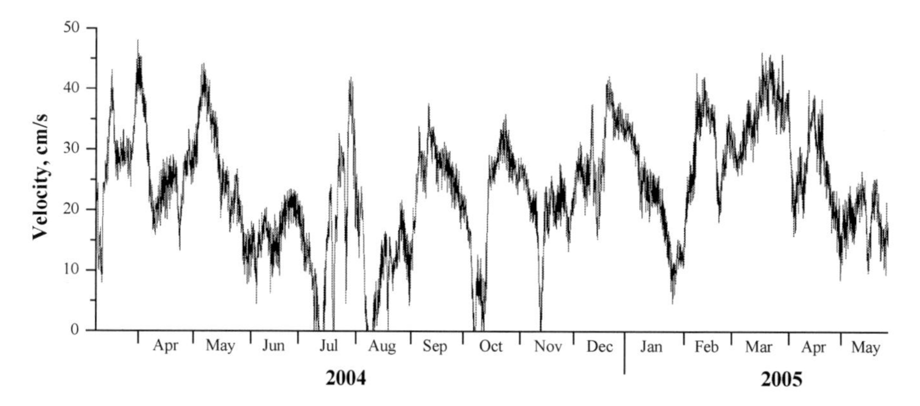

Fig. 4.44 Variations in northern velocity at 4,517 m (ocean depth 4,535 m) on mooring 438 deployed at 31°15.28′ S, 39°18.97′ W in the Vema Channel from March 2004 to May 2005

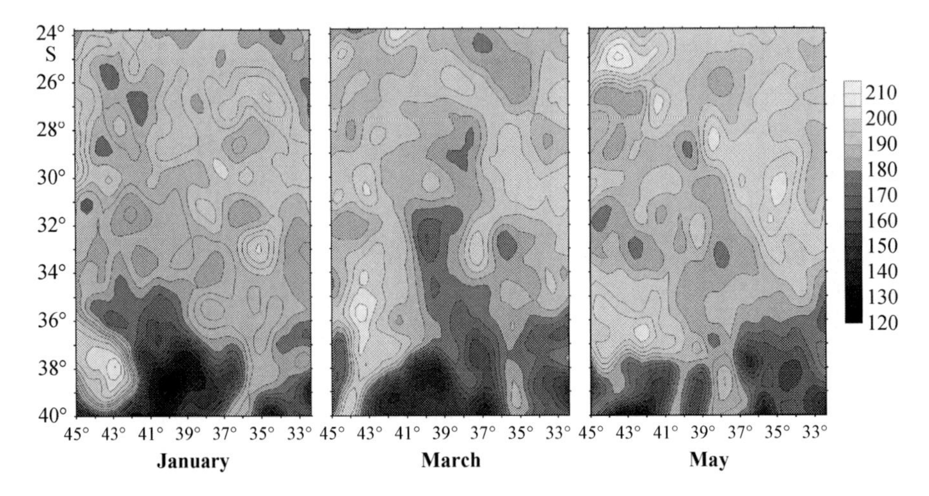

Fig. 4.45 Dynamic topography in the Vema region (dyn. cm) based on satellite TOPEX/POSEIDON data in January, March, and May 2005

4.2.12 Temperature Fluctuations and Current Shear in the Flow of Antarctic Bottom Water at the Vema Sill

This section describes the results of moored thermistor chain measurements supplemented by current meters deployed near the Vema Sill. The measurements continued for almost 2 years from 1998 to 2000. The lowest 500 m of the mooring line was equipped with a chain of thermistor temperature meters to monitor the processes in the benthic boundary layer (Zenk 2008).

A new method of observations with closely spaced instruments was tested to sample the lowest 500 m of the water column (Pätzold et al. 1999) using a thermistor line of sensors mounted on a mooring line. Mooring V389 was deployed on April 21, 1998, and recovered on March 8, 2000, near the eastern wall of the Vema Channel at 31°14.30′ S, 39°20.00′ W at a depth of 4,580 m.

The main objective of the experiment was to monitor the processes in the bottom boundary layer and vertical displacement of isotherms around $\theta = 0.2$°C at the interface between the two prevailing water masses: Weddell Sea Deep Water and Lower Circumpolar Water (Reid et al. 1977; Speer and Zenk 1993).

To monitor the persistence of a stable θ/S relation in the deepest part of the water column, we moored a high precision MicroCat CTD recorder with conductivity, temperature, and pressure sensors in the benthic boundary layer (48 m above the bottom). Above the MicroCat, we installed two 200-m-long Aanderaa thermistor chains attached to the mooring rope (490–290 and 267–67 m above the bottom). The mooring also supported two mechanical Aanderaa current meters, one in the bottom mixed layer (50 m above the bottom) and the other near the expected depth of the maximum current velocity (270 m above the bottom) (Hogg et al. 1999).

The mooring yielded a 687 day record from 23 levels with thermistors, two current-meter levels, and one level with simultaneous temperature and conductivity record (MicroCat). However, the conductivity measurements remain questionable. A scheme of the mooring is shown in Fig. 4.46.

The interval of thermistor sampling was 2 h. The original frequency of MicroCat's records was 20 min. The records were box-averaged over six values to get a 2-h sampling interval. During the deployment cruise, a supplementary CTD section was occupied across the middle part the channel. One of the stations was used as a reference point for the calibration of moored sensors. Likewise two additional temperature profiles obtained during the recovery cruise provided validation information at the end of the time series.

The mean pressure from the CTD recorder was (4,612.79 ± 0.73) dbar, which corresponds to 4,531.91 ± 0.72 m depth. This agrees well with the nominal depth of the recorder. We obtained a robust record mean in situ of the MicroCat temperature estimate of 0.2121 ± 0.0069°C, which equals a mean potential temperature of −0.1346 ± 0.066°C. This in situ temperature record was used as a prime reference for validation of the temperature series from the other thermometers on the mooring. The time-averaged temperature profile data showed a systematic decrease of the temperature variance toward the bottom. Such a behavior in long-term records

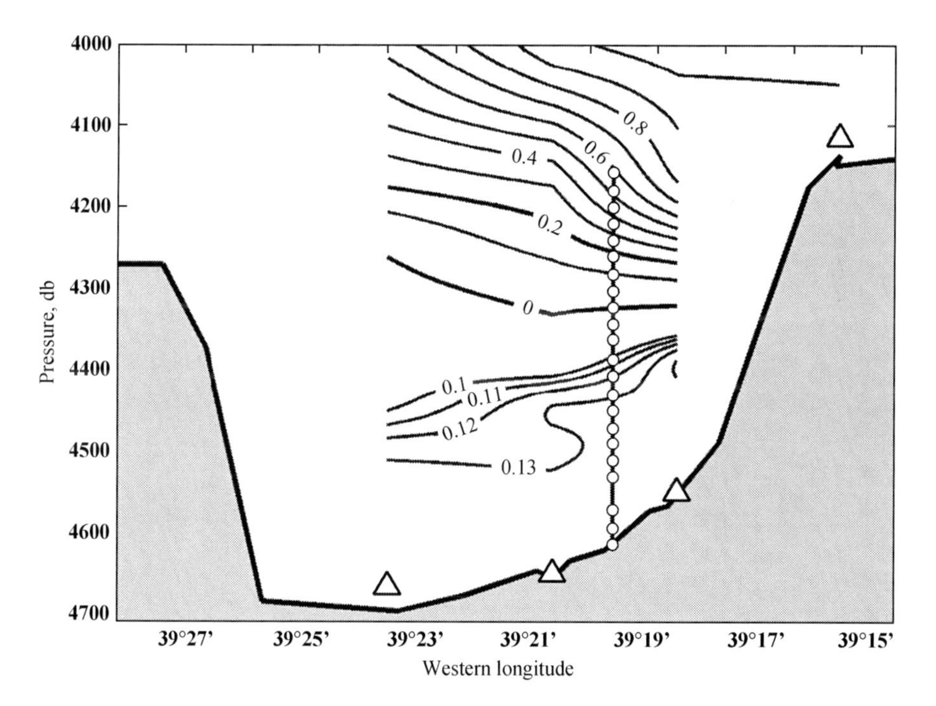

Fig. 4.46 Scheme of the mooring and section of potential temperature (θ, °C) across the Vema Sill in April 1998. *Triangles* show the deepest level reached by the CTD probe. The mooring was deployed about 2,000 m downstream from the shown CTD section. (Modified and redrawn from Zenk 2008)

was observed repeatedly in similar cases, e.g., the Hunter Channel (Zenk et al. 1999), nameless Equatorial Channel of the western Atlantic (Hall et al. 1997), and Romanche and Chain fracture zones (Mercier and Speer 1998).

Two time series of in situ temperature measurements between both current meters are shown in Fig. 4.47. For the succeeding analysis, potential temperatures were calculated from in situ temperatures by applying near-bottom salinity and pressure data from the CTD station on April 21, 1998. The upper current meter, 270 m above the ground, returned both vector velocity and temperature time series.

Let us analyze the temperature distribution across the sill. One can note three characteristics of the thermal stratification, which are typical for deep passages in the Southern Hemisphere away from the equator with enhanced bottom flow under frictional influence: (1) the gradient layer composed of weakly stratified Lower Circumpolar Water above 4,250 dbar with isotherms tilted down to the east; (2)

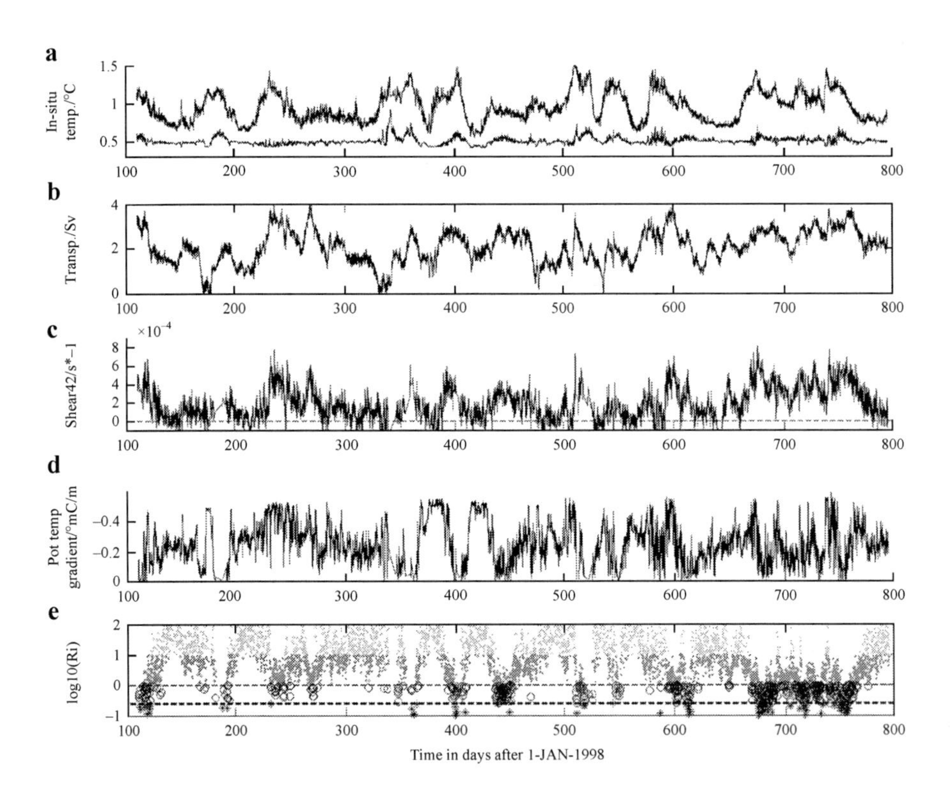

Fig. 4.47 Selected time series based on the moored current meters and thermistor chains: (**a**) in situ temperature series from the uppermost sensors in the thermistor chains, 490 and 267 m above the bottom; (**b**) equatorward transport of bottom water; (**c**) vertical current shear between the two current meters 270 and 50 m above the bottom; (**d**) least square fit approximation of the temperature gradient within the Weddell Sea Deep Water inferred from the lower thermistor chain, 67–267 m above the bottom; and (**e**) evolution of bulk Richardson numbers. Critical Ric values supporting vertical turbulence (Ri < 1) are highlighted by different markers and *darker gray* symbols. (Reproduced from Zenk 2008 with permission from Elsevier)

the pinching of isotherms along the eastern wall at ~4,300 dbar; and (3) the bottom mixed layer of Weddell Sea Deep Water core with a reversed slope of isotherms tilted upward to the east. From west to east, minimum potential temperature values recorded on the channel bed from the three deep stations were −0.133, −0.133, and −0.137°C.

Results of fluctuations in the thermal stratification based on low-pass filtered data are shown in Fig. 4.48. Besides the isotherms of potential temperature, the graph shows depth location of thermistors at both sides of the time axis. The lowest 500 m of the water column is divided into a benthic boundary layer carrying the pure Weddell Sea Deep Water with potential temperatures $\theta < -0.1°C$ and a gradient layer with the mixed Weddell Sea Deep Water and Lower Circumpolar Water. On average, the transition between both water masses is located near 4,330 m depth.

Contour intervals for potential temperatures $\theta > -0.1°C$ is 0.1°. The interval below is 0.01°. Note the differences in the stratification character separated by the interface at $\theta = 0.2°C$ between Weddell Sea Bottom Water and Lower Circumpolar Water.

Temperature fluctuations at the bottom of the benthic boundary layer are extremely small, which makes their resolution sensitive to instrument stability. Keeping in mind that our reference thermometer was located 48 m above the bottom, we consider that these small fluctuations are reliable. Standard deviations of potential

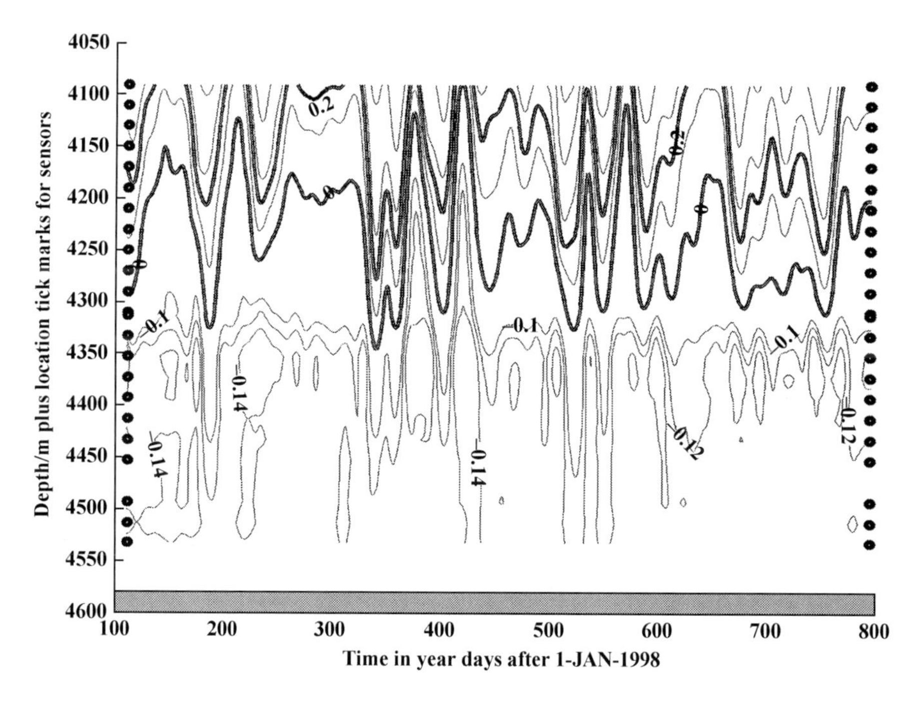

Fig. 4.48 Time evolution of the isotherms of potential temperature (°C) recorded over the Vema Sill. Water depth is 4,580 m. Depth location of thermistors is shown at both sides of the time axis for reference. (Reproduced from Zenk 2008 with permission from Elsevier)

temperature are about 0.02°C in the lowest 180 m above the seafloor. The lowest potential temperatures near the bottom vary between $-0.14 < \theta < -0.12$°C. Temperature variance increases monotonically in the direction from the bottom, from <0.002 to 0.040°C at 490 m above the bottom. A comparison of this integral quantity with the isotherm distribution in Fig. 4.48 shows again that quasi-homogeneity dominates the lower end of the mooring. Striking events of irregular vertical extent fluctuate significantly with time. Such perturbations of the interface position between Weddell Sea Deep Water and Lower Circumpolar Water make it difficult to define a traditional fixed level between these two distinct water masses.

During the total record, the 0°C isotherm underwent a vertical excursion of about 100 m. A similar displacement of the 0.2°C isotherm reveals an overall deepening rate of 60 m/1,000 days with increased tendency in the second half of the record (~140 m/1,000 days). The energy-preserving spectral estimate of the lowest temperature fluctuations (Fig. 4.49c) shows the only significant peak at the period of sub-mesoscale abyssal eddies (14–20 days). No fluctuating energy in the tidal band is present in the time series from the mixed benthic layer.

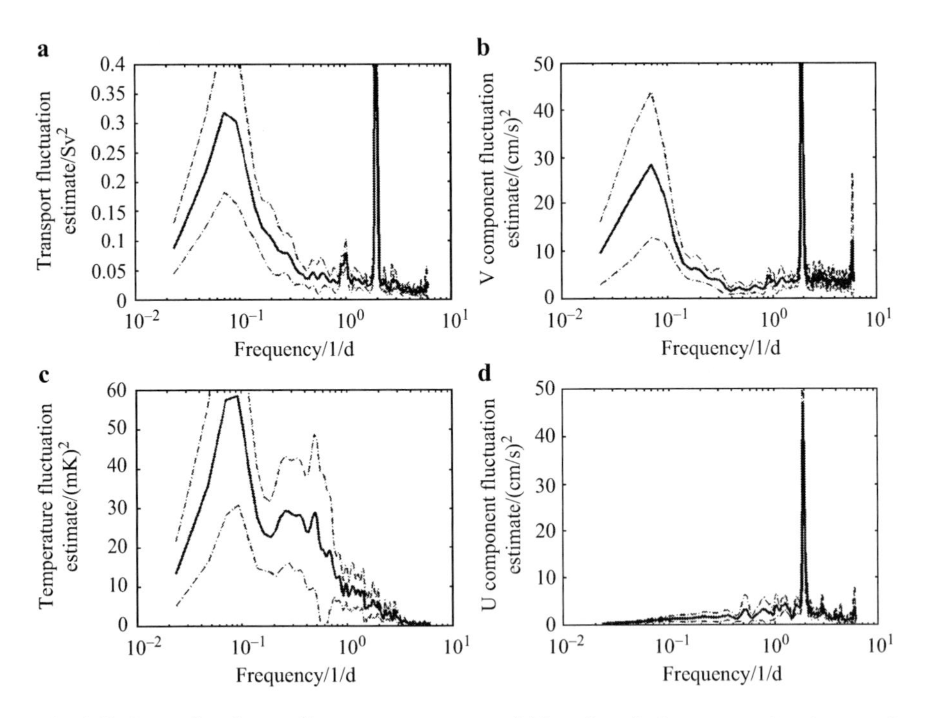

Fig. 4.49 Spectral analyses of bottom water transport (**a**) based on the lower current-meter record in the Weddell Sea Deep Water (**b**, **d**). The energy-preserving representation of spectral estimates indicates the domination of semidiurnal tides in the current records. At lower frequencies, one recognizes additional energy in the longer period bands, which can be caused by the activities of drifting eddies. This broad peak also dominates the variance in the lowest temperature record (**c**), in which no tidal signal could be identified. 95% confidence intervals are shown by *dashed lines*. (Reproduced from Zenk 2008 with the permission of Elsevier)

Table 4.6 Statistics of the velocity measurements for the two moored instruments on mooring V389 in the Vema Channel. Numbers were calculated from bi-hourly samples

Distance to the bottom, m	Mean V, cm/s over 687 days	Max V, cm/s	Min V, cm s^{-1}	MKE/EKE
270	22.14	45.59	14.90	5.87
50	18.16	35.59	16.61	8.09

V is the meridional current component. MKE is the energy of the time-averaged flow, EKE is energy of the fluctuations.

Let us analyze abyssal currents and estimate water transport. Along-channel current components exceeded the cross-stream components roughly by a factor of 8. The substantial confinement of the deep flow is also expressed by the unusually high ratio of the mean kinetic energy (MKE) to eddy kinetic energy (EKE) of 5.9 and 8.1 at depths of 4,310 and 4,530 m, respectively. Values smaller than unity are more typical for the abyssal flows in the open ocean (Mercier and Speer 1998). Statistical parameters of the observed currents are summarized in Table 4.6.

In fact, inequity of the across- and along-channel current components (designated U and V, respectively) is reproduced in Fig. 4.49b,d. In the energy-preserving presentation of current fluctuations, a rather narrow peak with only a small amount of energy (area under the spectral curve) is indicated at the semidiurnal tidal period of 12.4 h, at which the along-channel component surpasses the across-channel component by a factor of 1.8. A major portion of the high current variance is found in the sub-mesoscale eddy range exclusively in the along-channel current component V with a period of 14–20 days, i.e. somewhat longer than seen in the temperature record from the boundary layer (Fig. 4.49c).

The vertical shear between 4,310 and 4,530 m (Fig. 4.47c) is negative almost throughout the record (>90%), indicating clearly that the average value of currents measured at the higher level exceeded that of the near-bottom meter. This behavior suggests strong frictional control in the bottom layer, as has been observed previously at this location in Hogg et al. (1996, 1999). The corresponding potential temperature gradient is shown in Fig. 4.47d, which shows records of several episodes of significantly weakened thermal stratification. The available time series of along-channel currents in the Vema Channel suggest a rough transport time series for Weddell Sea Deep Water. Following an earlier study of the abyssal throughflow (Hogg et al. 1999), we assume the speed measured by the lower instrument to be representative for the lowest 100 m of the water column over the sill. The second instrument covers the depth range of 100–400 m above the bottom marked by the upper boundary of Weddell Sea Deep Water with potential temperatures $\theta \leq 0.2$°C, on the average. We assume that the channel width at the mooring site is 24 km. Under these assumptions, we get a mean transport estimate of 2.07 ± 0.73 Sv. The resulting time series is shown in Fig. 4.47b. This crude transport estimate lies in the expected range in comparison to the WOCE results reported in Hogg et al. (1999). They found a near-bottom transport of 2.40 Sv for the water with potential temperature below 0°C and 3.45 Sv below 0.8°C. Admittedly, their database, highly resolved in the vertical, was much more adequate for transport calculations of Weddell Sea Deep Water based on a large array of moored current meters and CTD stations across the southern boundary of the Brazil Basin.

Analysis of the isotherm diagram (Fig. 4.48) allows us to detect a series of stratification collapses in the gradient region (above ~4,330 m). If we take fluctuations of the 0°C isotherm as an indicator of advected temperature signals, we count approximately ten abrupt displacements. They are not at all restricted to warm Lower Circumpolar Water layers but could also be triggered by a near-bottom signal of the penetrating cold Weddell Sea Deep Water. Such episodic changes can be traced almost throughout the whole water column analyzed here. Their typical duration can be as long as 3–4 weeks. The number of warm and cold events is approximately the same.

Several dynamic effects, such as interface sharpening, tilt seiching (Lukas et al. 2001), and enhanced entrainment, may interact in a complex manner, making further interpretation without any long-term cross-channel section coverage impossible. The quasi-homogeneous waters of the deep boundary layer (in the lower part of Fig. 4.48) with temperature fluctuations of $\delta\theta < 20$ mK can last several weeks and cover a large vertical range.

To investigate the intrinsic mixing process in this layer, we follow Saunder's approach (2001) in the northern Iceland Basin. Although we have no highly resolved current profiles from the sill, we still can estimate time series of bulk Richardson numbers (Ri) for the Vema Channel over a depth interval of 220 m from our discrete current meter and temperature records; i.e., we approximate Richardson numbers by finite property differences across the boundary layer between Weddell Sea Deep Water and Lower Circumpolar Water centered at 160 m above the seafloor. A range of critical Richardson numbers Ri_c between 0.20 and 1.00 can be found in the meteorological and oceanographic literature. As a relevant example, we refer to Thorpe (1978) who observed instabilities due to vertical shear in a tank experiment even when $Ri_c > 1$. More recently, Canuto et al. (2001) came to a similar conclusion ($Ri_c \sim 1$) by a model study incorporating the latest advances in turbulent closure. The value of $Ri_c = 1/4$ is generally accepted as the canonical threshold value for the Kelvin–Helmholtz shear instability occurrence in fluid dynamics.

From the closest CTD station, we approximate a density versus potential temperature ratio of $\Delta\sigma_4/\Delta\theta = 93.9$ g m^{-3} K^{-1} for the deepest layer. The necessary temperature gradient series with a 2-h resolution were calculated from least square fits of the lower thermistor chain records between the current-meter pair (Fig. 4.47d). The resulting series of bulk Richardson numbers Ri is shown on the bottom of Fig. 4.47e. Symbols are given in different gray shades. Low Richardson numbers, (<1) shown in black and dark gray shades, indicate a turbulent regime due to weakening of the stratification, increasing vertical shear, or a combination of both. The much more frequent stable conditions with Ri > 1 are shown in lighter gray. Several occurrences of low Ri are seen during the almost 2-year record. The cumulative frequency distribution of Ri (Fig. 4.50) indicates that about 15% of the observations in the boundary layer carry the potential to maintain or boost turbulence in the abyssal flow regime at the Vema Sill. The Ri series suggest the possibility of occasional ac-

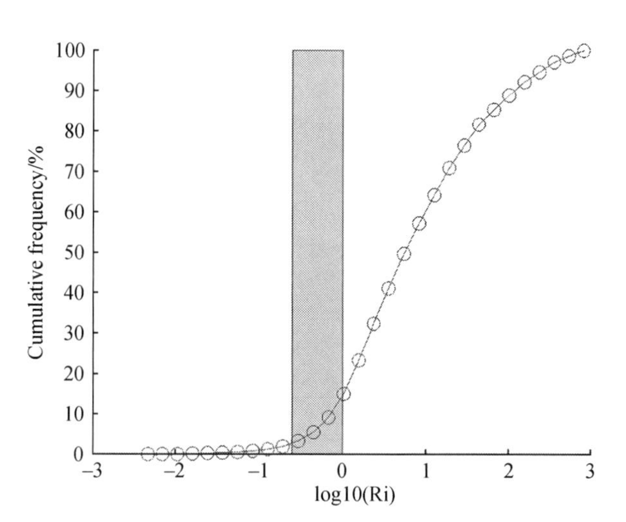

Fig. 4.50 Cumulative diagram of bulk Richardson numbers Ri from the Vema Sill. The range of critical Ri_c is labeled by a vertical bar. It reflects the transition zone between unstable and stable flow conditions. Note that smaller turbulence elements were lost due to the given large vertical range, over which vertical shear has been observed. (Reproduced from Zenk 2008 with the permission of Elsevier)

tive vertical mixing in the narrow Vema Channel. During 14 of the total 98 weeks, low Ri values implied the entrainment across the WSDW/CDW interface.

Figure 4.51 shows the dependence of the Richardson number on bottom water transport through the Vema Channel. Although it is difficult to infer absolute transport estimates from only two recording current meters, we recognize that the

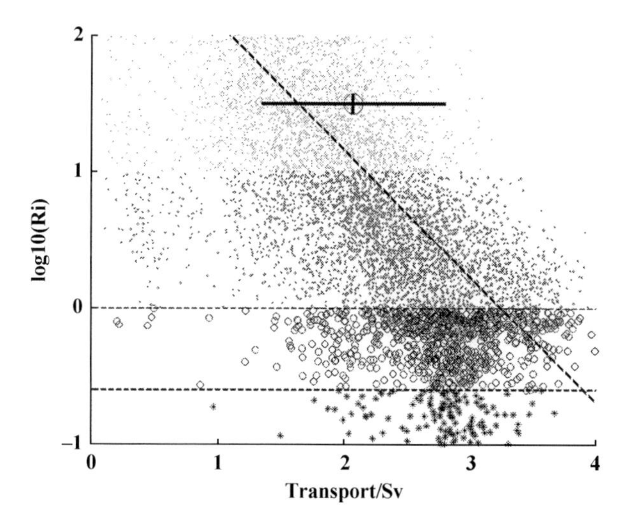

Fig. 4.51 Transport estimates of Weddell Sea Deep Water in the Vema Sill vs. bulk Richardson numbers Ri. The boundaries of critical thresholds of Ri (similar to Fig. 4.50) are shown with two *dashed lines*. They are highlighted by different markers and darker *gray* symbols. Averaged transport and its standard deviation are shown by a *horizontal bar* in the upper part of the graph. Note that the majority of low Ri numbers occurred at higher transports. A linear regression is suggested by the *broken line*. (Reproduced from Zenk 2008 with the permission of Elsevier)

majority of Ri_c is associated with higher transport numbers (>2.5 Sv). Similar to the overflow case described in Saunders (2001) in the Iceland Basin, the occurrence of unstable stratification seems to be associated with high transport and concurrent shear maxima. The linear fit in Fig. 4.51 appears to confirm this tendency to increased instability (lower Ri) with higher transport episodes.

Besides the shown potential for bottom water transformation due to vertical shear, a few cases of lateral property degradation are documented in the study site. A dual-divided bottom flow occurs farther to the south in the western Argentine Basin (Smythe-Wright and Boswell 1998). Supported by tracer analyses, they identified a well-developed deep western boundary current confined by an adjacent shear zone farther to the east. According to these investigators, the most dominant flow of abyssal water in the Argentine Basin is controlled by the local topography. The Zapiola Ridge represents the center of an anticyclone (anticlockwise) circulation cell at about 45° S, 45° W. On its western flank, the recirculating bottom water encounters the deep boundary current, thus forming an additional region of enhanced abyssal mixing. Smythe-Wright and Boswell (1998) concluded further from their observations that only a fraction of the original Weddell Sea Deep Water makes its way north into the Brazil Basin and beyond.

More recently, McDonagh et al. (2002) analyzed the WOCE observations. In a sequence of σ_4/S curves from meridionally aligned CTD stations, they documented significant modifications of abyssal water masses by lateral mixing within just a 100-km-long pathway between the northern Argentine Basin (WOCE Section A17) and the Rio Grande Rise (Section A10). Potential spots of enhanced mixing in the Romanche Fracture Zone are discussed in Polzin et al. (1996). They identified such locations more downstream of the main sill, where entrainment amplifies the transport of bottom water flow into the Sierra Leone Basin by a factor of 2.

Additional observations – more upstream and with higher vertical shear resolution above the sill – are necessary to clarify the dynamics of the observed event-like variability of the thermal stratification in the Vema Channel and to study the impact of the local vertical mixing in the Vema Channel, including the vigorous recirculation cell suggested in McDonagh et al. (2002, Fig. 9). Stability of the σ/S relation in the sill region and the adjacent eastern abyssal plain depends critically on the effects of potential internal surges signified in the highly resolved thermal distribution in Fig. 4.48.

Forcing mechanisms may include passing deep eddies and long baroclinic Rossby waves, the two most likely candidates that Lukas et al. (2001) identified in their study of overflow events across an unnamed sill between the Maui and Kauai Deeps of the subtropical North Pacific. The first suggested scenario, the passage of drifting abyssal eddies across the Vema Sill, was already discussed in Arhan et al. (2002a) during the analysis of a subset of the current-meter data used in this study. Rare episodes of rotating current vectors in records with a duration of 2–10 days seem rather suspicious. However, since all of them are reproduced in both independent current-meter records, they must be considered to be real. Arhan et al. (2002a) offer isolated vortices occasionally migrating northward through the abyssal Argentine Basin as a potential explanation. Due to its funnel property for bottom flow, the Vema Channel may attract such deep eddies where they appear as event-like perturbations in long-term records. One source region for abyssal eddies was identified in the southwest corner of the Argentine Basin based on a small population census of such deep Argentine eddies (DADDIES) (Arhan et al. 2002a).

Our direct observations of highly resolved near-bottom temperature fluctuations on the Vema Sill combined with high shear occurrence, suggest an additional potential for deep oceanic mixing. So far, this source of degradation of distinct properties along the equatorward pathway of Antarctic Bottom Water has been discussed less frequently in the literature devoted to the South Atlantic. Not only bottom roughness or entrained overflow plumes, but also deep channels with their frictionally induced instabilities can have a sustained effect on local mixing with a regional or even basin-wide impact on abyssal water mass properties.

4.2.13 Suspended Matter in the Channel and Adjacent Slopes of the Rio Grande Rise[1]

The present-day concept about suspended matter distribution in the bottom and deep waters of the World Ocean was formed in the 1980s (McCave 1986). A remarkable feature of the lower water column in most parts of the deep ocean is a high level of light scattering caused by the presence of increased amounts of suspended sediments. This part of the water column is called the bottom nepheloid layer. The increase in light scattering is perceived relative to the minimum values recorded at mid-water depths between 3,000 and 4,000 m. The global distribution of nepheloid layers is related (due to instrumental reasons) to the global distribution of fine particles (<2 μm). Their concentrations are high in the Southern Ocean and in the vicinity of deep western boundary currents.

The bottom nepheloid layer is up to 2,000 m thick (sometimes even more in trenches and channels) and generally has a basal uniform region, the bottom mixed nepheloid layer, corresponding quite closely to the bottom mixed layer defined by potential temperature. Above the bottom mixed nepheloid layer there is a more or less logarithmic fall-off in intensity of light scattering up to the clear-water minimum marking the top of the bottom nepheloid layer.

Another class of nepheloid layers found at continental margins is an intermediate nepheloid layer. These layers occur frequently at high levels off the upper slope and at the depth of the shelf-edge. From here they may spread out across the continental margin.

The nepheloid layers of both types are produced by resuspension of bottom sediments. Thus, their distribution indicates the dispersion of resuspended sediments in the ocean basins. Most concentrated nepheloid layers are located at deep continental margins (≤4,000 m depth) and thus the nepheloid layers indicate the regions of resuspension and deposition of sediment originally deposited in other regions.

[1] This section of Chapter 4, was written in cooperation with V. Sivkov, M. Kravchishina, and A. Klyuvitkin.

The properties and behavior inferred from nephelometer data (Biscaye and Eittreim 1977) are biased towards small particles, but larger particles are also present in the size spectrum and play a significant role in sedimentation. The Vema Channel, with a sill depth of approximately 4,620 m, is the passage for a strong Antarctic Bottom Water flow from the Argentine Basin to the Brazil Basin. In this section, we analyze the nepheloid layers in the Vema Channel based on the data on concentration of suspended particles with clay and fine silt size exceeding 2 μm.

A total of 18 hydrographic stations were occupied on the Rio Grande Rise along the Vema Channel and adjacent slopes of the Brazil and Argentine basins (Fig. 4.52) during cruise 17 of R/V *Akademik Sergei Vavilov* on November 1–17, 2003. During

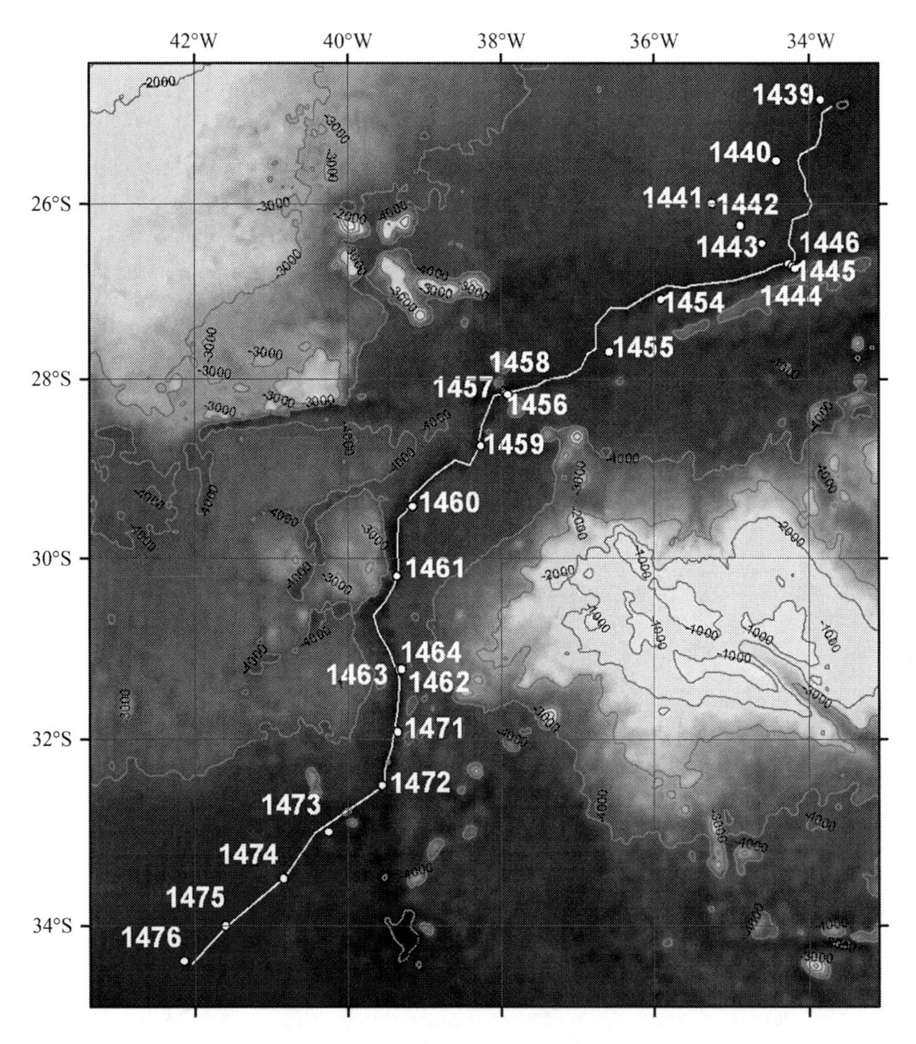

Fig. 4.52 Location of the sampling stations along the Vema Channel during cruise 17 of R/V *Akademik Sergei Vavilov*. Bathymetry is shown in meters

each hydrographic cast we sampled twelve 1.7-liter Niskin bottles for collecting suspended particulate matter. Concentrations of suspended particulate matter were determined using a Coulter Counter (Model Z_{BI}) applying the methods described in Sheldon and Parsons (1967) and modified by Richardson and Gardner (1985). The Coulter Counter data were recorded in terms of total volume of particles counted and subdivided into increasing size grades. A 70 μm aperture and 0.5 ml sample size were used in this study to measure the volume concentration (and size distribution) of particles with equivalent spherical diameters from 2 to 28 μm. A total of 282 samples of the suspended particulate matter were examined aboard the vessel.

We also applied a filtration method for collecting suspended particulate matter. The investigations were carried out during cruises 11 and 16 of R/V *Akademik Ioffe* and cruise 17 of R/V *Akademik Sergey Vavilov* in 2002–2004. A total of 84 samples of suspended particulate matter were collected. The mass concentrations (mg/l) of suspended particulate matter were determined using water filtration (3–12 l of water sample) through preliminary weighed nucleopore membrane filters 47 mm in diameter with a pore size of 0.45 μm under a vacuum of 0.4 atm. After filtration, the samples were washed with triple-distilled water, placed into Petri dishes, and dried at a temperature of about 50°C. The suspended particulate matter concentrations were determined later with an accuracy of 0.1 mg. A total of 36 samples of suspended particulate matter were examined using a JSM-U3 scanning electron microscope.

According to our data the thickness of the bottom nepheloid layer in the Vema Channel is generally up to 500 m (Fig. 4.53). This is much greater than the thickness of the bottom mixed layer, a fact which precludes the possibility of simple mixing by boundary turbulence being a sufficient mechanism. The typical values of suspended particulate matter volume concentration in this layer are in the range 0.1–0.3 ppm (mm^3/l).

It is known that a flow through the Vema Channel is well mixed in temperature and turbidity up to 400 m above the seafloor. Models of boundary layer development show that this cannot be caused only by vertical turbulent diffusion (Weatherly and Martin 1978; Richards 1982).

The central part of the Argentine Basin is a region of very high concentration of suspended particulate matter, which is not likely related to increased velocities near the bottom (Biscaye and Eittreim 1977). However, this high turbidity was caused by intermittently high velocities owing to high surface eddy kinetic energy propagated downwards, resulting in high abyssal eddy kinetic energy (McCave 1986). It is noteworthy that greater concentration of suspended material at depths exceeding 4,000 m can be accounted for by the fact that waters at these depths are in contact with a much greater area of seafloor than the same water volume in shallower parts of the oceans.

Very high concentrations of suspended matter (>100 μg/l) are found in near-bottom waters at the lower part of the continental slope in the northwestern Argentine Basin. The concentration decreases rapidly at depths 4,760–4,780 m (50–100 m above the seafloor). On the other hand, at GEOSECS station 59 (30°11′ S, 39°23′ W, depth 4,819 m, 1976) near the Vema Channel axis, the near-bottom values of par-

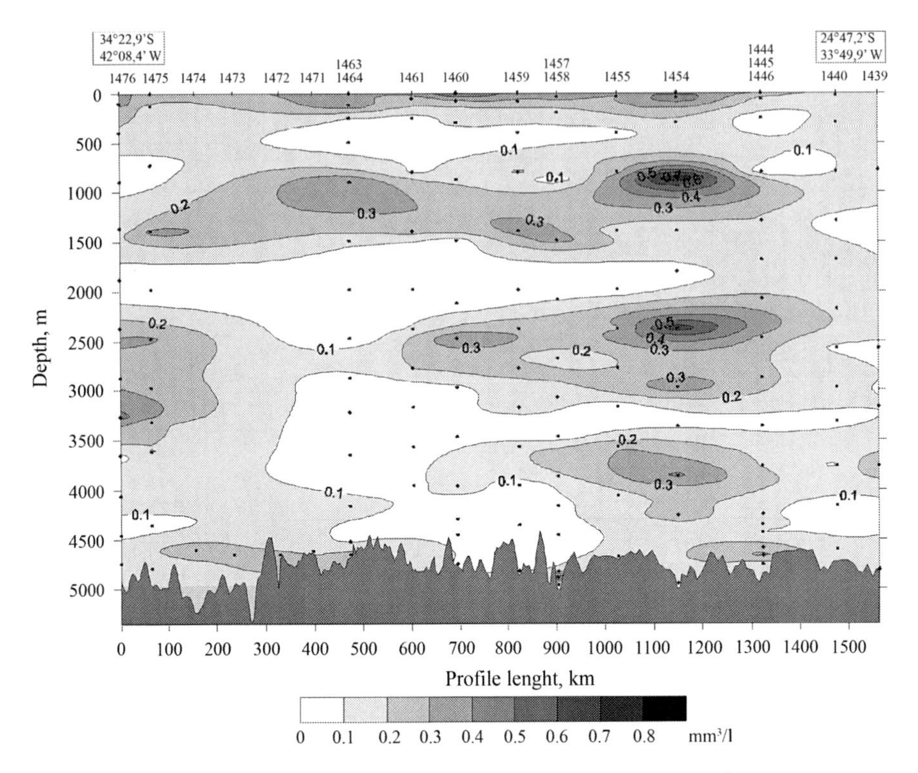

Fig. 4.53 Vertical distribution of suspended matter volume concentration (mm^3/l) along the Vema Channel (station numbers are shown at the *top*, depths of samples are denoted by *black dots*)

ticulate concentration are only ~30 μg/l, but the bottom mixed nepheloid layer is 400–500 m thick. At this station, the integrated mass of suspended matter per unit area in the lower 450 m (bottom mixed layer) is 1.12×10^{-3} g/cm^2. At the stations in the Argentine Basin, the integrated mass of suspended matter in the lower 450 m is $1.05–1.11 \times 10^{-3}$ g/cm^2. Remarkably close values of these estimates suggest that the local resuspension of bottom sediments within the Vema Channel may be relatively unimportant (Johnson et al. 1976).

Three intermediate nepheloid layers can be seen on the section along the channel in the depth intervals 3,500–4,300 m, 2,200–3,500 m and 800–1,500 m (see Fig. 4.53). In these layers, the volume concentration of suspended particulate matter reached 0.8 ppm. The major source areas for two lower layers are shallower regions (up to 2,000 m depth) where material is resuspended at the edges of terraces and slopes of the Rio Grande Rise and Santos Plateau, possibly due to tidal motions. The resuspended matter spreads seaward. According to Armi (1978) and Armi and D'Asaro (1980), who used the ideas put forward by Munk (1966), a vertical transport mechanism exists, which involves turbulent mixing in bottom layers 50-m thick, followed by the detachment of particles and lateral advection along isopycnal surfaces. The detachment occurs where isopycnals intersect the bottom and this is likely to occur in regions of steep topography, and also in regions of lower bottom slope gradient at benthic fronts where sloping isopycnals intersect the bottom. Inversions in the concentration of suspended matter are incompatible with turbulent mixing up to a few kilometers above the bed. However, these features are explicable if the layers were recently separated from the bottom. After the detachment, a decrease in turbulence and sedimentation of particles occurs in these layers.

The origin of the uppermost intermediate nepheloid layer (800–1,500 m, see Fig. 4.53) is still unclear. The source of the suspended particles is not completely clear. It may be suggested that the particles have a biological origin. It is possible that the muddy "trace" is a result of the influence of the Falkland Current frontal zone. This is an area with a high primary production, whose waters are transported to the north with Antarctic Intermediate Water. As this takes place, the influence of intense terrigenous flow from the Rio de La Plata cannot be excluded.

Thus, the Vema Channel is a pronounced example of the influence of the sides of a basin on the vertical structure of the suspended particles (2–28 μm) distribution which makes sandwiched nepheloid layers. Vertical profiling by nephelometer responds to the less fine particles and intermediate nepheloid layers are not revealed.

According to our data on suspended particulate matter in the Argentine Basin, cyclic and pinnate Diatomea algae are commonly found in the near-bottom water (about 4,000 m depth) (Fig. 4.54). The condition of the Diatomea cells is quite good. In the Vema Channel, only fragments of diatoms were found, mostly biogenic detritus, separate Coccoliths cells, and occasionally partly destroyed Coccolithophorids. Mineral particles were found in the samples as well. In addition, Peridinea algae were found in the sample at station 1464.

In the southern part of the Vema Channel, we found a direct relationship between volume and mass concentrations of suspended particulate matter ($R^2 = 0.94$ for 17 samples) based on our measurements during cruise 17 of R/V *Akademik Sergei Vavilov*. In the northern part of the channel this dependence disappeared. From south to north, we found a decrease in the concentration of terrigenous suspended particles from 49.2 to 9.9%.

The cross asymmetry of the near-bottom boundary layer is typical for the Vema Channel and is embodied in optical characteristics as well as values of the SPM mass concentration.

Light scattering and temperature profiles have striking cross-channel asymmetry. At the eastern side of the channel the top nepheloid layer is sharp and corresponds in position to the benthic thermocline. In the channel axis and in the western branch the gradients are lower and the transition zone is thicker (Johnson et al. 1976; Macrander, pers. com. 2006) (Fig. 4.55).

Measurements of suspended matter were made in different cruises by water samples filtering and further weighing. Samples were taken at a distance of 20 m above the bottom. The distribution of suspended matter over the standard section across the Vema Channel at 31°12′–31°14′ S shows that maximum concentration is found in the cold jet near the eastern slope of the channel. Changes in the concentration of suspended matter in different years are shown in Fig. 4.56.

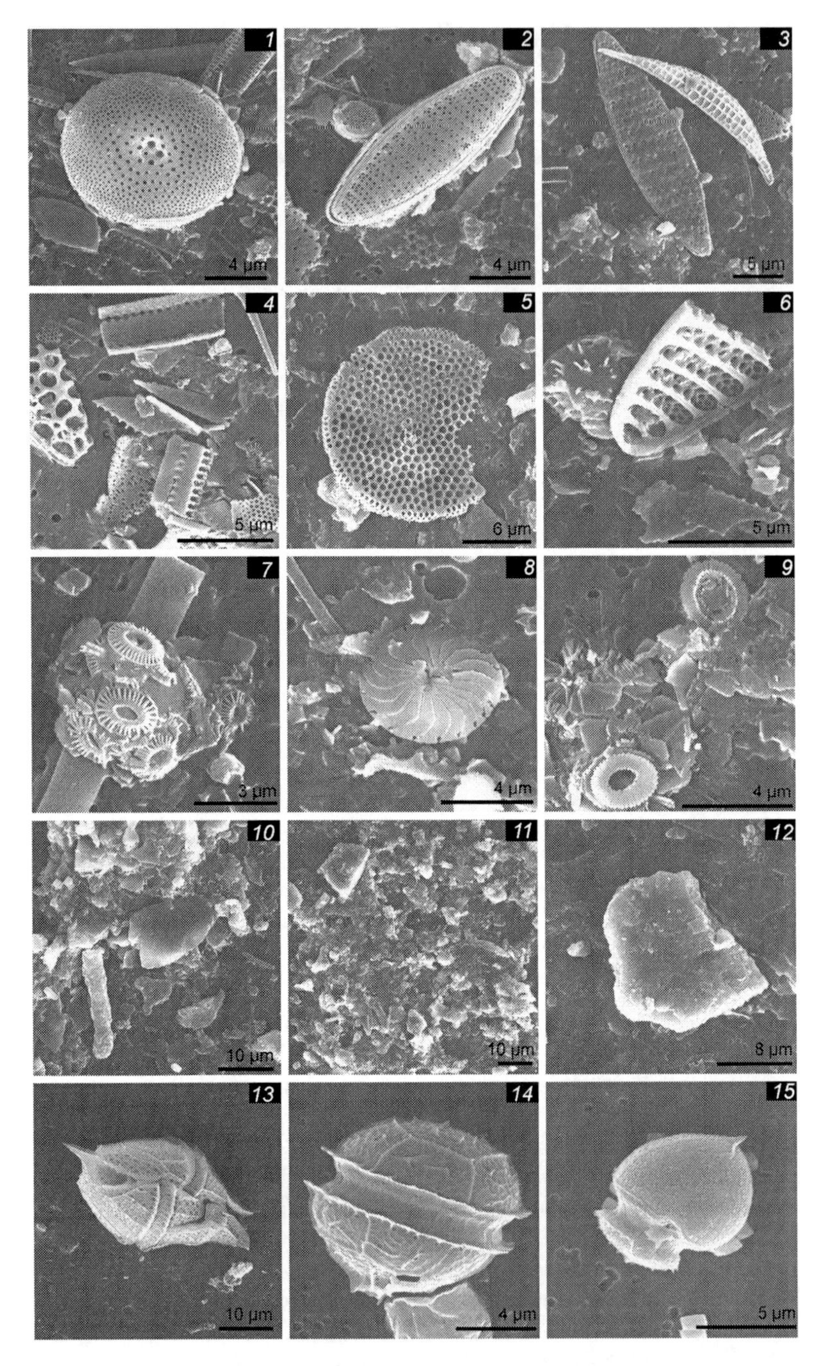

Fig. 4.54 The photomicrography of suspended matter samples from the Argentine Basin (*1–3*) and Vema Channel (*4–15*): (*1–3*) Diatomea algae (centric and pinnate forms) and their detritus; (*4–6*) fragments of Diatomea algae; (*7–9*) destroyed Coccoliths and coccosphere; (*10–12*) mineral particles and organic detritus; (*13–15*) Peridinea algae

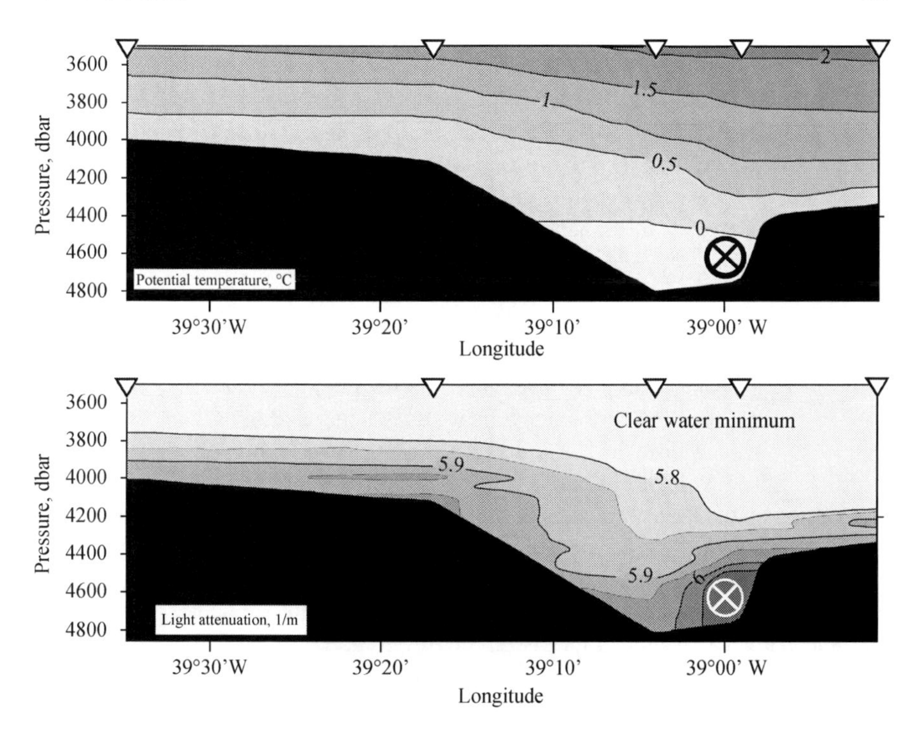

Fig. 4.55 Vertical distribution of abyssal potential temperature (*top*) and red light attenuation (*bottom*) on a zonal section between the Santos Plateau and the Vema Channel at approximately 29° S. The graph intersects the channel extension transporting cold (<2°C) Antarctic Bottom Water. Stations are indicated by *open triangles*; the *X*-symbols denote the area of the strongest near-bottom current (it coincides with the highest light attenuation). This correlation indicates the influence of the abyssal current in transporting the sediment

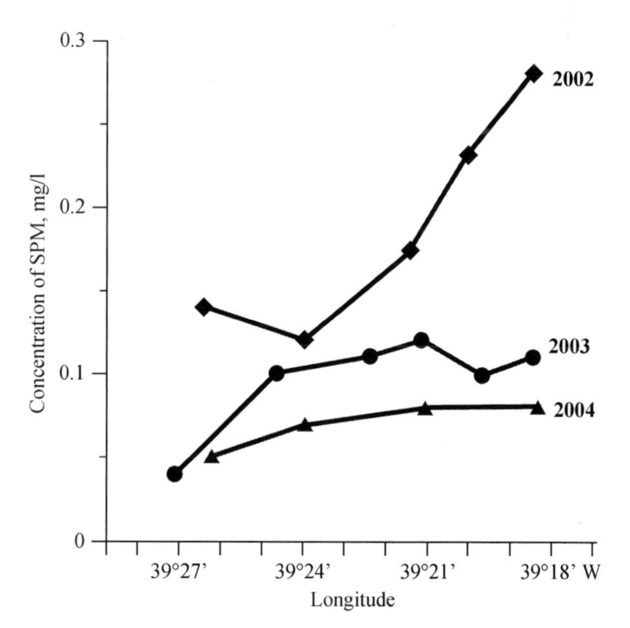

Fig. 4.56 Concentration of suspended particle matter across the Vema Channel at 31°12′ S in 2002, 2003, and 2004

4.3 Santos Plateau

The Santos Plateau is located between the continental slope of South America and the Vema Channel. The major part of the plateau lies at a depth of 4,000 m, which is well below the depth of the 2°C potential temperature isotherm. It might seem likely to provide the transport of at least the upper portion of Antarctic Bottom Water (Speer and Zenk 1993).

Geostrophic transport calculations based on the hydrographic section in 1991 (Speer and Zenk 1993), relative to the surface of zero motion coinciding with the 2°C potential temperature isotherm give a total northward transport below the 2°C isotherm of approximately 3 Sv and a total southward transport of 1 Sv, resulting in the net flow of 2 Sv. The flow can merge with the upper part of the flow through the Vema Channel and flow further to the north.

Much of our knowledge about water transport over the Santos Plateau comes from Hogg et al. (1999). The data include time series of currents measured at an array of moorings deployed across the southern boundary of the Brazil Basin in early 1991. The length of the time series exceeds 1.5 year. The array spanned from the lower continental slope of South America across the Santos Plateau, Vema Channel, and Hunter Channel. A quasi-zonal hydrographic section was occupied from the continental slope to the Rio Grande Rise between latitudes 35° S and 30° S. The measurements were made by cooperative efforts of scientists from the *Institut für Meereskunde* and the *Woods Hole Oceanographic Institution* (Moorings 1077, 1082, 1084, and 1088). Information about the moorings is given in Table 4.7.

In 2003, the Russian scientists deployed two moorings at the Santos Plateau with almost 1 year time series of measurements 20 and 30 m above the bottom (moorings 1066 and 1069).

Progressive diagrams for the near-bottom instrument at the six moorings are shown in Fig. 4.57. The diagrams cover a period of 3 months. The particle travel distances correspond to the scale of the figure.

Progressive diagrams both in 1991 and 2003 show that there is no net northward transport of bottom waters over the Santos Plateau. Similarly, the vectors of horizontal velocities demonstrate alternative currents to the north and to the south. Figure 4.58 shows the vectors based on the data of two moorings in 1991 and 1992 deployed on the Santos Plateau. The moored array data provided direct current

Table 4.7 Moorings in the Santos Plateau

Mooring number	Latitude	Longitude	Deepest instrument level	Ocean depth
1077	28°28′ S	44°28′ W	3,532 m	3,632 m
1082	29°03′ S	43°30′ W	3,850 m	3,950 m
1084	29°32′ S	42°42′ W	3,918 m	4,019 m
1088	30°05′ S	41°44′ W	3,714 m	3,814 m
1066	31°12′ S	41°51′ W	3,750 m	3,770 m
1069	30°39′ S	45°01′ W	3,460 m	3,490 m

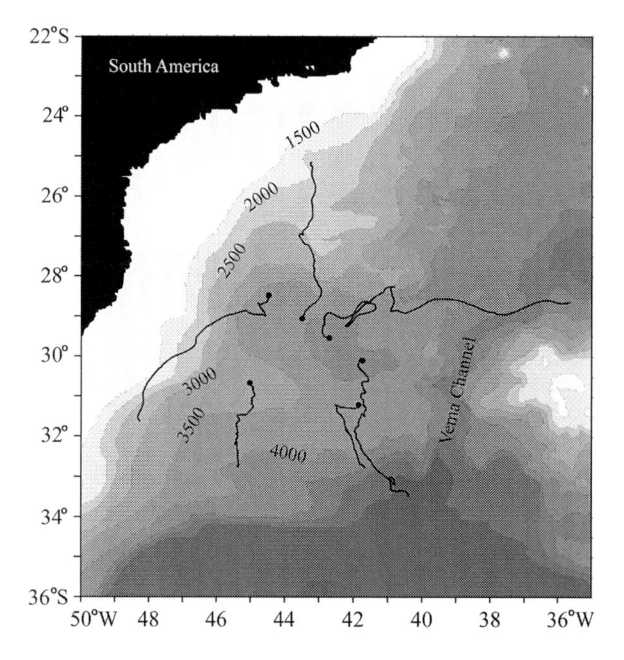

Fig. 4.57 Bottom topography of the Santos Plateau (meters) and 3-months-long progressive diagrams for the near-bottom current measurements on the Santos Plateau

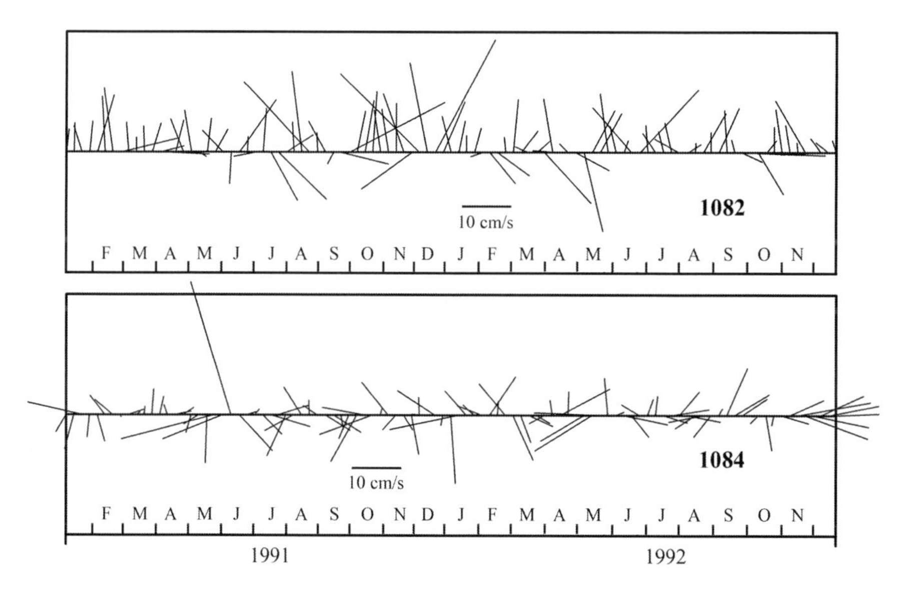

Fig. 4.58 Vectors of horizontal velocities as a function of time at moorings 1,082 (29°03′ S, 43°30′ W; instrument at 3,850 m) and 1084 (29°32′ S, 42°42′ W; instrument at 3,918 m). Currents were low-pass filtered to remove frequencies higher than 1/48 cyc/h. Vectors are shown with a week interval. *Letters* indicate months of the year

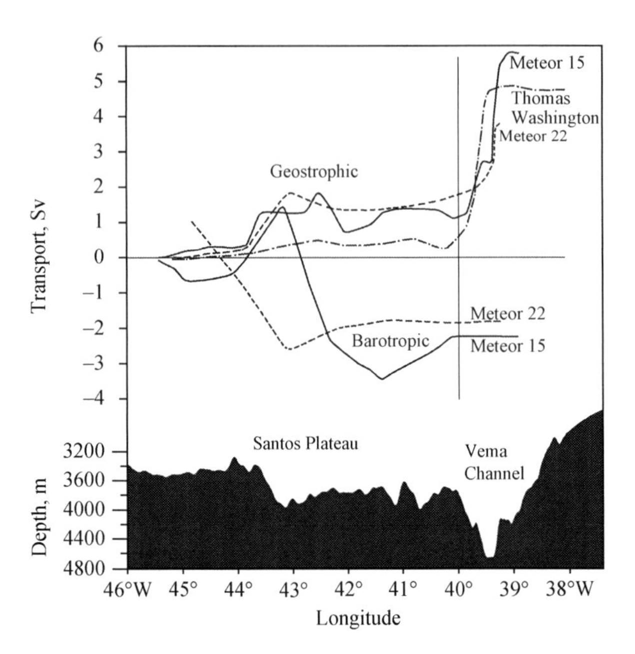

Fig. 4.59 Components of AABW transport over the Santos Plateau for the *Marathon* (R/V *Thomas Washington*), *Meteor* 15, and *Meteor* 22 cruises (cumulative curves). The reference velocity was selected at 2°C potential temperature isotherm. The *vertical line* at 40° W divides the Santos Plateau and Vema Channel longitudes. Bottom depth (meters) is shown in the *lower panel*. (Adapted and redrawn from Hogg et al. 1999)

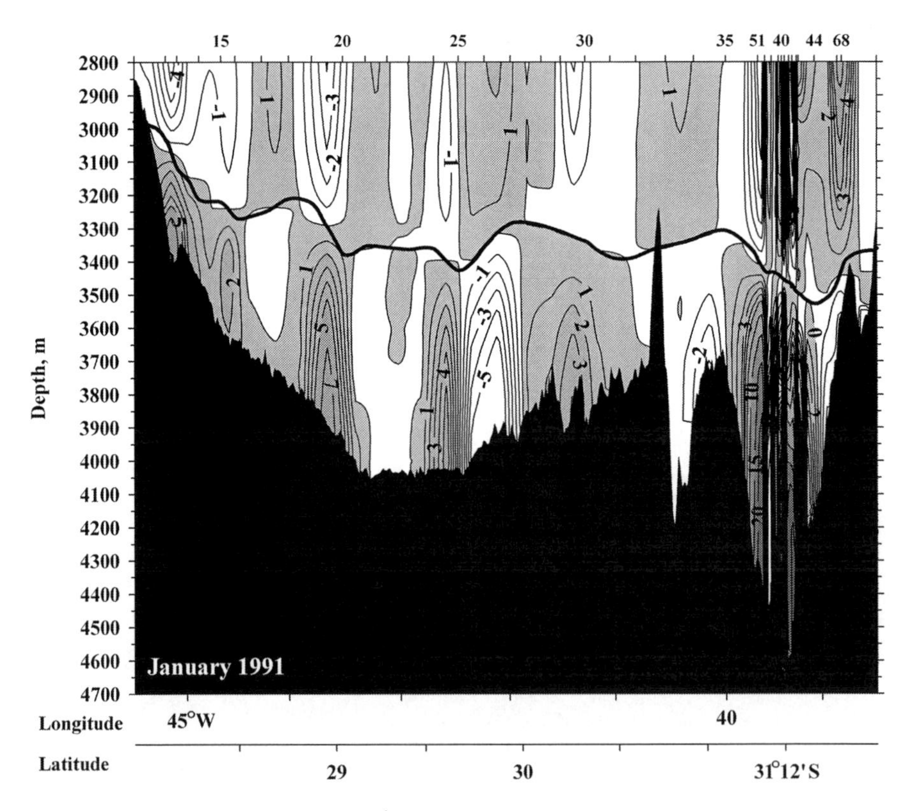

Fig. 4.60 Geostrophic velocities (cm s^{-1}) over section AR15C in 1991. The section includes the Vema Channel and its eastern flank. The *gray color* shows the northward flow

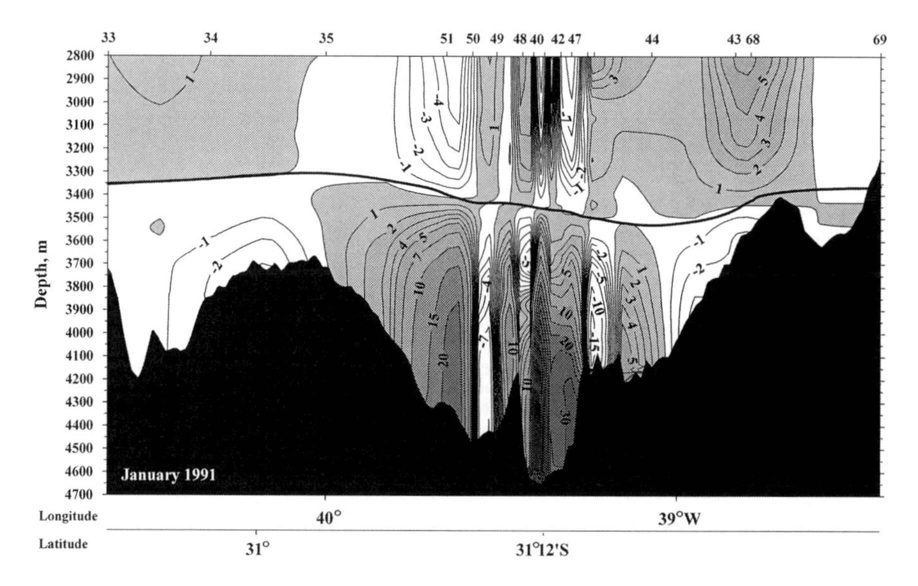

Fig. 4.61 Same as in Fig. 4.60, but without the western part of the section to show in more detail the geostrophic flow in the Vema Channel and its eastern flank. The *gray color* shows the northward flow

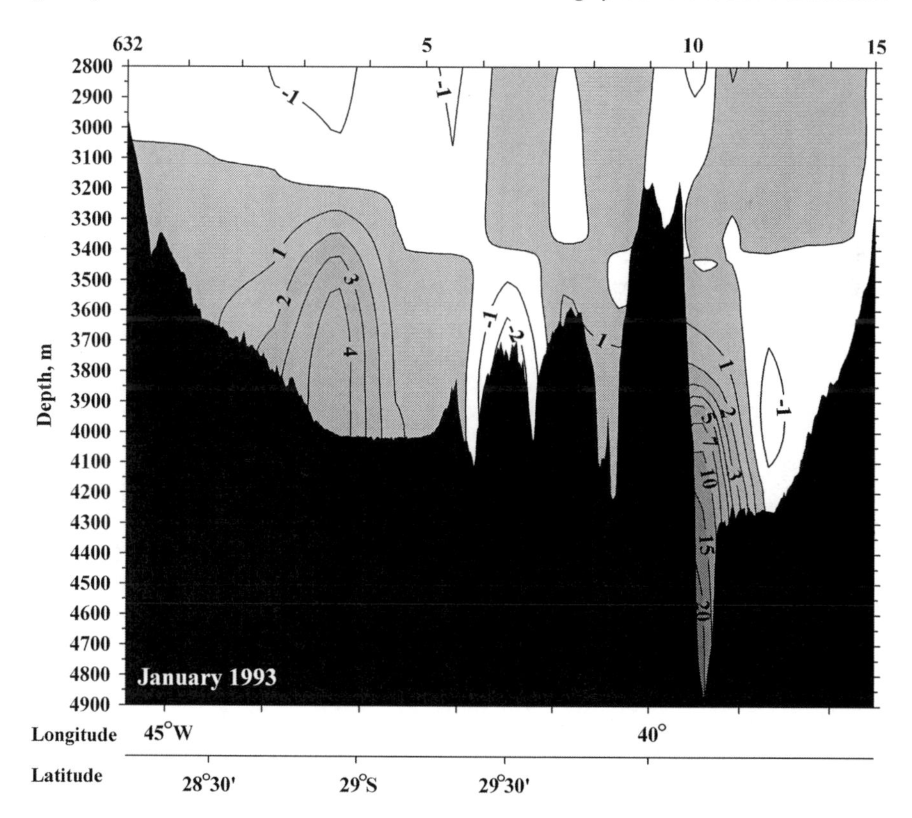

Fig. 4.62 Geostrophic velocities (cm s^{-1}) over section A10 in 1993. The *gray color* shows the northward flow

measurements to estimate the bottom water flux from the Argentine Basin to the Brazil Basin. Although the current-meter moorings were deployed sparsely, the direct measurements were combined with the dynamic computations of geostrophic velocities. While the geostrophic computations give a uniform northerly flow, current-meter data reveal two regions of southward flow.

Hogg et al. (1999) separated geostrophic and barotropic components of the flow on the basis of moored measurements between two CTD stations. The combination of moored measurements and geostrophic calculations during cruises 15 (1991) and 22 (1992) of R/V *Meteor* and the cruise of R/V *Thomas Washington* (1984) under the "Marathon" program resulted in the barotropic and geostrophic transport curves as functions of longitude (Fig. 4.59). Thus, the total transport of Antarctic Bottom Water is the sum of the barotropic and baroclinic contributions. The authors conclude that the northward geostrophic transport of 1.3–1.8 Sv calculated from the data of cruise 22 of R/V *Meteor* is nullified by a comparable southward transport. Upon emerging from the Vema Channel, the upper (shallow) components of Antarctic Bottom Water recirculate over the Santos Plateau and there is very little if any net northward flow over the plateau. This conclusion is

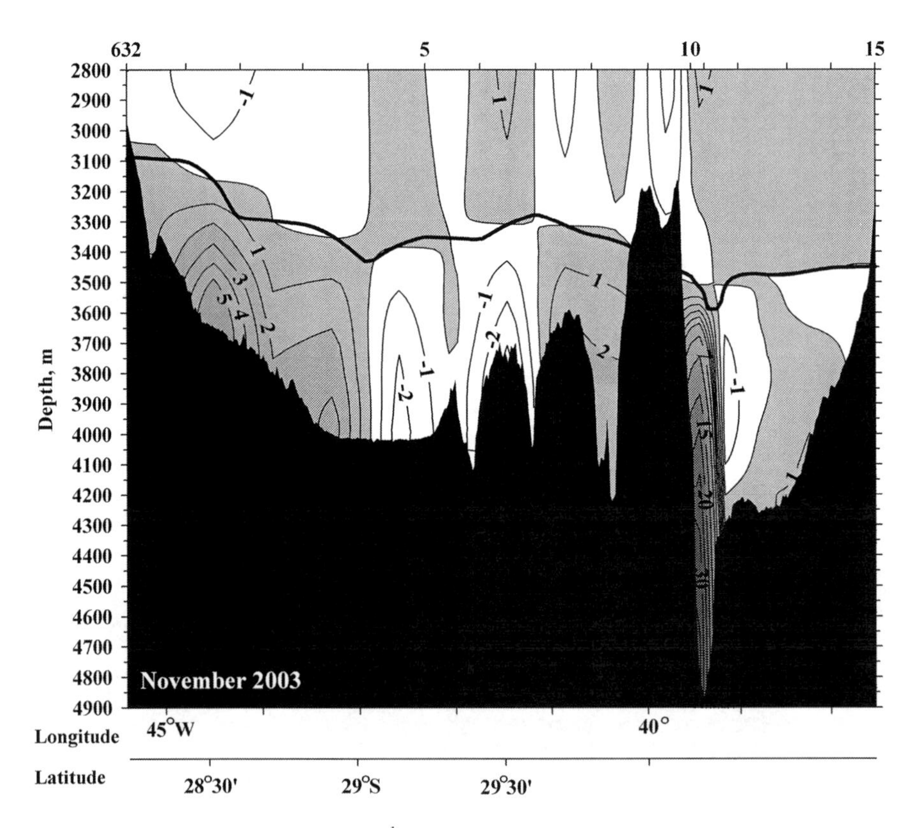

Fig. 4.63 Geostrophic velocities (cm s^{-1}) over section A10 in 2003. The *gray color* shows the northward flow

supported in McDonagh et al. (2002): near-zero net flow over the Santos Plateau results from a near-closed cyclonic circulation fed by the deep Vema Channel throughflow.

Geostrophic calculations based on the data of hydrographic sections across the Santos Plateau occupied in 1991 (AR15C), 1993 (A10), and 2003 (A10) resulted in compilation of baroclinic velocities and transport. The geostrophic velocity sections relative to the 2°C of potential temperature are shown in Figs. 4.60–4.63. The eastern part of the section in 1991 is shown separately in Fig. 4.61 to demonstrate the geostrophic flow structure in the Vema Channel and its eastern flank. All sections show a northward flow over the continental slope of South America below 3,000 m. Geostrophic calculations incorrectly show intensification of currents at the western slope of the Vema Channel and even southward flow at its eastern slope.

4.4 Hunter Channel

The Hunter Channel is a more than 200 km wide valley located between two ridges at 32–35° S. The maximum depth in the channel is approximately 4,300 m, which is only about 300 m shallower than the Vema Channel. The main sill is located at about 35°12′ S, 27°47′ W, i.e., approximately 70 km upstream of the CTD sections occupied in different years (Special researches aimed at the study of the flow in the channel were carried out in 1991, 1992, 1994, and 2002). Locations of hydrographic stations and moorings in the Hunter Channel are shown in Fig. 4.64.

The first standard hydrographic measurements were carried out in the Hunter Channel in 1959 using Nansen bottles and reversing thermometers. A total of seven stations were occupied along 32° S between the Rio Grande and Mid-Atlantic ridges (Fuglister 1960). During the WOCE field program, ten CTD stations were made across the narrow part of the Hunter Channel during cruise 15 of R/V *Meteor* in January 1991. Later, the region was visited in December 1992, during cruise 22 of R/V *Meteor* (four CTD stations) and then in May 1994; another section (five CTD stations) was occupied during cruise 28 of R/V *Meteor*. In November, 2002, a section (five CTD stations) was occupied during cruise 11 of R/V *Akademik Ioffe*. The most complete dataset was provided by cruise 15 of R/V *Meteor* when ten CTD stations were occupied in 1991.

During the measurements in 1991, the coldest bottom water with a potential temperature of about 0.2°C was found on the eastern side of the gap, which is similar but more prominent in the Vema Channel. At shallower depths in the water layer with potential temperatures between 0.5 and 2°C, the coolest water was found at the western boundary of the section. The isotherms slope downward to the east, indicating the northward flow of bottom water relative to the warmer surface. The salinity field varies mainly in the vertical direction and correlates with temperatures such that isopycnals are parallel to isotherms below 3,000 m.

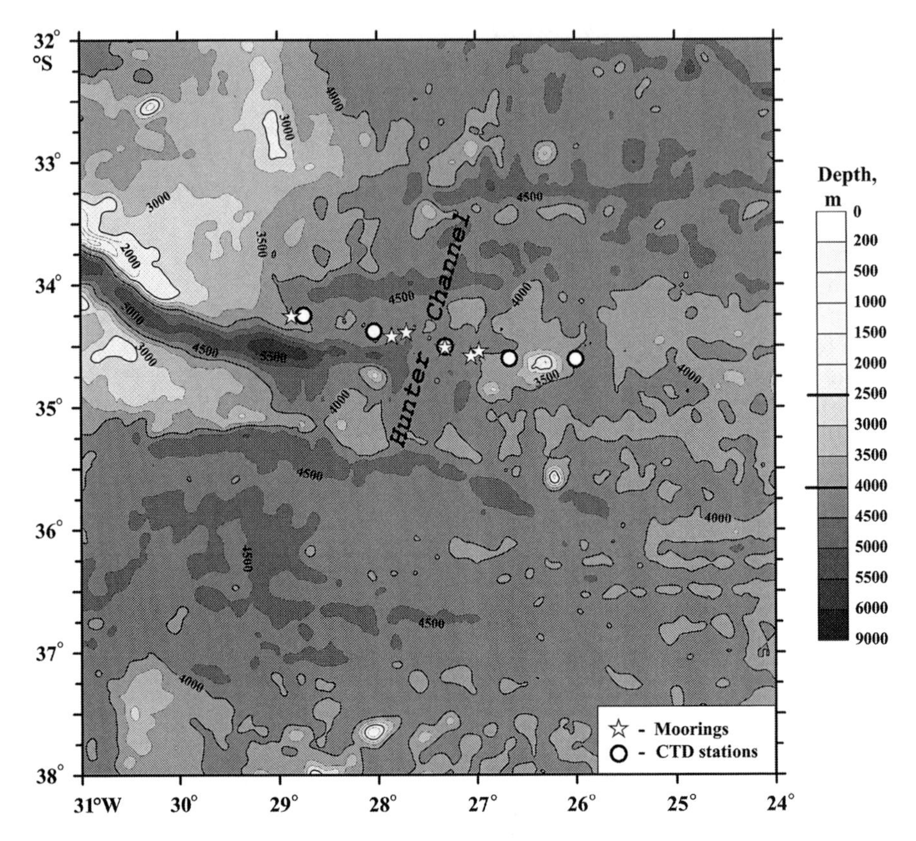

Fig. 4.64 Bottom topography (meters) of the Hunter Channel and locations of moorings and CTD-stations occupied in different years

The horizontal and vertical resolution of measurements in 1959 was not sufficient to distinguish a weak flow near the bottom. Reid (1989) was the first to estimate the deep transport through the Hunter Channel at 1.8 Sv. Two high-resolution geostrophic transport estimates of bottom-water overflow became available based on the sections in 1991 and 1994.

Using the *Meteor* (cruise15) data, Speer et al. (1992) calculated the geostrophic currents in the channel. Taking the flow to be zero at the 2°C surface, the bottom (geostrophic) northward velocities were typically 1–2 cm s^{-1}, and the total transport below 2°C was 0.7 Sv. The accuracy was estimated to be within ±50%.Geostrophic velocities were calculated relative to other surfaces of zero velocity. The results ranged within small variations near this value. It is considered that the above transport estimate, relative to the zero surface at 2°C, is accurate to better than 50%. The calculation was based on the assumption that the velocity below the deepest common level of station pairs is representative of the whole bottom triangle set up by these stations. Speer et al. (1992) supposed that a higher horizontal station resolution, revealing some bottom intensified flow, would raise the transport closer to 1 Sv. Northward bottom velocities were of approximately 1 cm s^{-1}. A weak countercurrent of about −0.5 cm s^{-1} was present at 27°30′ W.

The second CTD section in the Hunter Channel was occupied 3 years later during cruise 28 of R/V *Meteor* (May 1994). Again, the 2°C isotherm was chosen as an upper limit of bottom water and a reference level of zero velocity for geostrophic calculations. For the bottom triangles, a linear decrease of the current was assumed at the lowest data pair to zero over a distance of 1,000 m. The structure of the geostrophic current distribution appears as an eddy with the rotational velocity exceeding 0.5 cm s^{-1} in its core. The bottom water export to the Brazil Basin had almost ceased (<0.1 Sv).

Thermistor chain records are available from locations of moorings H4 (depth 4,336 m) and H5 (4,303 m), each carrying a 200-m near-bottom thermistor chain (see Fig. 4.64; coordinates of moorings H4: 34°30.8′ S, 27°19.2′ W and H5: 34°35.1′ S, 27°03.4′ W (Zenk et al. 1999)). The chains collected data during 3.5 years (1991–1994). Their separation was as high as 26 km. Thus, they cover 13% of the total width of the Hunter Channel. Daily averaged time series of temperature at these positions revealed that the short-term temperature fluctuations exceeding 0.2°C in magnitude are superimposed on a long-term trend. Fitting straight lines to curves in a least square sense provides temperature trends. The largest values (exceeding 0.5°C/1,000 days) are found in the upper portion of the water column at mooring H4. The chain at the other mooring (H5) delivered more uniform values between 0.2 and 0.3°C/1,000 days. Assuming a simple geostrophic balance in the trough between H4 and H5 (Fig. 4.64), we expect slope changes of the thermal stratification to be proportional to local changes of bottom-water throughflow. In the geostrophic sense, the first period corresponds to a strong northward bottom-water export situation; the second period, to a slower flow.

The difference in the depth of the Vema and Hunter channels eliminates water colder than 0°C from the Hunter Channel. Thus, the average potential temperature of the bottom current in the Hunter Channel is higher by 1°C. The warmer Hunter Channel flow supplies water to the warmer southeastern part of the Brazil Basin.

From December 1992 to May 1994, deep flow through the Hunter Channel was directly observed by an array of moored current meters and thermistor chains. Direct measurements of the Hunter Channel throughflow reveal a persistent northward transport of bottom waters. Current reversals exist in all records, with time scales of 1–2 weeks.

Highest speeds were observed on the eastern side of the Hunter Channel 15 m above the bottom. Both current components had typical values above 4 cm s^{-1}, resulting in an average speed of 7.1 cm s^{-1} and a maximum speed of 17.6 cm s^{-1}. The strong eastward flow at this site is caused by the local orientation of a deep northeast-oriented valley. Mean meridional components of about 0.9 cm s^{-1} are characteristic for the central and western sides of the Hunter Channel.

The first long-term moored current-meter observations from the Hunter Channel were reported in Pätzold et al. (1996). Six moorings were deployed in the Hunter Channel. The sampling rate was 2 h. The mean northerly transport in the channel was estimated from moored current meters at 2.92 Sv, which is significantly higher

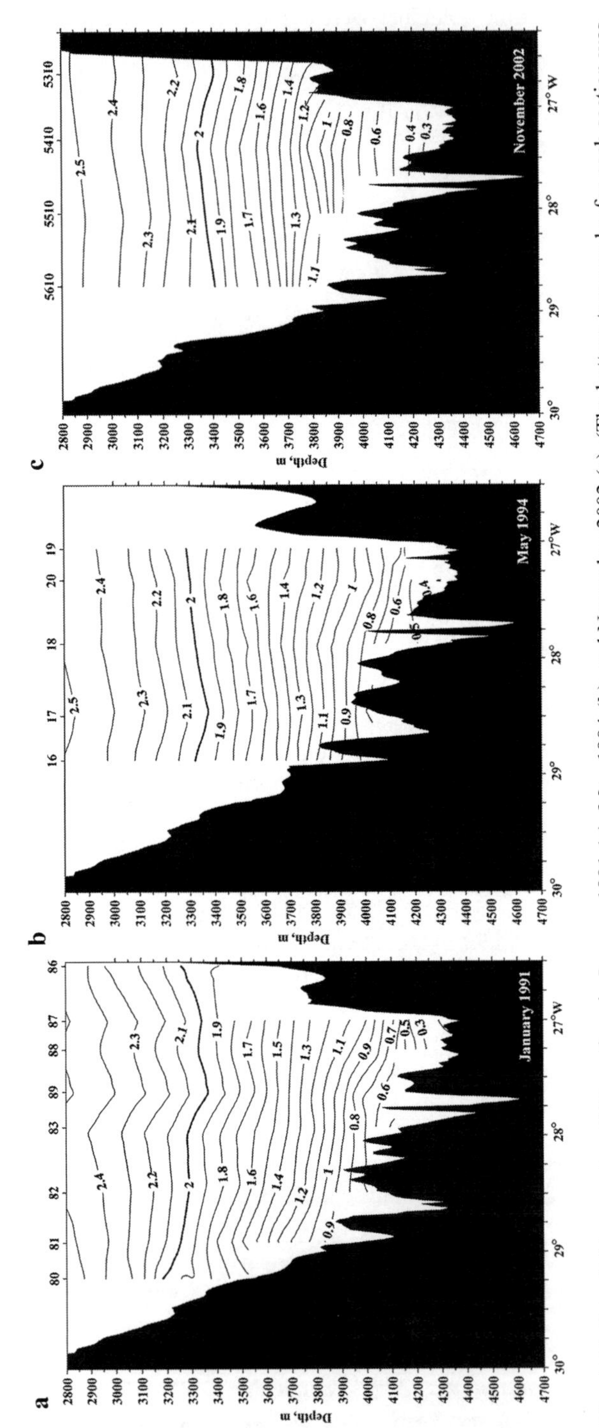

Fig. 4.65 Potential temperature (°C) sections in January 1991 (**a**), May 1994 (**b**), and November 2002 (**c**). (The bottom topography for each section was taken from Smith and Sandwell 1997)

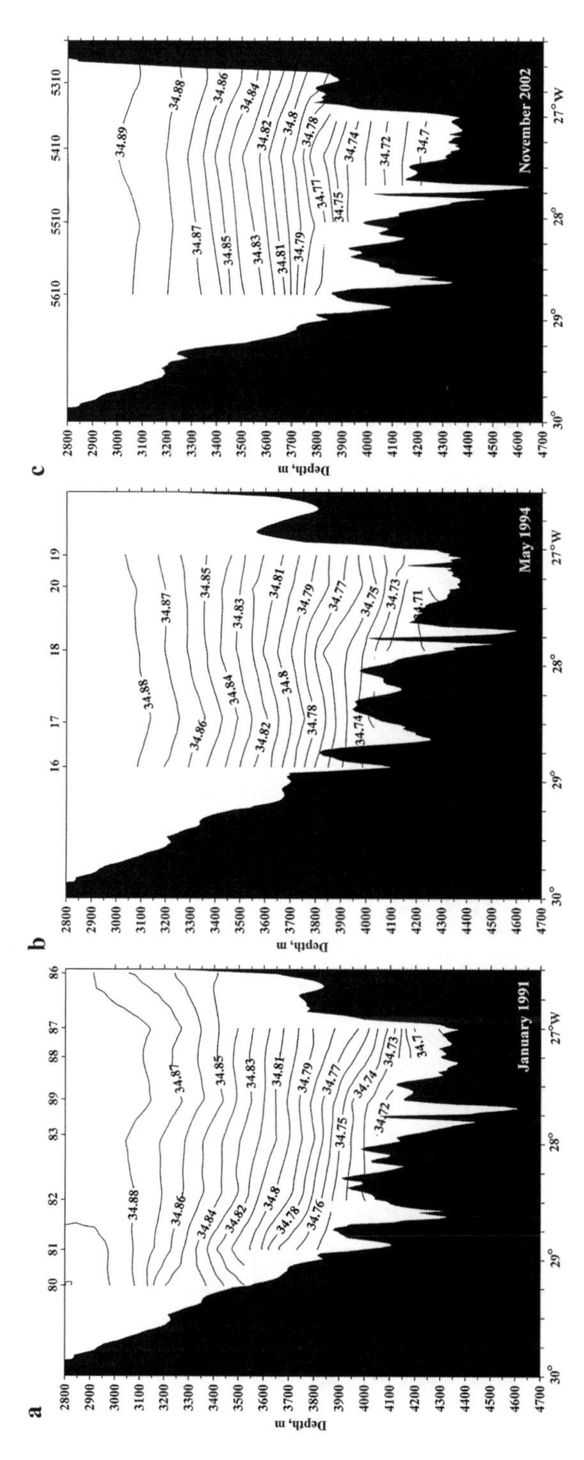

Fig. 4.66 Salinity sections (psu) in January 1991 (**a**), May 1994 (**b**), and November 2002 (**c**). (The bottom topography for each section was taken from Smith and Sandwell 1997)

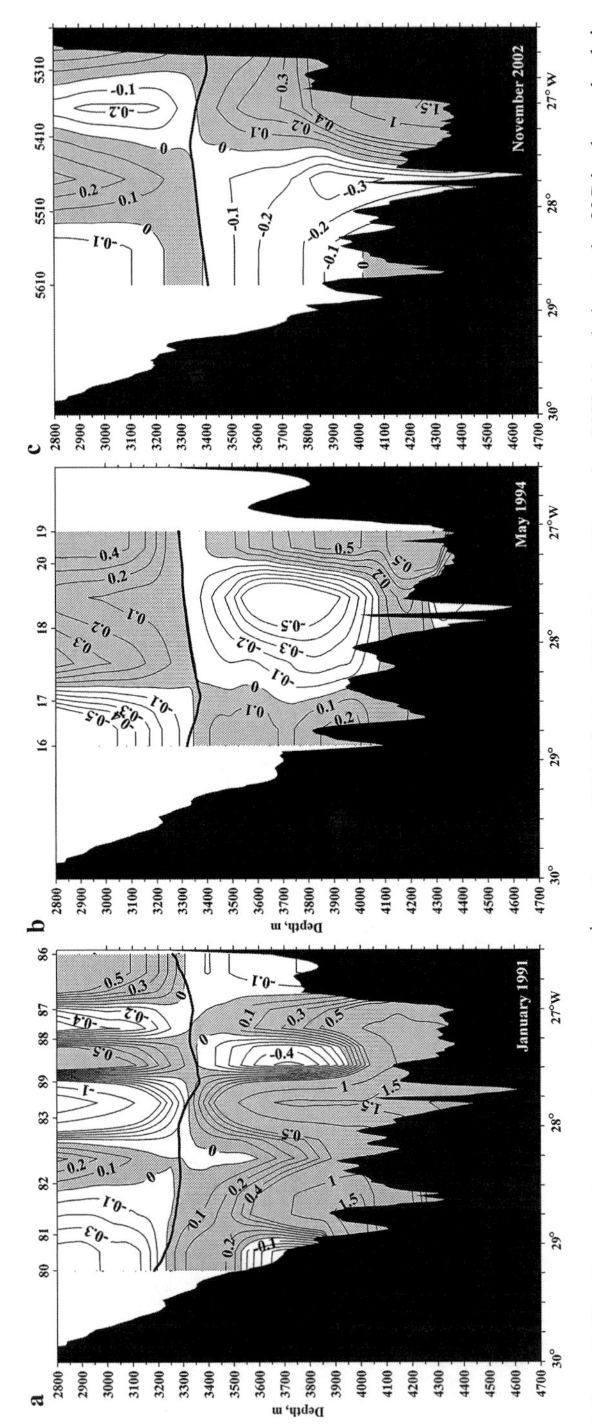

Fig. 4.67 Distribution of geostrophic velocity (cm s^{-1}) in January 1991 (**a**), May 1994 (**b**), and November 2002 (**c**) relative to the 2°C isotherm depth in Fig. 4.65. The *gray color* indicates the northward transport. (The bottom topography for each section was taken from Smith and Sandwell 1997)

than earlier estimates based on geostrophic calculations. The maximum transport exceeding 11 Sv was recorded in June 1993. Its impact on the mean transport was compensated only 6 weeks later by the minimum transport of −5 Sv during a current reversal event of only a few days duration.

A tendency toward increased bottom-water temperatures was observed during the WOCE observational period. The lowest potential temperature in 1991 was 0.223°C. In 1994, the minimum bottom temperatures barely reached below 0.301°C, and the lowest potential temperature measured in 2002 was 0.219°C. Lowest salinities observed in 1991 were 34.693 psu. In 1994, they were replaced by values above 34.695 psu. In 2002, the lowest salinities at the bottom were 34.689 psu.

Graphs of three sections of potential temperature measured in 1991, 1994, and 2002 are shown in Fig. 4.65. All graphs were calculated and plotted using the original data and the same programming code and graphic program. The graphs are similar, but lower temperatures were not found in 2002 at the eastern wall of the channel. However, only four stations were located in the gap in 2002. The eastern station (at 26°40′ W) was not occupied in the deepest place of the eastern part of the gap.

Distributions of salinities do not show strong differences among three measurements (Fig. 4.66). The geostrophic velocities relative to the depth of the 2°C potential temperature isotherm are shown in Fig. 4.67. They result in lower velocities than measured at moored buoys in 1994. The data of three sections were processed using the same approach. Note a good coincidence between the velocity distributions in 1991 and 1994 published in Zenk et al. (1999). Geostrophic transports of water layer below the 2°C isotherm in 1991, 1994, and 2002 are 0.75 Sv to the north, 0.10 Sv to the south, and 0.13 Sv to the north, respectively. The latter two transports are close to zero with a strong reverse southward flow in a part of the channel. Mean velocities and water transport measured in 1992–1994 by current meters on buoys (2.92 ± 1.24 Sv), which include the barotropic component, are notably greater.

Chapter 5
Further Propagation of Antarctic Bottom Water from the Brazil Basin

In this chapter we assume Wüst's (1936) sense of Antarctic Bottom Water propagation. In other words, this is the bottom water of the Antarctic rather than North Atlantic origin. However there is no generally accepted isotherm or value of any characteristics in literature for the upper boundary of Antarctic Bottom Water in the Equatorial and North Atlantic. Therefore, it is essential that quantitative estimates of Antarctic Bottom Water properties and transport depend on the choice of this boundary.

5.1 Brazil Basin

The Brazil Basin is a major transit region for the propagation of Antarctic waters in the abyssal depths. After the outflow from the Vema Channel, intensity of the bottom flow and vertical mixing decreases strongly in this large and deep basin. In the southern part, the basin is separated from the eastern basin of the Atlantic by the relatively high Mid-Atlantic Ridge without fractures that are crucial for the bottom water transport to the east. In the northern part, the ridge is crosscut by the Romanche and Chain Fracture zones.

Not all publications support the division of Antarctic Bottom Water in the Brazil Basin into several components. Many authors consider that Weddell Sea Deep Water disappears in the northern equatorial part of the basin (Reid 1989; Sandoval and Weatherly 2001). However, Weddell Sea Deep Water is distinguished at 5° S based on the measurements of CFC concentration in Fischer et al. (1996) and Andrie et al. (1998). The distinguishing indicators of Lower Circumpolar Water – the minimum of oxygen concentration (Reid et al. 1977) and local maximum of silicates (Mantyla and Reid 1983) – disappear in the Brazil Basin because relatively high oxygen concentration in the bottom layer of the Argentine Basin changes to a bottom minimum in the Brazil Basin (Mantyla and Reid 1983).

E. G. Morozov et al., *Abyssal Channels in the Atlantic Ocean*,
DOI 10.1007/978-90-481-9358-5_5, © Springer Science+Business Media B.V. 2010

Part of the Antarctic Bottom Water propagating from the Argentine Basin through the passages to the Brazil Basin continues its flow mainly along the western slope of the basin (Stommel and Arons 1960; Rhein et al. 1995; Sandoval and Weatherly 2001). A number of elevations located east of the continental slope of South America form obstacles for the propagation of these waters, and the flow turns around these topographic features. However, according to the scheme in Hogg and Owens (1999), these obstacles are possibly not very significant. The bottom waters propagate along the western slope of the Brazil Basin approximately to 5° S. According to De Madron and Weatherly (1994), a branch of the northerly Antarctic Bottom Water flow turns to the south in this region. This flow reaches the southern boundary of the Brazil Basin, then again turns to the north, and flows along the Mid-Atlantic Ridge. These authors consider that a cyclonic eddy is formed in the 4–11° S region near the ridge. This eddy entrains the branch propagating along the Mid-Atlantic Ridge and the branch flowing here directly from the western boundary region. Stephens and Marshall (2000) consider that waters from this eddy propagate further through the Romanche and Chain fractures into the eastern basin. Based on the results of geostrophic calculations over different sections, Sandoval and Weatherly (2001) suggest another scheme of Antarctic Bottom Water propagation in the bottom layer of the Brazil Basin. They distinguish two main flows of these waters, which they traced from the Vema Channel. The flows are located at 150–200 km from each other at depths of 4,700 and 4,250 m. They flow almost parallel to each other in the 16° S region and turn together around the Vitoria-Trindade ridge of seamounts. The lower flow propagates around the Bahia Mountains and then turns to the east to the Romanche and Chain fractures in the 4° S region. The upper flow propagates along the western coast of the ocean and further to the nameless Equatorial Channel. Thus, the flows are separated into two branches (one flowing to the Equatorial Channel and the other flowing to the eastern basin) immediately after the Vema Channel. According to Larque et al. (1997), Lower Circumpolar Water (in terms of Reid et al. (1977)) and Weddell Sea Deep Water in the Brazil Basin make up several quasi-meridional jets moving from the Vema and Hunter channels almost directly to the north.

Many estimates of integral water transport in the Brazil Basin are available in the literature (Onken 1995; and others). Transport of Antarctic Bottom Water decreases strongly from 5–7 to 3 Sv due to vertical mixing as it flows in the Brazil Basin. According to McCartney and Curry (1993), decrease in water transport can be caused by weak upwelling flow of water due to diapycnal mixing. Estimates of the transport of Antarctic Bottom Water in the bottom layer presented by researchers differ strongly not only owing to the difference in the definition of the upper boundary of these waters. Choice of different reference surfaces is another cause. Below, we present the available estimates of the position of reference surfaces in the northern part of the Brazil Basin: $\sigma_{1.5} = 32.15$ (Stramma 1991); $\sigma_4 = 45.90$ (Rhein et al. 1998); 2,000–3,000 m (Friedrichs and Hall 1993); 2,000 and 4,000 dbar (Fu 1981); and $\theta = 4.7°C$ (Molinari et al. 1992). A summary of bottom water transport

Table 5.1 Transport in the Southern part of the Brazil Basin (28–32° S, section A10 along 30° S)

Authors	AABW boundary	Transport (Sv)	Comments
Sverdrup et al. (1942)		3	Transport across 30° S, geostrophic calculation
Wright (1970)	$\Theta = 2$	5.1	32° S
Fu (1981)	$\sigma_4 = 45.92$	−1.5	Inverse model based on “Meteor” data at 28° S; reference surface = 2,000 dbar
		−5.5 … 1.4	Inverse model based on IGY section at 32° S; reference surfaces at 2,000 and 4,000 dbar
Macdonald (1993)	$\sigma_4 = 45.895$	6	
Speer and Zenk (1993)	$\Theta = 2$	6.6	Geostrophic transport, Meteor 15, reference surface $\theta = 2$; total transport from three sources
Saunders and Thompson (1993)	3,000 m	−4.5	FRAM model
De Madron and Weatherly (1994)		7.2	SAVE, 25° S
Cai and Greatbach (1995)	3,700 m	0.3	OGCM model
Schlitzer (1996)	$\sigma_4 = 45.9$	4.1–4.2	Numerical model
Hogg et al. (1999)		6.7	Total transport from three sources
Barnier et al. (1998)		1–2	Numerical model, annual cycle
Holfort and Siedler (2001)	$\sigma_4 = 45.93$	6.8	
Vanicek and Siedler (2002)	$\gamma^n = 28.12$	7 ± 2	Inverse model; transport of LCDW and AABW across A10 section
McCartney and Curry (1993)		7.0	
Ganachaud (2003)	$\gamma^n = 28.11$	6 ± 1.3	Inverse model

in the Brazil Basin is given in Tables 5.1, 5.2, and 5.3. A scheme of circulation is shown in Fig. 5.1.

The major part of estimates given in these tables were obtained using the data of inverse models (or box-models) for large-scale basins of the ocean limited by quasi-latitudinal sections. Although the authors used practically the same sections, the results of Antarctic Bottom Water transport sometimes differ even in sign. The transport can be even directed to the south, for example, in Fu (1981) and Roemmich (1983). This is likely explained by the assumption about zero total transport and the small proportion of Antarctic Bottom Water in the general mass transport. It is also noteworthy that the differences between the most recent publications became much smaller (1–2 Sv), which is within the errors presented in papers (Vanicek and Siedler 2002; Ganachaud 2003). Calculations of geostrophic velocities in Roemmich (1983) estimate the transport of Antarctic Bottom Water below $\sigma_4 = 45.92$ as 0.6 Sv at 24° S and 0.2 Sv at 8° S.

Table 5.2 Transport in the Central part of the Brazil Basin (18–24° S)

Authors	AABW boundary	Transport (Sv)	Comments
Sections along 18–19° S			
Fu (1981)	$\sigma_4 = 45.92$	5.2	Inverse model based on IGY section at 24° S, reference surfaces 2,000 and 4,000 dbar
Roemmich (1983)	$\sigma_4 = 45.92$	3.5	Inverse model based on section at 24° S
McCartney and Curry (1993)		6.7	Oceanus 1983
Speer and Zenk (1993)	$\theta = 2$	5.0	24° S
Macdonald (1998)	$\Theta = 1.7$	6 ± 1	Inverse model based on IGY section at 23° S, Oceanus 1983
		5 ± 1	Inverse model based on IGY section at 27° S, Knorr 1988 (SAVE). Transport 0 ± 1 in the eastern basin
Vanicek and Siedler (2002)	$\gamma^n = 28.12$	7 ± 2	Inverse model. Transport of LCDW and AABW across sections SAVE 3 and 4 at 25° S
		7 ± 2	Inverse model. Transport of LCDW and AABW across sections at 23° S; Oceanus
15–21° S, section A09 along 19° S			
Wright (1970)	$\theta = 2$	2.3	16° S
Fu (1981)	$\sigma_4 = 45.92$	−0.9	Inverse model based of "Meteor" section data at 21° S with reference level at 2,000 dbar
		−6.1 … −0.5	Inverse model based on IGY section at 16° S; reference surface 2,000 and 4,000 dbar
		−2.8	Inverse model based of the Meteor data at 15° S, reference surface 2,000 dbar
Speer and Zenk (1993)		4.5	19° S
De Madron and Weatherly (1994)		7.6	SAVE, 18° S
Vanicek and Siedler (2002)	$\gamma^n = 28.12$	7 ± 2	Inverse model. Transport of LCDW and AABW across A09 section
Ganachaud (2003)	$\gamma^n = 28.11$	3.4 ± 1.1	Inverse model
Weatherly et al. (2000)		4	Direct measurements at 18° S; seasonal variation ~1 Sv

Table 5.3 Transport in the Northern part of the Brazil Basin

Authors	AABW boundary	Transport (Sv)	Comments
Section A08 along 11° S			
Roemmich (1983)		5	
Friedrichs and Hall (1993)		4.9	
Macdonald (1993)	$\sigma_4 = 45.895$	4	
McCartney and Curry (1993)	$\theta = 1.9°$	5.5	Oceanus 1983
Speer and Zenk (1993)	$\theta = 2°$	3.1	11° S
Cai and Greatbach (1995)	3,700	3.7	OGCM, 11° S
Saunders and King (1995)	$\sigma_4 = 45.95$	4.4	
Macdonald (1998)	$\theta = 1.8°$	4 ± 1	Inverse model based on section at 11° S; Oceanus 1983
Sloyan and Rintoul (2001a)	$\gamma'' = 28.30$	0	Inverse model
Vanicek and Siedler (2002)	$\gamma'' = 28.12$	4 ± 3	Inverse model. Transport of LCDW and AABW across section at A08
		5 ± 2	Inverse model. Transport of LCDW and AABW across section at 11° S; Oceanus
Ganachaud (2003)	$\gamma'' = 28.11$	3.1 ± 1.5	Inverse model
Section A07 at 4.5° S			
Wright (1970)	$\theta = 2°$	2.8	8° S, IGY
Fu (1981)	$\sigma_4 = 45.92$	−2.7	Inverse model based on Meteor section data at 8° S; reference surface 2,000 dbar
		−5.3 … −0.5	Inverse model based on IGY section data at 8° S; reference surface 2,000 and 4,000 dbar
Vanicek and Siedler (2002)	$\gamma'' = 28.12$	3 ± 2	Inverse model. Transport of LCDW and AABW across section A08
Ganachaud (2003)	$\gamma'' = 28.11$	3.4 ± 1.2	Inverse model
Schott et al. (2003)	$\sigma_4 = 45.90$	1.3	Direct LADCP measurements at 5° S

The results of our geostrophic calculations based on the WOCE data with different reference surfaces yielded the following values of Antarctic Bottom Water transport to the north (isopycnal surfaces $\sigma_2 = 37.10–37.11$ were assumed as the upper boundary of Antarctic Bottom Water layer depending on latitude): 1.5–5.7 Sv at 30° S, 1.9–5.6 at 19° S, 1.6–3.3 Sv at 11° S, 0.5–3.0 at 4° S, and 0.3–0.9 Sv at 8° S. The square of the Antarctic Bottom Water layer section decreases from 16% at 30° S to 4% at 8° S. The flow of Antarctic Bottom Water in the Brazil Basin decreases from south to north due to mixing with overlying waters. Mixing coefficients within the Antarctic Bottom Water layer are in the range $k \sim 1–5 \times 10^{-4}$ m^2/s (Morris et al. 2001).

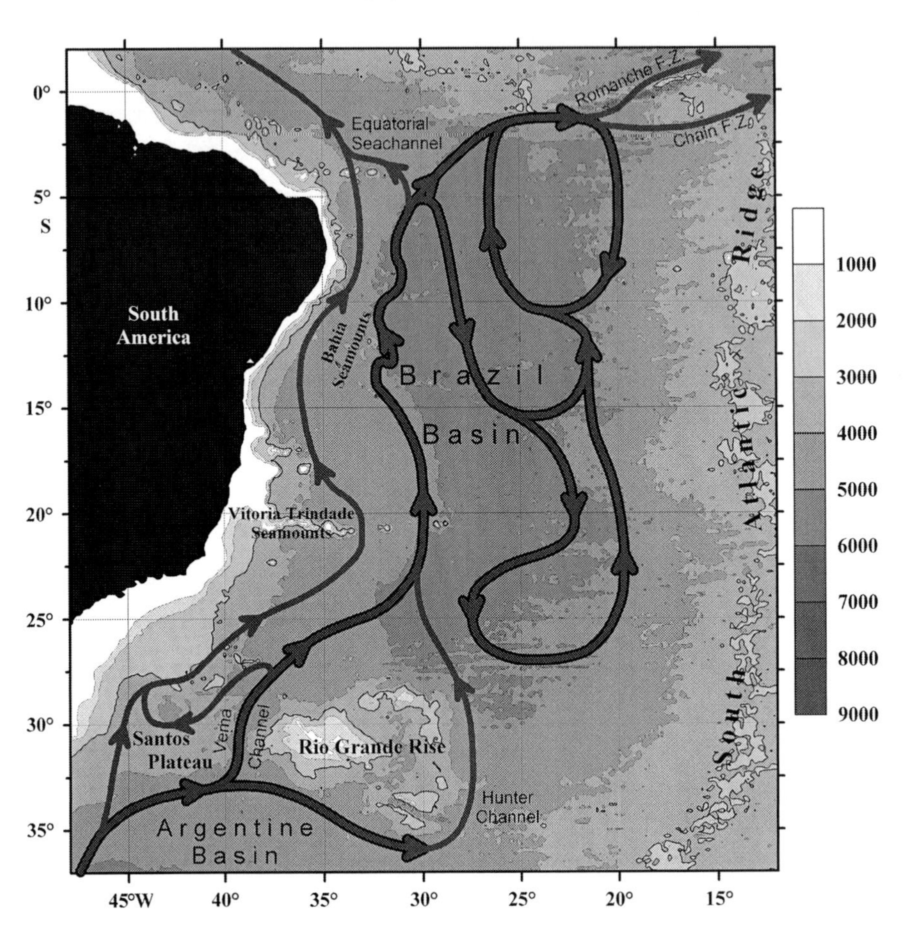

Fig. 5.1 Bottom circulation in the Brazil Basin (De Madron and Weatherly 1994; Sandoval and Weatherly 2001; Morozov et al. 2008). *Thick lines* show circulation of Weddell Sea Deep Water and Lower Circumpolar Water. *Thin lines* show circulation only of Lower Circumpolar Water

5.2 Flow in the Guiana Basin and Westward Equatorial Channels

Intensity of Antarctic water transport in the abyssal ocean decreases strongly north of the Brazil Basin. These waters are found here only in strongly transformed form. A large portion is transported to the eastern basin mainly through the Romanche, Chain, and Vema fracture zones.

Antarctic Bottom Water propagates to the western part of the North Atlantic through the nameless Equatorial Channel located north of the Brazil Basin. The width of this channel is approximately 300 km, and the maximum depth is 4,500 m. According to Hall et al. (2004), the Equatorial Channel starts near the Parnaiba Ridge (2° S) of the region of the Saint Paul Fracture Zone (1.5° N). Whitehead and

Table 5.4 Antarctic Bottom Water transport in the Equatorial Channel

Authors	AABW boundary	Transport (Sv)	Comments
Sverdrup et al. (1942)		1	
McCartney (1992)	$\theta = 1.9°$	2.7–4.3	Transport across section at 4° N with data extrapolation in bottom triangles
McCartney and Curry (1993)	$\theta = 1.9°$	2.7–4.3	Geostrophic calculation at the equator; reference surface 4,000 dbar; different extrapolation in bottom triangles
Hall et al. (1997)	$\theta = 1.8°$	2.1	Current meters
		0.2–3.1	Annual cycle of transport (February–September)
Rhein et al. (1995)		2.6 ± 2.0	
Schlitzer (1996)	$\sigma_4 = 45.9$	3.2–3.7	Equatorial model
Rhein et al. (1998)	$\sigma_4 = 45.90$	1.4–2.2	Transport across different parts of the channel; combination of direct measurements and calculations; reference surfaces $\sigma_4 = 45.90$ and $\sigma_4 = 32.15$
Limeburner et al. (2005)		2 ± 0.9	Array of current meters crossing the channel

Worthington (1982) write that the densest part of Antarctic waters with potential temperature less than 1°C cannot propagate through this channel.

According to Mantyla and Reid (1983), the coolest and least saline Antarctic water (among the bottom waters of all oceans) propagates in the western part of the Atlantic Ocean. The properties of this water change after it passes the equator. The formerly coolest and least saline water becomes warmer and more saline than in the other oceans. Mantyla and Reid (1983) conclude on the basis of radiocarbon concentration that Antarctic waters in the North Atlantic are younger compared to similar waters in the North Pacific. All these facts provide evidence that Atlantic Antarctic waters are strongly transformed under the influence of North Atlantic Deep Water.

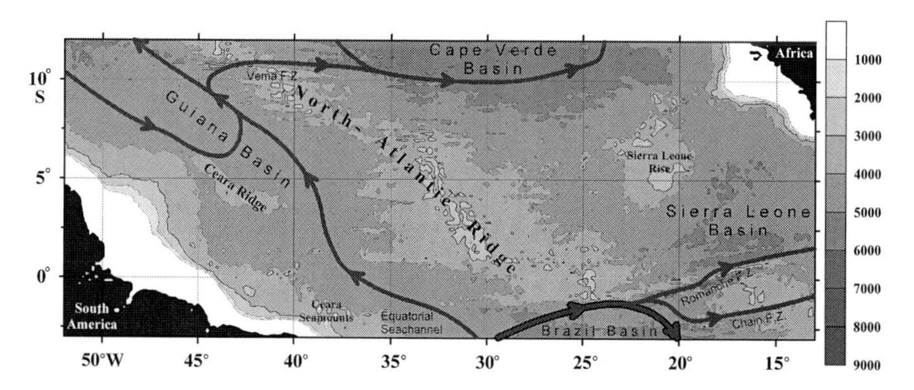

Fig. 5.2 Circulation of Antarctic Bottom Water in the Equatorial Channel and Guyana Basin and eastward flows through the Mid-Atlantic Ridge (Schmitz 1996a; Friedrichs and Hall 1993; McCartney et al. 1991). *Thick lines* show circulation of Weddell Sea Deep Water and Lower Circumpolar Water. *Thin lines* show circulation only of Lower Circumpolar Water

Since the Equatorial Channel is located close to the equator, the estimates of Antarctic Bottom Water transport in the channel are mainly the data of direct instrumental measurements of currents. Publications dedicated to the interpretation of these data provide evidence that a strong seasonal variability exists from 0.2 Sv in February and March to 3.1 Sv in September and October (Hall et al. 1997) (Table 5.1). The authors of this paper also note that the deep thermocline between the Antarctic and North Atlantic waters ascends and descends synchronously with the intensity of the Antarctic waters flow. All available estimates of transports from Table 5.4 fall practically within the range of seasonal variation presented above.

According to the scheme of circulation, Antarctic Bottom Water propagates in the Equatorial Channel and further in the Guiana Basin along the western slope of the North Atlantic Ridge (Friedrichs and Hall 1993) (Fig. 5.2). The Ceara Ridge is a high obstacle for the further propagation of these waters further to the north. A quasi-stationary cyclonic eddy existing north of the Ceara Ridge extends up to 17° N. This eddy transports approximately 3 Sv of bottom waters (Friedrichs and Hall 1993; Friedrichs et al. 1994). The flow of Antarctic Bottom Water splits in the Vema Fracture Zone region (11° N): part of this water flows through the Vema

Table 5.5 Antarctic Bottom Water transport in the Guyana Basin (8–13° N, section A06)

Authors	AABW boundary	Transport (Sv)	Comments
Wright (1970)	$\theta = 2$	2.7	8° S; IGY
Whitehead and Worthington (1982)	$\theta = 1.9$	2–2.3	
Roemmich (1983)	$\sigma_4 = 45.92$	−2.7…0.5	8° N; the first value is based on zero total transport
McCartney (1992)	$\theta = 1.9$	3.5–4.5	Transport across section at 13° N with data extrapolation in bottom triangles
Molinari et al. (1992)		2.2 ± 0.9	Geostrophic transport; reference surface $\theta = 4.7°$
Friedrichs and Hall (1993)	$\Theta = 1.8$	2.1	At 11° N
Luyten et al. (1993)		3.2–4.0	Calculation based on data in (Whitehead and Worthington 1982) with variations in reference surface
McCartney and Curry (1993)	$\theta = 1.9$	4.0	Calculation based on data in Whitehead and Worthington (1982); reference surface $\theta = 1.9°$, at 4° N
			Calculation with other method of determining bottom triangles
Klein et al. (1995)		0.69	Transport at 8° N
		0.31	Transport at 14.5° N
Macdonald 1998	$\theta = 1.8$	1 ± 1	Inverse model based on the data of section along 11° N; Oceanus 1989.
			Transport 0 ± 1 Sv in the western basin
Ganachaud (2003)	$\gamma^n = 28.11$	3.6 ± 1.2	Inverse model

Fracture Zone to the east, while the other part flows to the north of the Guiana Basin. The estimate in Rhein et al. (1998) indicates that 1.1 Sv of the total 1.4 Sv of bottom waters transported through the Equatorial Channel flow to the eastern part the Atlantic and only 0.3 Sv remains in the western basin. However, other estimates presented in Table 5.5 indicate that the remaining part is much greater.

5.3 North American Basin

After passing the Equatorial Channel and Guiana Basin, Antarctic Bottom Water flows to the North American Basin. Many researchers (Wunsch 1984; Stephens and Marshall 2000) consider that Antarctic Bottom Water propagates mainly near the western slope of the Mid-Atlantic Ridge, and, generally, the circulation in the basin is cyclonic (Stephens and Marshall 2000; Weatherly and Kelley 1982; Lavin et al. 2003) (Fig. 5.3).

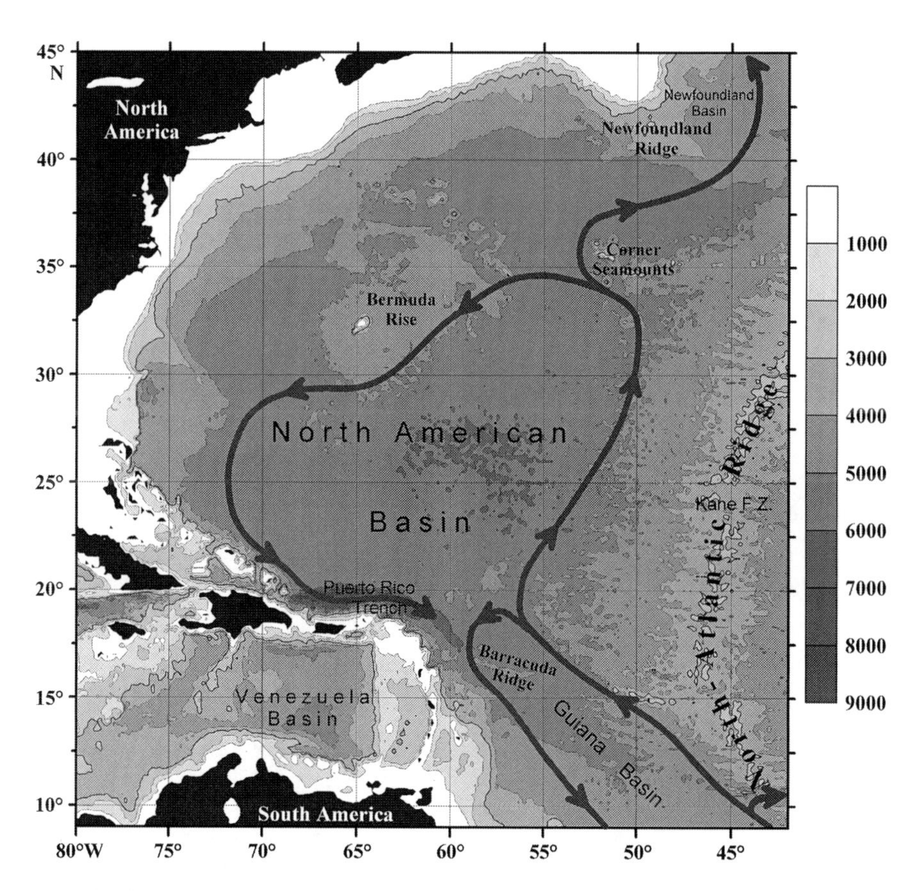

Fig. 5.3 Circulation of Antarctic Bottom Water (Lower Circumpolar Water) in the Northwest Atlantic. (Schmitz 1996a; Friedrichs and Hall 1993)

The authors of many publications consider that the Newfoundland Bank region (42–53° N) is the limit for the propagation of waters of Antarctic origin (Mantyla and Reid 1983; Mamayev 1992; Dobrovolsky and Zalogin 1992). The isotherm $\theta = 2.0°C$ is located at the bottom in this region. According to the classification in Wüst (1936), this isotherm is the northern boundary of Antarctic Bottom Water propagation. It is the opinion of Schmitz (1996a) that only 1 Sv of this water of the total 2 Sv, which pass the 25° N region, reaches the Newfoundland Bank. Traces of Antarctic Bottom Water were also found in the bottom layer at 50° N (Clarke and Gascard 1983).

Other authors consider that only one-half of Antarctic Bottom Water from the Guiana Basin flows to the Vema Fracture Zone, and its transport is estimated at 2.1–2.3 Sv (McCartney et al. 1991). The calculation of geostrophic velocities made in Molinari et al. (1992) provide evidence that 2.2 ± 0.9 Sv of Antarctic Bottom Water is transported to the Guiana Basin. Other authors (Klein et al. 1995) estimate that the transport of Antarctic Bottom Waters north of the equator is much smaller: 0.69 Sv at 8° N and only 0.31 Sv at 14.5° N (Table 5.6). A scheme of bottom water circulation in the Northwest Atlantic is shown in Fig. 5.3.

Table 5.6 Antarctic Bottom Water transport in the North American Basin

Authors	AABW boundary	Transport (Sv)	Comments
24° N section A05			
Wright (1970)	$\theta = 2$	1.4	16° N; IGY
Tucholke et al. (1973)	$\theta = 1.84$	0.5	Current meters and geostrophic calculation; reference surface 4,700 m
Schlitzer (1996)	$\sigma_4 = 45.9$	0.7–1.1	Model; 30° N
Macdonald (1998)	$\theta = 1.9$	−4 ± 2	Inverse model based on Atlantis II section at 24° N in 1981; transport only in the Western basin
Lavin et al. (2003)	4,900 м	1.3	Transport at 24° N; −1.1 Sv in the DWBC and 0.4 Sv in the Eastern Basin; calculation of geostrophic velocities
Ganachaud (2003)	$\gamma^n = 28.11$	1 ± 1	Inverse model
Marsh et al. (2005)		3	Numerical model; section A05
36° N; section A03			
Weatherly and Kelley (1982)	$\theta = 1.82$	− 0.5	Current meters only in the DWBC at 40° N
McCartney et al. (1991)	$\theta = 2$	− 0.37	Geostrophic transport at 36° N; reference surface 2–2.5°C
Macdonald (1998)	$\theta = 1.9$	−7 ± 1	Inverse model based on the data of section at 36° N; Atlantis II, 1981; transport only in the Western Basin
48° N, section A02			
Macdonald (1998)	$\theta = 2$	−10 ± 1	Inverse model based on the data of section at 48° N; Hudson 1982; transport only in the Western Basin
Ganachaud (2003)	$\gamma^n = 28.11$	0	Inverse model

5.4 Eastward Equatorial Channels. The Romanche and Chain Fracture Zones

5.4.1 Research History

The authors of many papers demonstrated that bottom topography of the Atlantic Ocean allows propagation of Antarctic Bottom Water to the northern latitudes of the East Atlantic only through the following deep channels: the Romanche and Chain fracture zones (at the equator) and the Vema Fracture Zone (11° N) (Messias et al. 1999; Wüst 1936; Leontieva 1985; McCartney et al. 1991). Other small and shallower passages are less significant.

The Romanche Fracture Zone is a deep passage in the Mid-Atlantic Ridge 800 km long and 10–40 km wide (Fig. 5). Together with the Chain Fracture Zone they form an equatorial pathway for Antarctic Bottom Water to the East Atlantic. The main sill across the Romanche Fracture Zone is located at a depth of 4,359 m (13°40′ W). The Vema Deep (7,856 m) is the deepest place in the Romanche Fracture Zone. The Chain Fracture Zone is located south of the equator, 200 km south of the Romanche Fracture Zone. The main sill of the Chain Fracture Zone is located at a depth of 4,050 m (12°22′ W). Both fracture zones are corridors for water flow between the Brazil and Guinea basins.

The history of hydrographic research in the Romanche Fracture Zone is longer than a century. The first measurements in this region were made in 1901. Later, the fracture zone was studied during the International Geophysical Year (1958) and during Russian expeditions in the 1960s and 1970s, in particular, in 1967 onboard R/V *Akademik Kurchatov*. The Romanche and Chain fracture zones were intensely studied in 1991–1994 during the French expeditions within the Romanche-I, Romanche-II, and Romanche-III projects, which were parts of the International WOCE program. Many CTD stations were occupied in the region and moorings were deployed for long-term operation. These measurements with modern instruments were the most precise ones. A CTD-section across the Romanche Fracture Zone at 16°03′ W was occupied by Russian scientists in 2005 and repeated in 2009 (cruises 19 and 29 of R/V *Akademik Ioffe*).

The first hydrographic measurements in the region of the Chain Fracture Zone were carried out in the western part of the channel in 1903 and 1927. Unfortunately, they were not made in the axial part of the fracture. The high quality of measurements during the *Meteor* expedition in 1927 is notable. Their measurements are close to the data obtained in 1991. Much work in the region was conducted in 1961. A number of transform faults including the Chain transform fault were found during cruise 17 of R/V *Chain* in 1961. They are described in Metcalf et al. (1964) and Heezen et al. (1964b). It is likely that the first station in this cruise was occupied in the axial part of the fracture zone at 13°10′ W (depth 4,823 m). The bottom potential temperature recorded was equal to 0.806°C. The Chain Fracture Zone was studied in 1991–1994 during the French expeditions together with the Romanche Fracture Zone. In 2009, a Russian expedition during cruise 27 of R/V *Akademik Ioffe* occupied two CTD stations near the main sill of the Chain Fracture Zone. All CTD casts in 2005 and 2009 were accompanied with LADCP profiling.

5.4.2 Moored and LADCP Measurements of Currents

In 1992–1994, a series of measurements was carried out in the Romanche and Chain fracture zones to obtain reliable estimates of the eastward bottom water transport and its variability (Mercier and Speer 1998). The program included moored velocity and temperature measurements over a period of 2 years from November 1992 to November 1994. Four moorings were deployed in each of the fracture zones.

In both fracture zones, the sites of mooring deployment were located near the first sill, where strong bottom water modification was observed. The moorings were deployed upstream of the shallow sills in the fracture zones. A total of 44 current and temperature meters were set in the bottom layer 500 m thick. The total return rate of the data was about 90%.

Direct measurements of velocities presented in Mercier and Speer (1998) revealed easterly transport of the Antarctic Bottom Water at a rate of 1.22 Sv through both fracture zones (Table 5.7). The transport share of the two fracture zones is 0.66 and 0.56 Sv in the Romanche and Chain fracture zones, respectively. Previous studies resulted in greater transports. According to the estimates in Schlitzer (1987), Warren and Speer (1991), and Polzin et al. (1996), transport in zones deeper than 4,000 m is equal to 2 Sv (Table 5.7).

Table 5.7 Propagation of Antarctic Bottom Water to the eastern basin of the Atlantic Ocean: Romanche and Chain fracture zones

Authors	AABW boundary	Transport (Sv)	Comments
Schlitzer (1987)	$\theta = 2.07$	2.6–5.1	Box model
Warren and Speer (1991)	4,000 m	2	
Polzin et al. (1996)		1.4	Profiling
Mercier and Speer (1998)	$\theta = 1.9$	1.22 ± 0.25	Current meter. Joint transport in Romanche and Chain FZ
		0.3	Minimal transport through the Romanche FZ over 2 years based on current meter data
		0.3	Minimal transport through the Chain FZ over 2 years based on current meter data
		0.66 ± 0.14	Mean transport through the Romanche FZ over 2 years based on current meter data
		0.56 ± 0.17	Mean transport through the Chain FZ over 2 years based on current meter data
Stephens and Marshall (2000)	$\theta = 1.8$	~1	
Demidov et al. (2006)	$\theta = 1.8$–2	0.15–0.78	LADCP

The maximum transport of Antarctic Bottom Water over a 2-year-long period is as large as 0.9 Sv in the Romanche Fracture Zone and 0.8 Sv in the Chain Fracture Zone (Mercier and Speer 1998). The minimum transport of Antarctic Bottom Water, which was observed in November, is 0.3 Sv in both fracture zones. The apparent variability period is approximately 1 year. This variability is observed mainly in the warmer (1.38–2.18°C) layers.

The maximum velocities averaged over the two-year-long period in the Antarctic Bottom Water layer ($\theta < 2.0°C$) were more than 21 cm s^{-1} in the Romanche Fracture Zone and as high as 14 cm s^{-1} in the Chain Fracture zone (Fig. 5.4a, b). The lowest mean potential temperature θ was recorded in the Chain Fracture Zone (0.83°C) compared to 0.89°C in the Romanche fracture Zone. In both fracture zones, the speeds are highest in the middle of the section of channels and decay toward the walls and bottom.

The cores of maximum velocities based on moored current measurements were found at 4,200 m in the Romanche Fracture Zone and 4,000 m in the Chain Fracture Zone. Both cores are located approximately 200–300 m above the bottom. In the Chain Fracture Zone, the core of the bottom water current was found at a warmer potential temperature θ (1.88°C) than in the Romanche Fracture Zone (1.38°C).

Sections of mean velocities over the period of measurements (2 years) in both fracture zones are shown in Fig. 5.3. The figures are plotted on the basis of data in Mercier and Speer (1998). The moorings in the Romanche Fracture Zone were located in the deep northern channel of the fracture. A lesser portion of bottom water could propagate through the southern passage.

In 2005 and 2009, Russian CTD stations were accompanied with LADCP measurements. According to the LADCP measurements in 2005 and 2009 (16°03′ W), the core of the Antarctic Bottom Water current in the Romanche Fracture Zone is located at a depth of 3,900–4,000 m. The maximum velocities of the flow exceed 25 cm s^{-1}. The potential temperature isotherm 2°C, which is considered the upper boundary of the bottom water flow, is located approximately at 3,800 m.

The section of along-channel velocity based on the LADCP data in 2009 is shown in Fig. 5.5. The transport estimates of Antarctic Bottom Water flow are 0.35 Sv in 2005 and 0.88 Sv in 2009. The maximum velocities recorded in 2009 exceed 25 cm s^{-1}, whereas in 2005 they were only slightly greater than 10 cm s^{-1}. Transports of bottom water in the Romanche Fracture Zone are given in Table 5.8.

Measurements of currents using LADCP in the Romanche Fracture Zone demonstrated that currents in the periphery of the flow core, including the near-bottom layer, are directed to the west (Fig. 5.5). Most likely, this is caused by tidal motions. Tidal motions are clearly seen from the moored current measurements from 1992 to 1994. The diurnal and semidiurnal tidal cycles are superimposed in the records and the currents near the bottom periodically change direction to the west reaching a few centimeters per second. Band filtering of current records showed that the amplitudes of the diurnal and semidiurnal cycle in the currents are 1–2 cm s^{-1} for each tidal component.

The Russian expedition in 2009 measured velocities at two stations in the Chain Fracture Zone using LADCP. Figures 5.6 and 5.7 show profiles of potential tem-

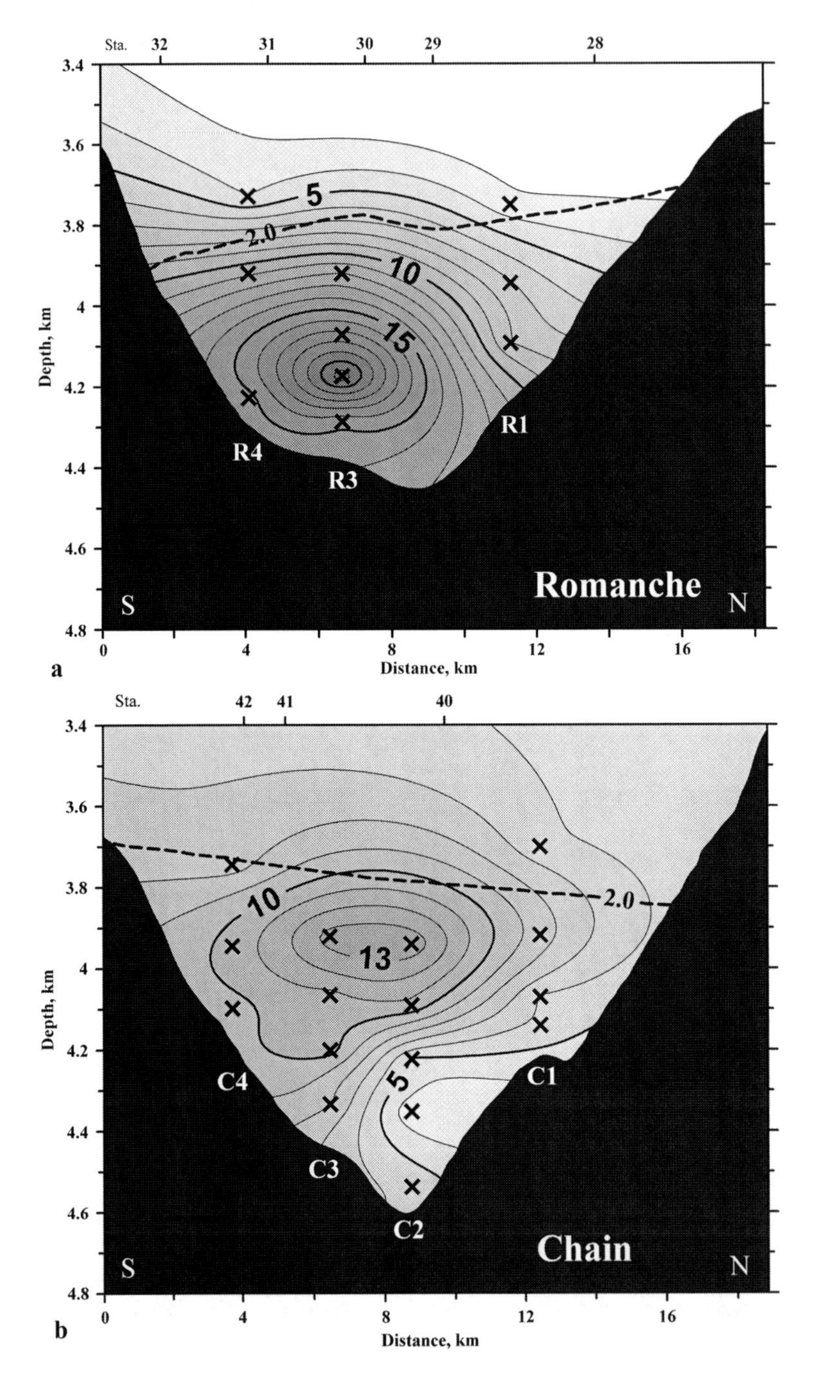

Fig. 5.4 Mean along-channel velocities (cm s^{-1}) in the Romanche (**a**) and Chain (**b**) fracture zones based on moored measurements during the French expeditions in 1992–1994. Potential temperature isotherm 2.0°C based on CTD section in the same expedition is shown

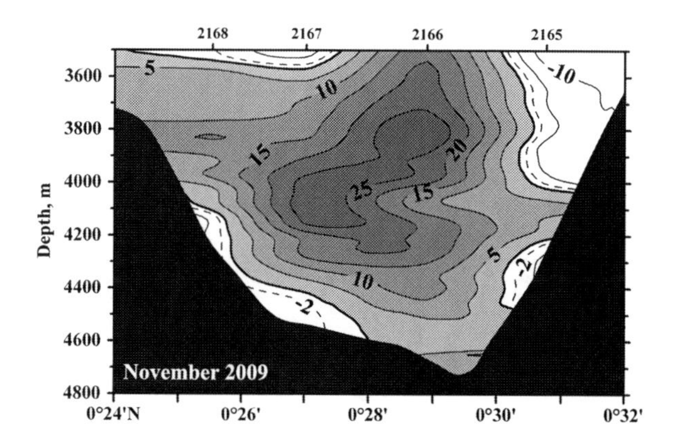

Fig. 5.5 Along-channel velocity (cm s^{-1}) based on LADCP data in November 2009

perature and the along -channel velocity component in the fracture zone measured at these stations. Greater velocities (up to 20 cm s^{-1}) were recorded at the station over the main sill. Lower velocities up to 10 cm s^{-1} were recorded at the deeper station downstream, which can be explained by the fact that the fracture widens almost twice. It is seen from the figures that the zones of maximum velocities are confined to increased temperature and density gradients. We used the topography from the database in Smith and Sandwell (1997) to construct the bottom section normal to the flow and to calculate the transport of bottom waters and Lower North Atlantic Deep Water in the channel. If we accept the upper boundary of bottom waters at $\theta = 1.9°C$ and the boundary of Lower North Atlantic Deep Water at $\theta = 2.1°C$ as in Mercier and Speer (1998), transport of the bottom water and Lower North Atlantic Deep Water would be equal to 0.11 and 0.26 Sv, respectively. The estimate of Lower North Atlantic Deep Water transport equal to 0.26 Sv is close to 0.22 Sv estimated in Mercier and Speer (1998). If we take isotherm $\theta = 2°C$ as the upper boundary of bottom waters, the transport will increase up to 0.17 Sv.

A scheme of the bottom for calculations based on LADCP measurements is shown in Fig. 5.7.

Thus, bottom water transport in the main sill region of the Chain Fracture Zone can be estimated at 0.11–0.17 Sv, which is less than 0.3 Sv estimated in Mercier and Speer (1998). However, the French moorings in 1991–1992 (13°30′ W) were

Table 5.8 Transport of Antarctic Bottom Water (Sv) at different selection of the upper boundary in the Romanche Fracture Zone; measurements in 2005 and 2009

Water mass	Upper boundary, θ(°C)	Transport in 2005	Transport in 2009
AABW	Below $\theta = 2.0$	0.35	0.88
	Below $\theta = 1.9$	0.33	0.69
	Below $\theta = 1.8$	0.29	0.62
	Below $\theta = 1.5$	0.23	0.43

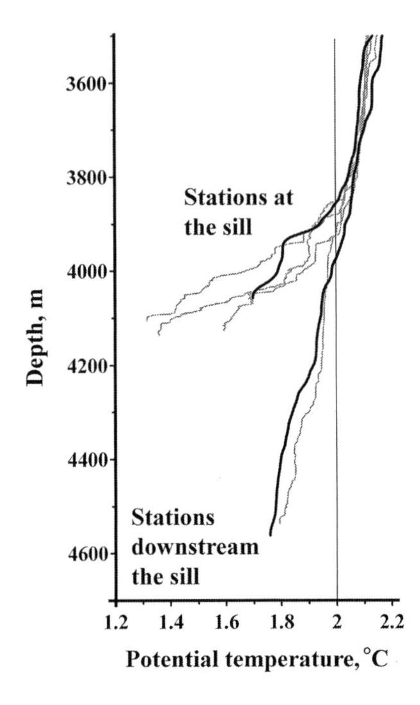

Fig. 5.6 Vertical distribution of potential temperature (°C) in different years at stations near the main sill of the Chain Fracture Zone. *Thick lines* denote measurements in 2009; *thin lines* are related to the measurements in 1991–1992

deployed west of the Russian LADCP measurements in 2009 (12°20′ W). We can admit that a flow exists between two fracture zones (Romanche and Chain) and between the region of moorings deployment and the main sill where LADCP measurements were made. This flow can pass along a series of quasi-meridional valleys that cut the northern slope of the Chain Fracture Zone.

Near 12°30′–13° W, the northern slope of the Chain Fracture Zone is broken by a number of valleys. Possible inflows and outflows through these valleys between two fracture zones are expected. Also part of the water flow can reach the Guinea Basin through these valleys by-passing the main channels of the fracture zones. Scarcity of CTD casts in these channels unaccompanied by LADCP measurements

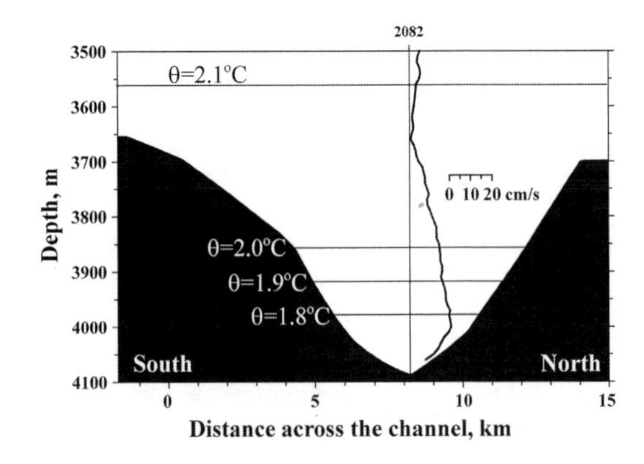

Fig. 5.7 Vertical profile of velocity (cm s^{-1}) along the channel measured in 2009. Bottom boundaries for the calculation of transport from LADCP measurements in the Chain Fracture Zone (Smith and Sandwell 1997) at 12°23.9′ W. The width of the channel in kilometers is laid along the x axis. Levels of different potential temperatures are indicated

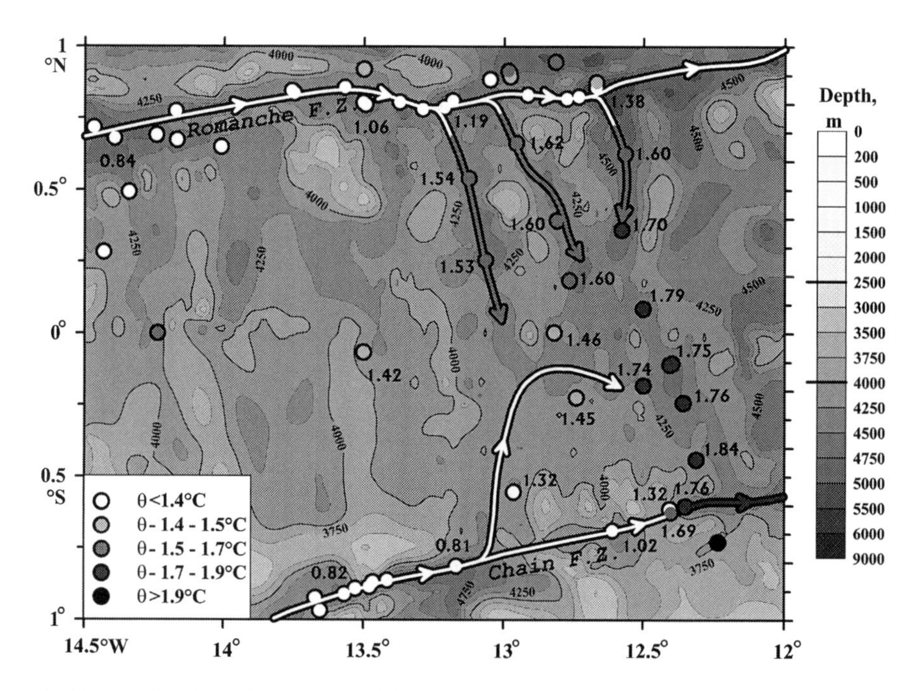

Fig. 5.8 Distribution of bottom potential temperatures (°C) in the region of the Romanche and Chain fracture zones. Bottom topography is shown according to Smith and Sandwell (1997). *Arrows* indicate possible directions of flows between the fracture zones based on the decrease of temperature along the flow

does not allow us to estimate the volume flux in these valleys. However, we think that the main transport is concentrated along the axes of the fracture zones as the deepest channels.

A series of CTD casts in 1994 in the region between the Romanche and Chain fracture zones near the main sill of the Chain Fracture Zone allowed us to plot the bottom temperature field between the fractures and suggest a scheme of the main flows between the fractures shown in Fig. 5.8.

5.4.3 CTD-Sections Along and Across the Fracture Zones

Our analysis is based on the results of French expeditions in 1991–1994 and Russian measurements in 2005 and 2009.

The coldest waters with a potential temperature of 0.68°C were found on the western side of the Mid-Atlantic Ridge near the Romanche and Chain Fracture zones (McCartney et al. 1991; WODB-2005 data). It is likely that bottom waters propagate freely over the southern ridge that limits the Romanche Fracture Zone near the Vema Deep (the deepest part of the fracture zone).

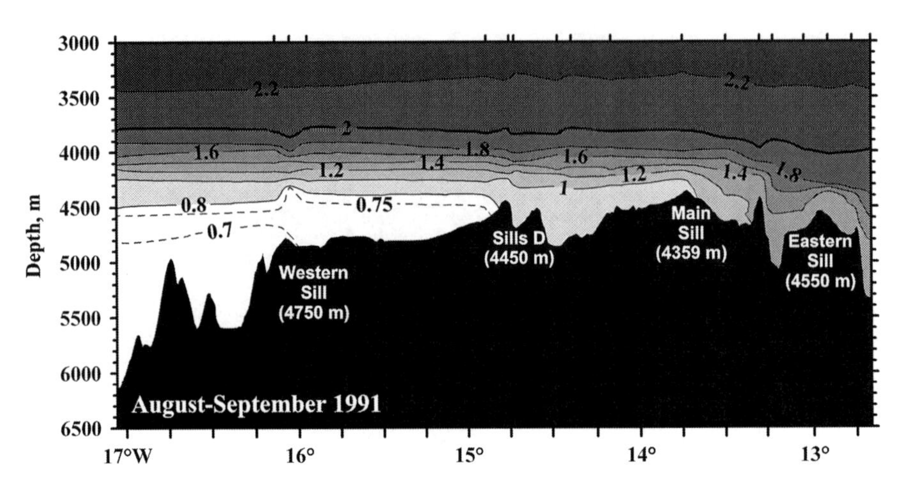

Fig. 5.9 Potential temperature section (°C) along the Romanche Fracture Zone based on the data in 1991

Bottom waters are intensely transformed during their propagation in the Romanche and Chain fracture zones. Horizontal temperature gradients in the fracture zones are as high as 0.15°C/100 km and 0.6°C/100 km, respectively. Due to different depths of the main sill and different intensities of water mixing in the fracture zones, potential temperatures at the main sills and outflow from the channels are much warmer than at the inflow on the western slopes of the Mid-Atlantic Ridge. The potential temperatures measured at the main sills of the Romanche and Chain fracture zones are 0.92 and 1.32–1.69°C, respectively (WODB-2005). The depths of the sills are 4,350 and 4,050 m, respectively (Messias et al. 1999). Potential temperatures at the outflow from the channels are 1.66°C (Romanche) (McCartney et al. 1991), and 1.76°C (Chain Fracture Zone, based on 2009 data). Here we also give for comparison the value of potential temperature at the outflow from the Vema Fracture Zone (11° N), which is 1.69°C.

Mixing along the channels is also responsible for variations in other parameters: salinity changes along the Romanche Fracture Zone from 34.75 to 34.82 psu, and density from $\sigma_2 = 37.15$ to 37.10 (or from $\sigma_4 = 46.00$ to 45.91). The oxygen concentration changes from 5.4 to 5.8 ml/l; the silicate concentration, from ~100 to ~70 μmol/kg. In addition, the densest waters of the flow cannot overflow numerous sills in the fracture zones.

Figure 5.9 shows the distribution of potential temperature over the section along the Romanche Fracture Zone east of the Vema Deep based on the data of the French expedition in 1991. This interval of the fracture length includes several sills that serve as topographic obstacles on the pathway of Antarctic Bottom Water propagation to the east. Only the upper parts of these waters pass the obstacles, which are subsequently transformed due to mixing with the overlying Lower North Atlantic Deep Water.

Potential temperatures at the bottom as function of longitude along the Romanche Fracture Zone are shown in Fig. 5.10. The temperature increases sharply only

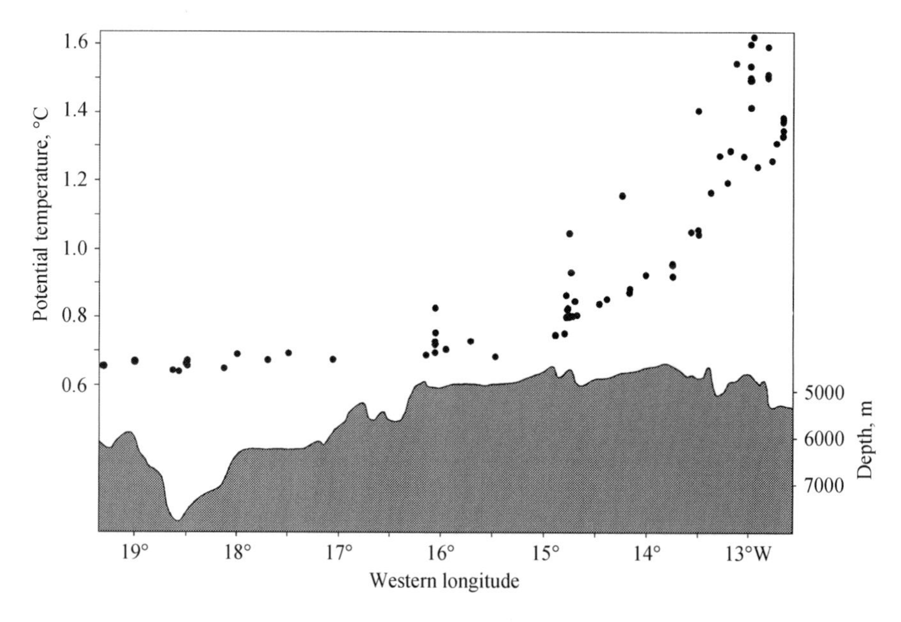

Fig. 5.10 Bottom potential temperature (°C) vs. longitude in the Romanche Fracture Zone based on measurements in different years

near the main sill in the eastern part of the fracture zone. Such variations in temperature are associated with intense mixing near the main sill. The mixing is possibly related to tidal motions. Strong mixing in the Romanche Fracture Zone was reported in Ferron et al. (1998).

Bottom potential temperatures as a function of longitude along the axial zone of the Chain Fracture Zone are shown in Fig. 5.11. Strong temperature increase due to mixing with the overlying layers is found only near the main sill. Salinity variation

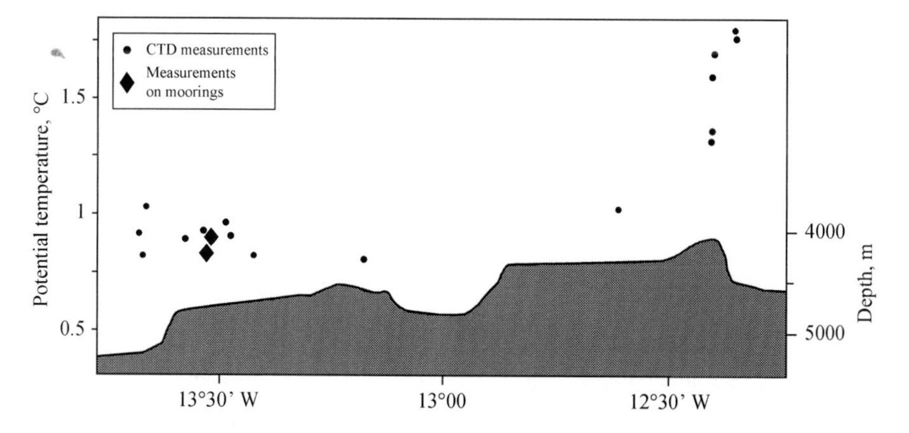

Fig. 5.11 Bottom potential temperature (°C) variation with longitude in the Chain Fracture Zone based on measurements in different years. *Diamonds* denote measurements by temperature sensors of current meters in 1992–1994

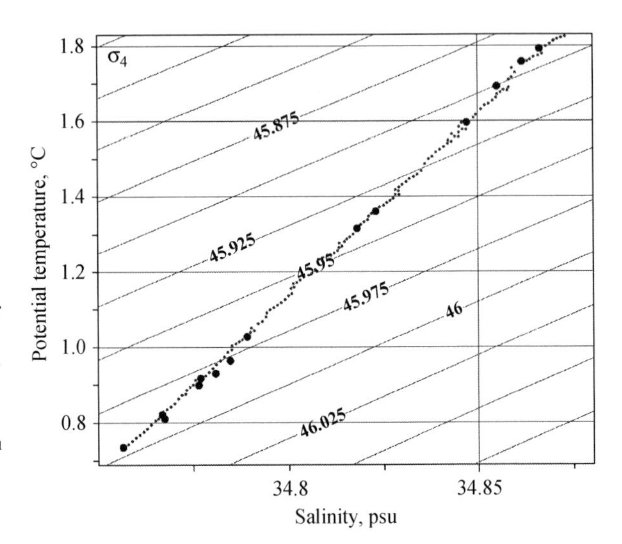

Fig. 5.12 The θ/S-relations for bottom measurements along the Chain Fracture Zone demonstrating that geothermal warming of bottom waters in the fracture zone is negligible. The line of *small dots* represents the θ/S-relation for a station in the extreme western part of the fracture zone. Other larger *dots* falling on the same line are related to the bottom measurements at points located along the fracture axis. Isopycnals referenced to 4,000 m (σ_4 in kg/m^3) are shown

is similar. Such variations in temperature indicate that strong mixing occurs near the main sill, possibly, due to tidal forcing.

Since the geothermal flux in the Romanche and Chain fracture zones is rather high, it could influence the bottom water temperature. We shall consider this problem using the data along the Chain Fracture Zone. Variations in the bottom temperature/salinity ratio as a function of longitude in the Chain Fracture Zone are shown in Fig. 5.12. The θ/S-curve for a station in the western part of the Chain Fracture Zone is shown with a line of small dots. The values of θ/S-relation at the bottom for other stations along the axis of the fracture are shown with large dots. One can see that all of them fall on one line. Only the minimum temperature and salinity at each station increase with longitude. This demonstrates negligible effect of the geothermal flux on warming of the bottom water layer. If the warming caused by the geothermal flux was strong, the dots related to the station along the fracture would be elevated relative to the θ/S-curve of the western station. Variations in temperature and salinity as functions of longitude are similar, suggesting that these variations are caused only by mixing.

5.4.4 Time Variations in Temperature and Salinity

Let us consider distributions of temperature and salinity over a section across the Romanche Fracture Zone and their variations in time since 1991. This analysis is based on the data of French CTD section in 1991, which was repeated by the Russian expeditions in 2005 and 2009. The section crossed the local sill of the fracture along 16°03′ W (the shallowest depth is 4,750 m).

Fourteen and eighteen years passed between the French measurements in 1991 and Russian expeditions in 2005 and 2009. Comparison of measurements in 1991,

2005, and 2009 shows that no principal changes occurred in the structure of the flow. Equatorial location of the fracture zone leads to symmetrical distribution of temperature and salinity with respect to the axis of the channel. No significant displacements of isothermal and isohaline surfaces occurred during the years between the expeditions (Fig. 5.13). The potential temperature section across the Romanche Fracture Zone measured in 2009 is shown in Fig. 5.14. The salinity section is shown in Fig. 5.15. Here we show only the sections in 2009 because all three of them are very similar.

The minimum potential temperature at the bottom in 2009 (0.752°C) was higher by 0.057°C than in 1991 (0.695°C) and by 0.022°C than in 2005 (0.730°C). Salinity minimum at the bottom almost did not change.

The depths of both fracture zones allow the flow of North Atlantic Deep Water over the bottom water. Both flows are directed to the east. Mean temperature in

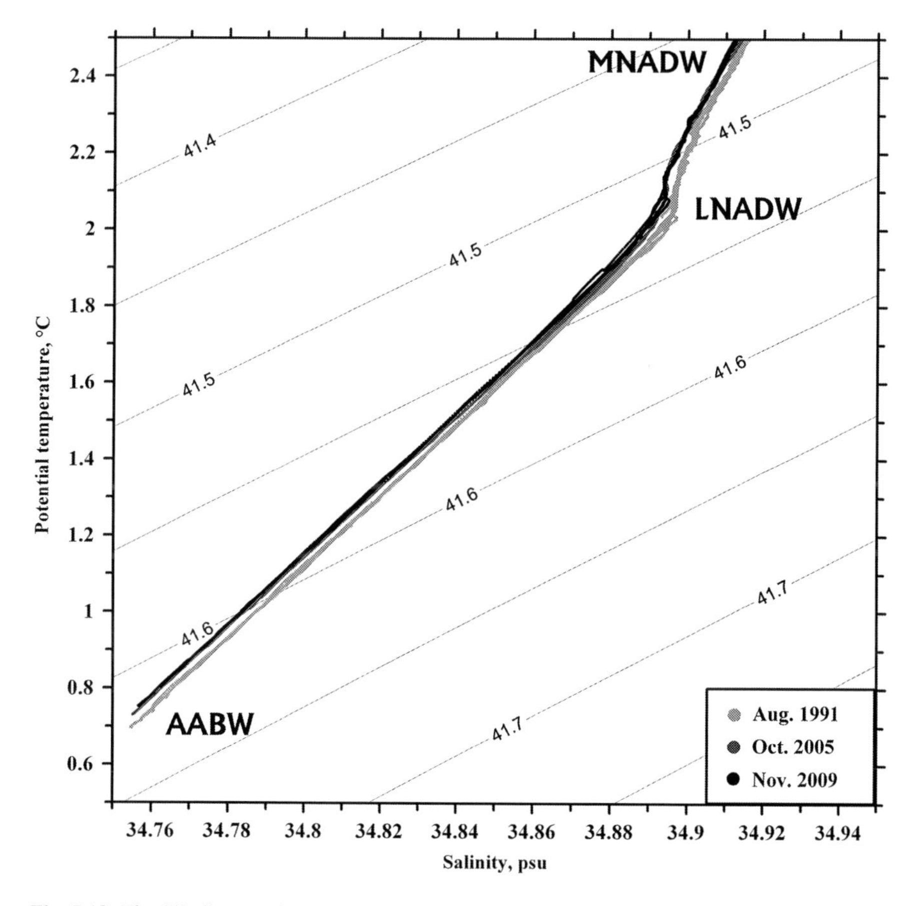

Fig. 5.13 The θ/S-diagram for deep waters based on the data of the deepest stations in the expeditions of 1991, 2005, and 2009. Names of water masses are indicated. Contour lines of potential density σ_3 are shown

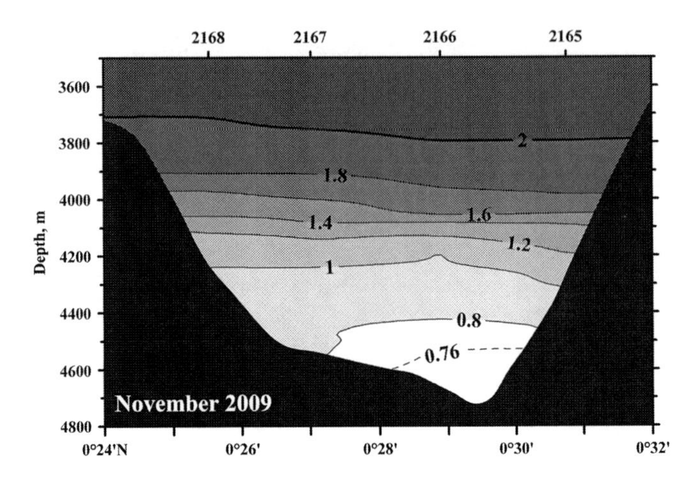

Fig. 5.14 Potential temperature section (°C) across the Romanche Fracture Zone based on the data in 2009

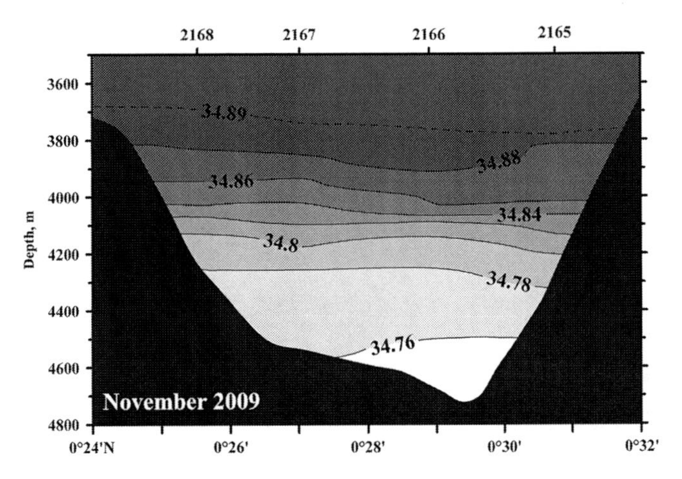

Fig. 5.15 Salinity section (psu) across the Romanche Fracture Zone based on the data in 2009

the layer of North Atlantic Deep Water (2,000–3,000 dbar) increased by 0.075°C from 1991 to 2009. Mean salinity of this layer decreased by 0.03 psu. Warming (0.075°C) of the North Atlantic Deep Water layer (2,000–3,000 dbar) from 1991 to 2009 can be related to the warming in the source of this water. Since the initial formation of characteristics of deep and bottom ocean waters occurs in the upper layer, there are grounds to think that they can conserve climatic signals in their temperature field for a long time during their gradual sinking to the depths. The observed warming of North Atlantic Deep Water in the Romanche Fracture Zone is related to the warming of these waters in the tropical North Atlantic (You 2005). Even more, this can be related to the temperature increase in the formation region of North Atlantic Deep Water (Labrador Basin), which has actually been observed from the beginning of the 1970s (Koltermann et al. 1999). The travel time of such a climatic signal from the source to the equator is estimated at 25–40 years (Koltermann et al. 1999). Hence, we can suppose that in the Romanche Fracture Zone we observe a remote response to warming of North Atlantic Deep Water in its source.

Time changes in temperature and salinity at the outflow from the Chain Fracture Zone is not at all stable compared to the variations in the middle section of the Romanche Fracture Zone. According to the WODB-2005 data, three stations were occupied near 12°24′ W during the French expedition in 1991–1994. Coordinates of two stations were the same and the depth difference was only 4 m. However, the difference of potential temperatures at the bottom was as high as 0.24°C (!). This fact indicates that short-period variations in potential temperature are significant in the main sill region (Fig. 5.6). This fact also provides evidence that the flow is not stationary and can be perturbed by tidal forcing that causes the oscillatory transport of bottom water along the channel axis near the main sill.

The measurements in 2009 in the Chain Fracture Zone revealed that potential temperature increased by 0.1°C compared to the measurements in 1991–1994 if we consider the greatest value measured in 1994. The difference with the lowest temperature is 0.34°C. Measurements at greater depths east of the main sill in 2009 did not reveal strong differences in potential temperature. Bottom potential temperature at the station downstream the sill decreased by 0.034°C. Measurements at greater depths east of the main sill in 1991 and 2009 did not demonstrate strong temperature variations. Additional considerations of the tidal influence on mixing and variations in temperature and salinity at the outflow from the Romanche and Chain fracture zones will be discussed in Sect. 5.6.2.

Considering the time variations in temperature near the sills of the Romanche Fracture Zone, it is useful to analyze additional materials of earlier observations. Although these observations had a lower accuracy as compared with the modern data, they can still be applied to determine the dominating tendency of water temperature change at abyssal depths. Therefore, we present Table 5.9 with a time series of instrumental measurements of potential temperature θ in the Vema Deep, the deepest part of the Romanche Fracture Zone within the layers of North Atlantic Deep Water (4,000 m) and Antarctic Bottom Water (7,000 m) based on the data of several known expeditions (Leontieva 1985).

The data presented in the table demonstrate significant fluctuations of temperature in the North Atlantic Deep Water layer, but no gradual secular climatic trend in North Atlantic Deep Water can be seen. At the same time, warming of Antarctic

Table 5.9 Potential temperature (°C) of deep and bottom waters in the Romanche Fracture Zone (historical data)

Level	Expedition and year						
	Albatross, 1948	Crawford, 1958	Chain, 1965	Meteor, 1965	Akademik Kurchatov, 1967	Akademik Vernadsky, 1969	Mikhail Lomonosov, 1972
4,000 m	1.75	1.66	1.61	1.64	1.77	1.57	1.54
7,000 m	0.63	0.64	0.62	0.60	0.65	0.66	0.67

Bottom Water by 0.04°C has undoubtedly been observed starting approximately from the second half of the twenteeith century.

5.4.5 *Underwater Cataract in the Chain Fracture Zone*

Let us consider the flow in the Chain Fracture Zone, keeping in mind a sharp deepening of the bottom to the east near the main sill. The bottom slope here is approximately 30–40°. Such water flow is a cataract (Worthington 1969; Whitehead 1989), and the water flow between these two stations descends by 500 m. A bathymetry chart is shown in Fig. 5.16. We occupied two CTD stations with LADCP measurements here. One station was located at the main sill, the other one 3 miles to the east downstream from the bottom flow at a depth greater by 500 m than the first station. The stations were located at 00°37.4′ S, 12°23.9′ W (4,067 m) and 00°36.5′ S, 12°21.0′ W (4,573 m).

We compare the bottom water flow down the slope in the Chain Fracture Zone over a height of 500 m with laboratory experiments of V. Liapidevsky. He made experiments with two layers of fluid of different densities in a laboratory setup. Water with a density greater than water in both layers is let to flow down the slope, while both layers are at rest. The flow splits into two branches. Part of the flow that mixed with the fluid at rest continues its motion in the horizontal direction almost along the interface between two layers due to inertia. The densest part of the fluid flows down the slope due to the gravity force (Fig. 5.17).

Similar effects are observed in ocean hydrodynamics when fluids of different densities (Antarctic Bottom Water and Lower North Atlantic Water) flow down the slope (Fig. 5.18). The intermediate layer is approximately at a depth of 4,000 m. A thin intermediate mixed layer exists between these fluids above the sill. Its thickness is not more than 50 m. Neither layer is homogeneous, but they are stratified by temperature and salinity. The lower layer of Antarctic Bottom Water is approximately 150 m thick. Its velocity is greater than the velocity of the Lower North Atlantic Deep Water layer. The velocity is 18 cm s^{-1} above the sill. Velocity de-

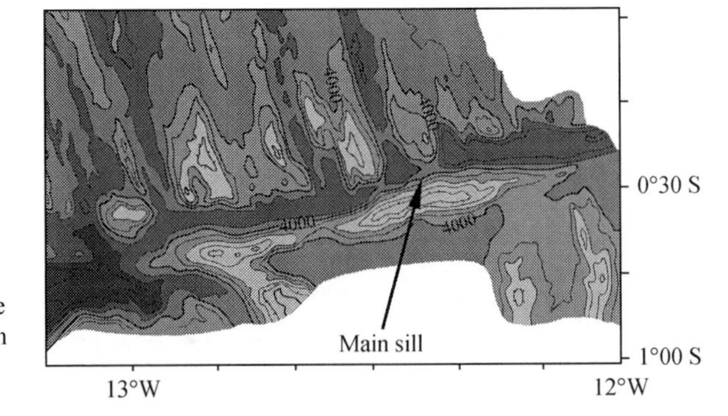

Fig. 5.16 Bathymetry in the main sill region of the Chain Fracture Zone. Depths are shown in meters

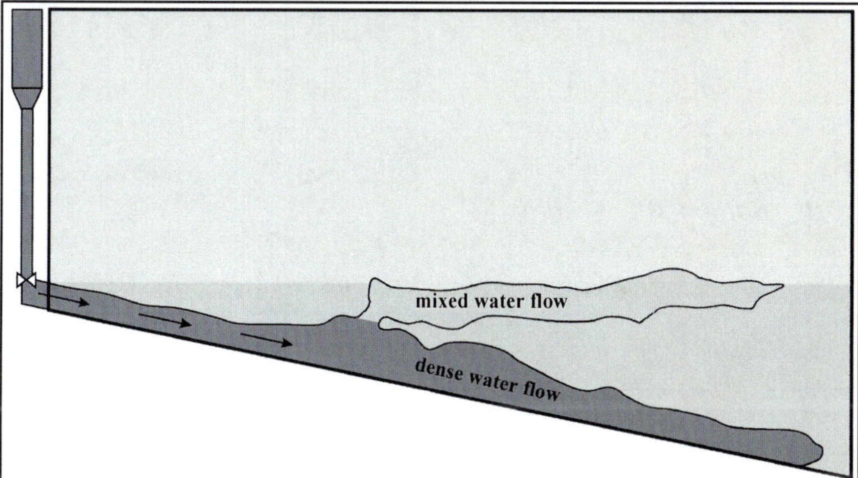

Fig. 5.17 Scheme of laboratory experiments by Liapidevsky at the Institute of Hydrodynamics (Siberian Branch of the Russian Academy of Sciences). Splitting of water flow is observed as it descends down a sloping bottom. Water layers with different densities are shown in different color

creases sharply in the bottom boundary layer 50 m thick. The velocity of Lower North Atlantic Deep Water is estimated at 8 cm/c. The Antarctic Bottom Water layer slides down the thin boundary layer with minor changes in its properties.

Below the cataract, when the Antarctic Bottom Water layer descends by 500 m, the thickness of the intermediate layer increases due to the shear between layers. The second horizontal current appears along the depth of 4,000 m. Velocities in the flow decrease, because the width of the channel below the sill increases almost twice.

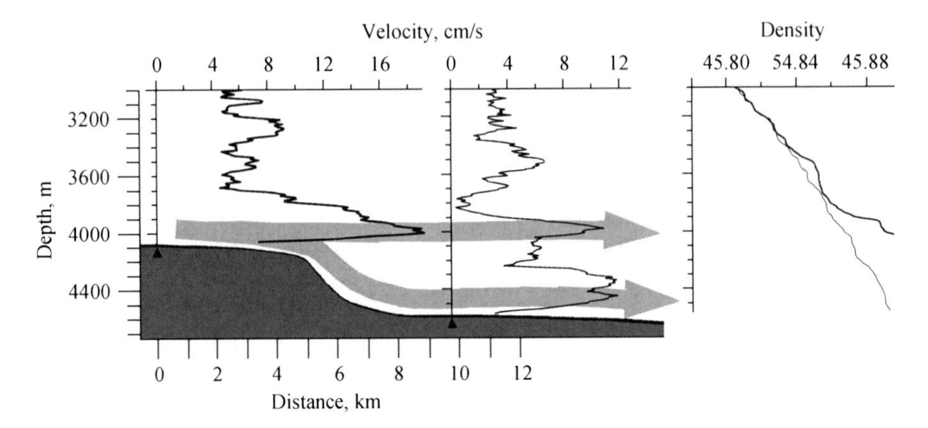

Fig. 5.18 Profiles of velocity (cm s^{-1}) along the channel from the LADCP data at two stations with respect to bottom topography shown in *dark gray*. *Light gray arrows* show a scheme of the water flow. The *right panel* shows density profiles (σ_4) at two stations in the Chain Fracture Zone at the main sill (*thick line*) and east of the sill below the sill level by almost 500 m (*thin line*)

5.5 Vema Fracture Zone

5.5.1 *Bottom Topography*

During the German expedition in 1927–1929 on R/V *Meteor*, Wüst (1936) found cold abyssal waters with potential temperature θ = 1.74°C at the bottom of the Gambia Abyssal Plain. He suggested that these waters were of Antarctic origin. He attributed their existence to the penetration through the Romanche Fracture Zone (known already by that time) and a fracture in the Mid-Atlantic Ridge presumptively located near 7–13° N. Only in 1956, deep sounding from R/V *Vema* revealed a deep fracture near 11° N, which was later called the Vema Fracture Zone after the name of the ship (Heezen et al. 1964a). Later, the morphology of the Vema Fracture Zone was studied in detail in Vangriesheim (1980). The fracture occurs between 43.5 and 41° W and connects the Demerara and Gambia abyssal plains. The width of the fracture zone is 8–10 km and the maximum depth is approximately 5,200 m.

Vangriesheim (1980) indicated the existence of three main sills of the fracture at 41°02′, 40°55′, and 40°53′ W, respectively. According to her data, sills are located at depths of 4,690, 4,650, and 4,710 m. During the geophysical expedition in 1998 onboard R/V *Akademik Nikolay Strakhov*, the existence of these sills was confirmed, but the measured depths were reported as 4,690, 4,660, and 4,660 m, respectively. Thus, the first and second sills seem to be located at the same depth. Based on data in Smith and Sandwell (1997), the main sill of the fracture zone is located at 41°01′ W at a depth of 4,571 m. A chart of the Vema Fracture Zone bottom topography based on Smith and Sandwell (1997) is shown in Fig. 4.

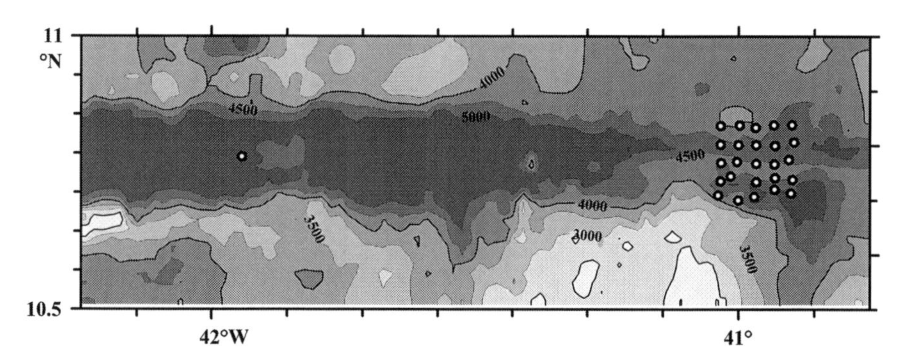

Fig. 5.19 Bathymetric chart of the Vema Fracture Zone and locations of CTD stations occupied in 2006 (*white circles*). Depth is given in meters. The remote western station at 10°47′ N, 41°56′ W is shown

In 2006, an expedition with CTD and LADCP measurements onboard R/V *Akademik Ioffe* visited the region of the main sills. The depth measurements using an ELAC LAZ 470 echo sounder and an Atlas Parasound parametric echo sounder during cruise 22 of R/V *Akademik Ioffe* in 2006 introduced more details in the bottom topography data in the region by combining the previous and new bottom soundings. The resulting bathymetric chart is shown in Fig. 5.19 and the study region is shown in more detail in Fig. 5.20.

5.5.2 *Measurements*

According to the WODB-2005 database, the first temperature and salinity measurements in the Vema Fracture Zone were made in 1959. One measurement was made at a depth of 5,165 m in the fracture, and the measured potential temperature was $\theta = 1.79$°C. In 1977, a series of CTD-stations was occupied along the axial line (10°48′ N) of the fracture mainly in its eastern part (40°03′–40°55′ W). In the 1970s–1980s, the fracture zone was studied intensely. Stations in the fracture zone were made within the TTO program from R/V *Knorr*, *Robert Conrad*, *Oceanus*, and others. In 1979, a section along the fracture was occupied at a distance from the fracture axis. In 1994, a series of three stations across the fracture near its western end was occupied from R/V *Meteor* within the WOCE program.

During the expedition onboard R/V *Akademik Ioffe* in 2006, a survey of 25 CTD-stations (SBE-25 Sea-logger) was carried out in the study region with a size of 16 × 16 km over the main sill of the Vema Fracture Zone at 40°30′–41°00′ W and a station west of this region (at 41°56′) repeating the *Meteor* station in 1994. A survey of 5 × 5 stations with a distance of ~4 km between stations was located in the main channel of the fracture and a depression separated by a sill from the main channel of the fracture. Currents were measured using an LADCP profiler (RDI Workhorse Sentinel) at 18 stations including 5 stations along the main chan-

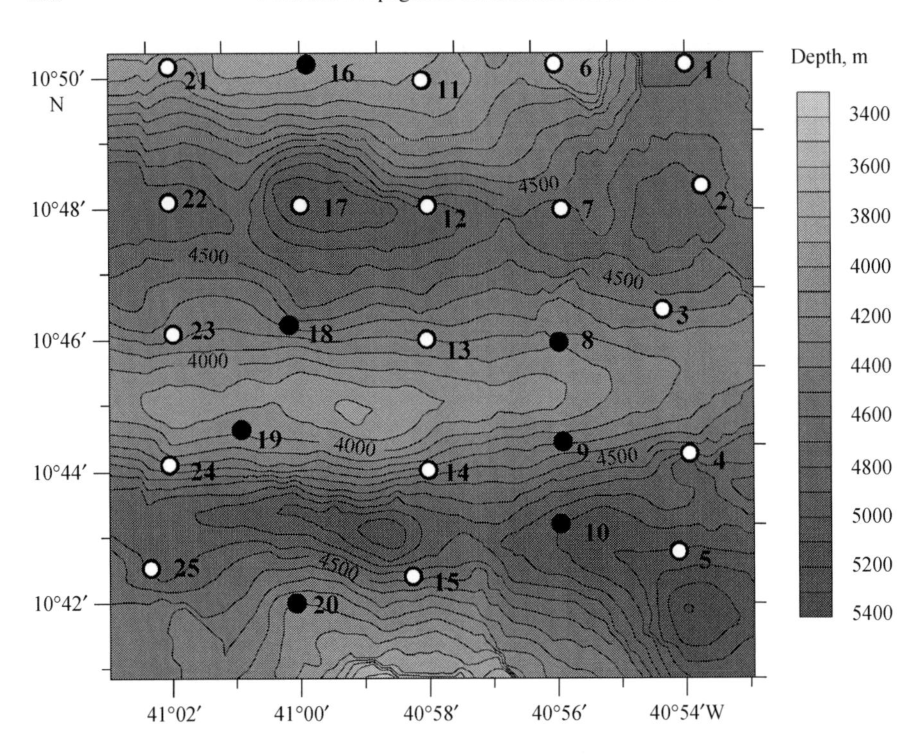

Fig. 5.20 Bathymetric chart (m) of the study region of the Vema Fracture Zone (main sill) based on the measurements in cruise 22 of R/V *Akademik Ioffe*. All *dots* indicate the locations of stations when the instrument was at the bottom. Dots indicating points with LADCP measurements are highlighted by the *white color* in the middle of the dots

nel (Fig. 5.20). Thus, in 2006, the total number of stations in the fracture zone increased from 21 to 47.

5.5.3 *Structure of Bottom Flow*

The deepest part of the Vema Fracture Zone is occupied by Antarctic Bottom Water and North Atlantic Deep Water. In many places, the Mid-Atlantic Ridge serves as an obstacle to the interbasin exchange of Antarctic Bottom Water and lower part of North Atlantic Deep Water, which can propagate only through some of the transform fractures, of which the Vema Fracture Zone is the deepest. The lower boundary of Lower North Atlantic Deep Water is located here within the layer approximately from 3,700 to 4,100 dbar (Demidov et al. 2007a).

According to Eittreim et al. (1983), Antarctic Bottom Water propagates in the Vema Fracture Zone as a homogeneous layer, which can be as thick as 700 m. The flow is well mixed due to bottom and lateral friction. The authors distinguish two gradient zones ('benthal thermocline' and 'transition zone') at the Antarctic Bottom Water boundary with Lower North Atlantic Deep Water. These zones differ by the vertical temperature gradients equal to 0.0012 and 0.0004°C/m, respectively. The maximum thickness of the transition zone can reach 300 m in the eastern part of the fracture zone. The benthal (bottom) thermocline is located approximately 1,000 m above the bottom. The potential temperature jump in the benthal thermocline is between 1.5 and 2.0°C. In Eittreim et al. (1983), its origin is related to the waters of two-degree discontinuity distinguished in Broecker et al. (1976), which are located at a significant distance south of the study region. Following the concept described in Broecker et al. (1976), Eittreim et al. (1983) considered that these waters are formed in the Rio Grande Rise region due to diffuse interaction between Antarctic Bottom Water and North Atlantic Deep Water, alternative intensification, or weakening of the inflow of North Atlantic Deep Water (or Antarctic Bottom Water) revealed in Broecker et al. (1976). Researchers have no common opinion whether the transition zone and benthal thermocline are included in the Antarctic Bottom Water layer. In rare cases, the transition zone is divided into several parts based on different estimates of the location of boundaries, which are within the range of potential temperature $\theta = 1.5–2°C$ and density $\sigma_4 = 45.895–45.920$.

During the cruise of R/V *Akademik Ioffe* in 2006, a structure similar to the one described in Eittreim et al. (1983) was found. A quasi-homogeneous layer of bottom water with a thickness of 400–500 m and potential temperature $\theta = 1.34°C$ and salinity 34.829 psu at the bottom was recorded at the distant western station (Fig. 5.21). While propagating to the east, the flow reaches the main sill region at station 22 (Fig. 5.20) after it passes the western sill and the region with a relatively smooth bottom topography. At station 22, the flow is approximately 100 m thick and its potential temperature $\theta = 1.36°C$ and salinity lower than 34.835 psu (Fig. 5.22).

By the time the water in the main channel reaches station 17, the thickness of the bottom mixed layer increases to 200 m and θ increases to 1.395°C, which probably indicates strong mixing in the bottom layer. Higher temperature at point 7 (in the second sill region) than at point 12 provides evidence of spatial inhomogeneity of the flow and mixing over the sill at 40°57′ W. Based on the bottom temperature distribution at a station in the depression (point 5) located south of the southern wall of the main channel (Fig. 5.20), we can conclude that the part of the above-mentioned Antarctic Bottom Water jet, which is characterized by lowest temperatures, does not get into the depression basin. The temperature at the bottom of the depression is close to the temperature in the upper part of the transition zone in the main channel. This temperature corresponds to the water that overflows the deepest crest over the elevation between the main channel and depression.

A strong spatial variability of water mass characteristics was found in the study region over a scale of a few miles, which is seen well in Figs. 5.21 and 5.22. Bottom water becomes warmer and warmer as it overflows the sill in the main channel. However, temperature increase is not monotonic, which can be explained by the fact that we could not follow the main jet of the flow. The θ/S-structure of water is similar to the structure found by Eittreim et al. (1983), which was described above (Fig. 5.23).

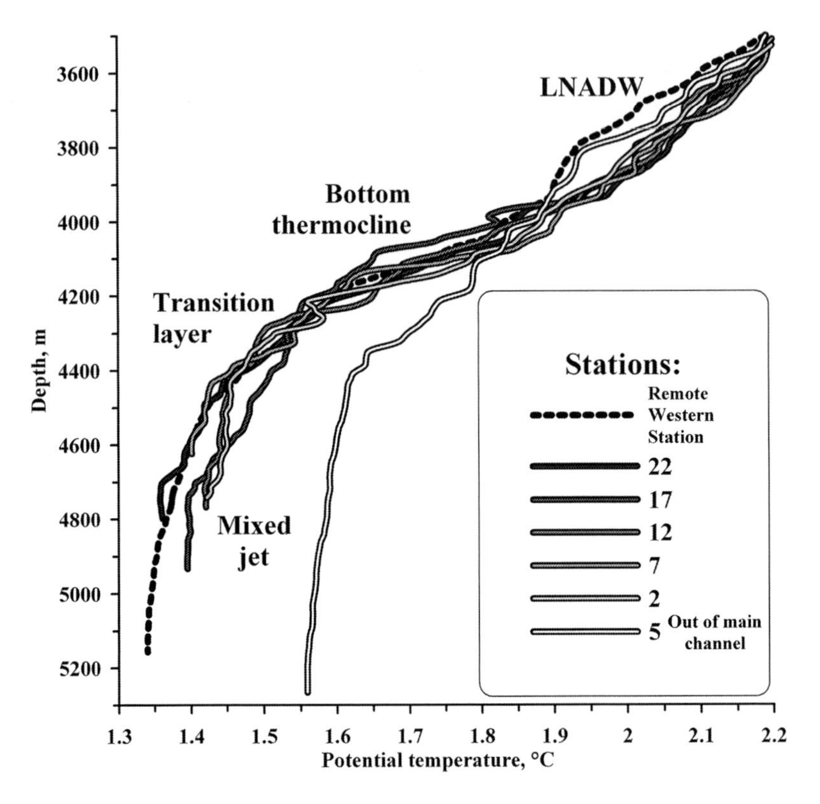

Fig. 5.21 Depth variation of potential temperature (°C) at several stations in the Vema Fracture Zone. Numbers of stations according to Fig. 5.20 are indicated. Line 5 shows a profile in the depression south of the main channel (see also Fig. 5.19). *Straight lines* indicate intervals of different vertical gradients of potential temperature

As to the time variability of water properties at the stations located in the region of the sills, no notable long-term temperature variations were found while comparing the measurements in 1977 and 2006 (Fig. 5.23). Maximum temperature variations exceeding 0.06°C were found near the 4,000-m isobath. The salinity differences reaching 0.05 psu are probably caused by insufficient accuracy of measurements in 1977. Potential temperatures at the bottom measured in 1977 were within 1.406–1.440°C (up to 1.441 in the mixed jet), which coincides with the range of values at the bottom at four nearest stations in 2006 (1.399–1.440°C).

An increase in potential temperature θ by 0.027°C and salinity by 0.009 psu in the Antarctic Bottom Water core was found by comparing the measurements at the distant station in 2006 and the measurements of R/V *Meteor* in 1994, which were made at the same place. The depth of the station in 2006 appeared 100 m greater than in 1994. We explain this fact as different positions of the instrument relative to the position of the vessel in different expeditions and steep slopes of the fracture walls. (The coordinates of the station are assumed as the position of the ship; however, the coordinates of the instrument at the bottom can be different).

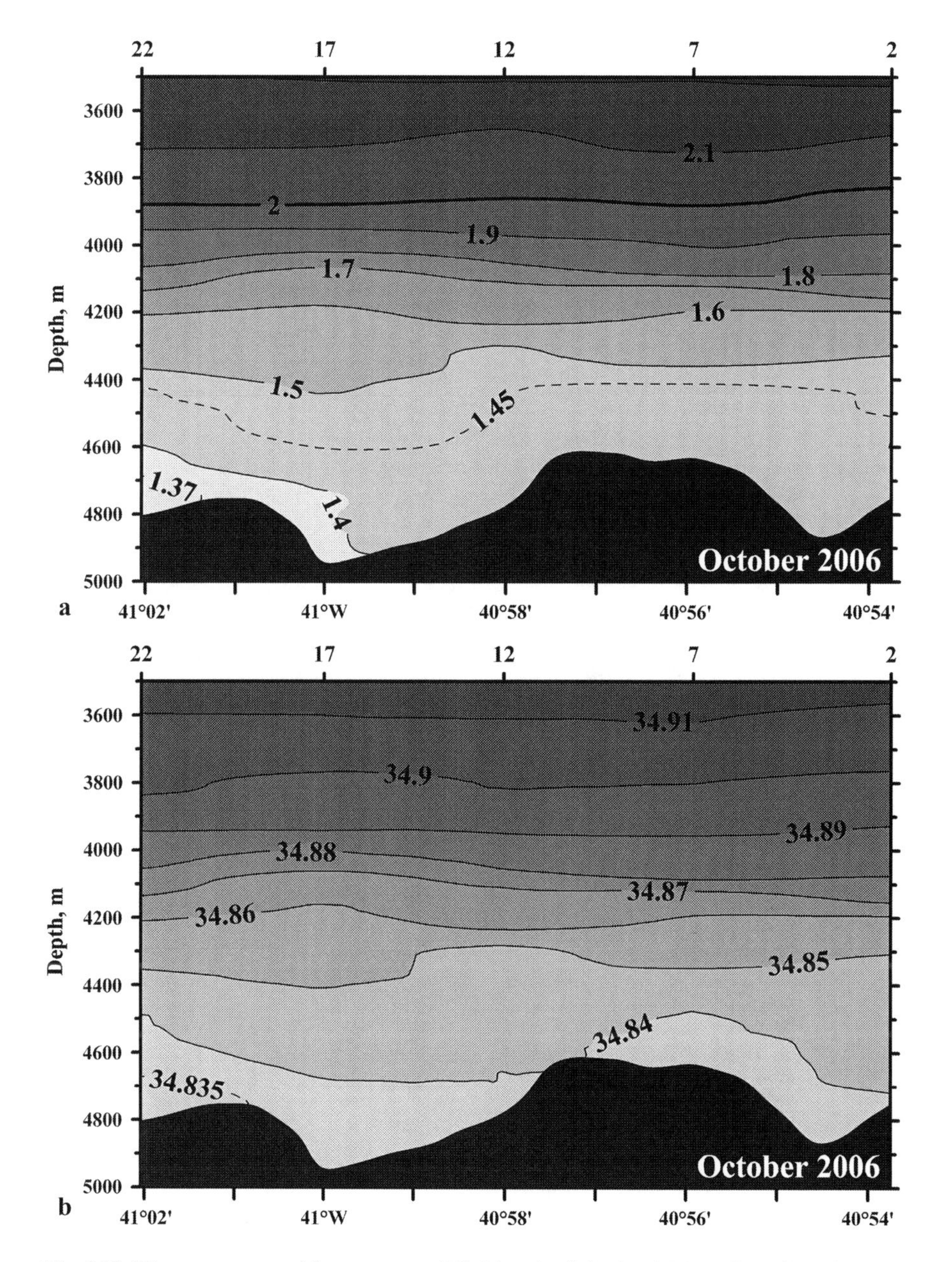

Fig. 5.22 West to east potential temperature (°C) (**a**) and salinity (psu) (**b**) sections along the main channel in the study region (Vema Fracture Zone)

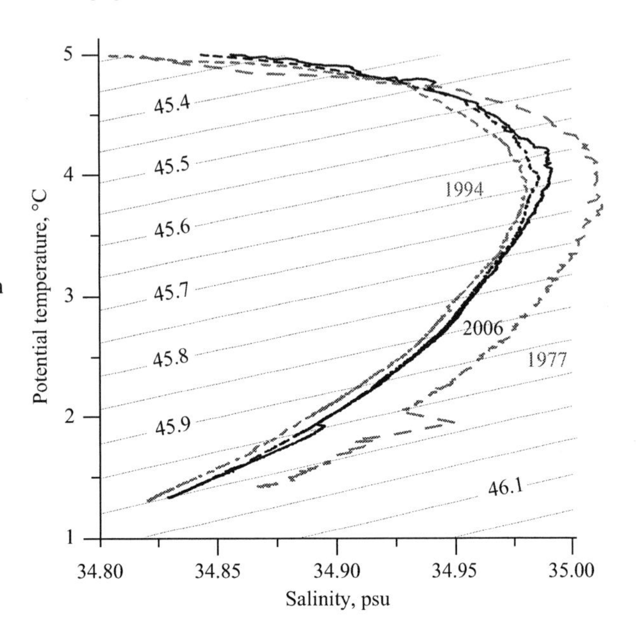

Fig. 5.23 The θ/S-diagram for the bottom layer waters in the Vema Fracture Zone based on the data of 1977 and 1994 with the nearest stations in 2006 (shown in *black color*; among them station 2 of the study region is shown with a *dashed line* and a remote station is shown with a *solid line*). Isopycnals referenced to 4,000 m (σ_4 in kg/m^3) (σ_4 = 45.4–46.1) are shown

Figure 5.24 demonstrates the distribution of potential temperature at the bottom as a function of the distance from the main sill of the Vema Fracture Zone. The figure presents combined data from Arhan et al. (1998) and the data of the new expeditions. In addition to the differences with the measurements in 1977 and 1994, transformation of properties of Antarctic Bottom Water in the study region caused

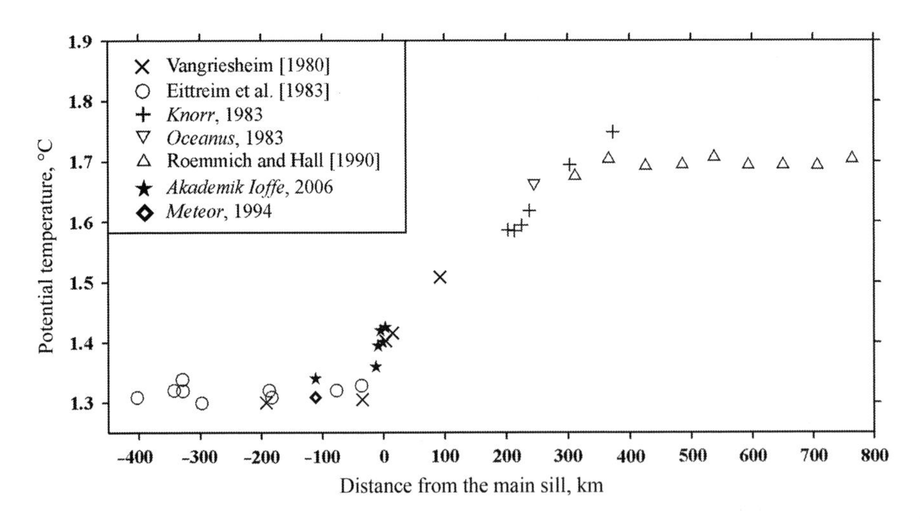

Fig. 5.24 Potential temperature (°C) at the bottom as a function of the distance from the main sill of the Vema Fracture Zone. Combined data from Arhan et al. (1998) and the data of new expeditions are presented

by the elevation of bottom topography is also shown in the figure. Only the upper part of Antarctic Bottom Water can overflow this elevation. The greatest longitudinal gradients of potential temperature are found in the region of two sills, where the study site was located. This is caused by the peculiarities of the bottom water overflow (the densest waters can overflow only occasionally) and possible intensification of mixing in this region due to steep slopes of the fracture walls. This figure also shows that increase in the water temperature and mixing is strongest near the slopes of the Mid-Atlantic Ridge and in the Vema Fracture Zone, mainly near the sills. After this, temperature changes only slightly during the further motion in the deep basin.

A local salinity maximum (Fig. 5.23) was found at a depth of 3,850 m (σ_4 = 45.95) at the station west of the study region. This maximum indirectly confirms the opinion in Friedrichs and Hall (1993) about the existence of a return northerly flow of Lower North Atlantic Deep Water in the western basin of the Atlantic (Fig. 2.7). A difference in the structure of the Lower North Atlantic Deep Water layer, which is manifested in salinity increase in the southern part of the study region, makes it possible to suppose the following pattern: after propagating into the fracture zone, this water flows to the east and shifts to the southern slope of the fracture. In previous studies, the existence of the above-mentioned local maximum of salinity in the Lower North Atlantic Deep Water layer was not mentioned.

5.5.4 Bottom Water Transport

The first estimate of Antarctic Bottom Water transport through the Vema Fracture Zone was obtained in Vangriesheim (1980) based on the data of two current meters west of the main sill (40°54′ W) at a depth of approximately 5,040 m (Table 5.7). The current meters operated during 26 and 9 days. These data showed that the flow direction varied strongly and the mean velocities were approximately 2.9 and 3.7 cm s^{-1}, while the maximum observed velocities reached 33 cm s^{-1}. Eittreim et al. (1983) found an easterly current at a depth of 4,790 m with maximum velocity equal to 3 cm s^{-1} (mean velocity 0.07 cm s^{-1}) observed during 11 days. These authors measured the easterly flow not only in the sill region, but also east of the region. However, underwater photography did not record any strong currents at the bottom (Eittreim et al. 1983). The calculation of geostrophic velocities in McCartney et al. (1991) based on the data of R/V *Knorr* (section along 35° W) allowed them to get an estimate of velocity equal to 20 cm s^{-1} in the bottom layer.

Velocity measurements using an LADCP from R/V *Meteor* (Fischer et al. 1996) indicated the presence of maximum velocities equal to 20 cm s^{-1} at a depth of 4,200 m in the middle of the channel, which is close to the geostrophic estimates. The total transport was calculated by combining the direct measurements and the geostrophic transport from the reference level with zero velocities found from the LADCP data located at a depth of 3,640 m. The results of calculation of the Antarctic Bottom Water transport are given in Table 5.10. It was already discussed above

Table 5.10 Propagation of Antarctic waters to the eastern basin of the Atlantic Ocean through the Vema Fracture Zone

Authors	AABW boundary	Transport (Sv)	Comments
Vangriesheim (1980)	$\theta = 1.5°$	0.05–0.46	Moored current meter
Schlitzer (1987)		0.0–0.7	Box model
McCartney et al. (1991)	$\theta = 2°$ $\theta = 1.5°$	2.08–2.24 0.46	Geostrophic transport, reference surface, $\theta =$ 2.17–2.43°C
Fischer et al. (1996)	$\theta = 2°$ $\sigma_4 = 45.9$	1.8–2.0 2.1–2.4	Combination of geostrophic transport and LADCP data at 42° W
Rhein et al. (1998)		1.1	Inclusion of part of LNADW into the AABW layer
Stephens and Marshall (2000)	$\theta = 1.8°$	2	
Demidov et al. (2007a)	$\theta = 1.8–2°$	0.12–0.64	LADCP measurements

that the transport estimates depend strongly on the choice of the upper boundary of Antarctic Bottom Water. For example, in the calculation of the geostrophic transport made in McCartney et al. (1991), the total transport increased more than four times (!) after they changed the upper boundary of Antarctic Bottom Water from $\theta = 1.5°C$ to $\theta = 2°C$ (Table 5.10).

In 2006, we repeated the station made in 1994 that allowed us to make a comparison of currents. Comparison of the velocity distribution by depth obtained during direct measurements of currents in 1994 and 2006 showed a decrease in the transport in the bottom part of the channel in 2006 (Fig. 5.25). However, the vertical distribution of velocities is quite similar, and even more this distribution is close to the distribution of geostrophic velocities based on the data in 1994.

The measurements conducted in 2006 showed that the core of bottom water flow in the channel of the Vema Fracture Zone was observed not at the bottom but in the layer between 3,700 and 4,000 m (maximum LADCP velocities), which is close to the upper boundary of Antarctic Bottom Water (Fig. 5.25). This is a similar property of bottom flows in many abyssal channels (see Fig. 10).

While estimating the transport of Antarctic Bottom Water we took into account the following:

1. The estimates of transport differ significantly depending on methods of determination of the section square and extrapolation of the current velocities to the section triangles, where the bottom waters exist but the measurements are lacking. Calculations taking into account the transport beyond the main channel within the study region, where there is no direct flow of bottom water, were carried out separately.
2. It is crucial that current velocity measured using an LADCP also includes the influence of tides and other periodical motions. Finally, the values of transports over the meridional sections differ by a factor greater than 1.5, which required the presentation of the final estimates in this paper only in the form of intervals.

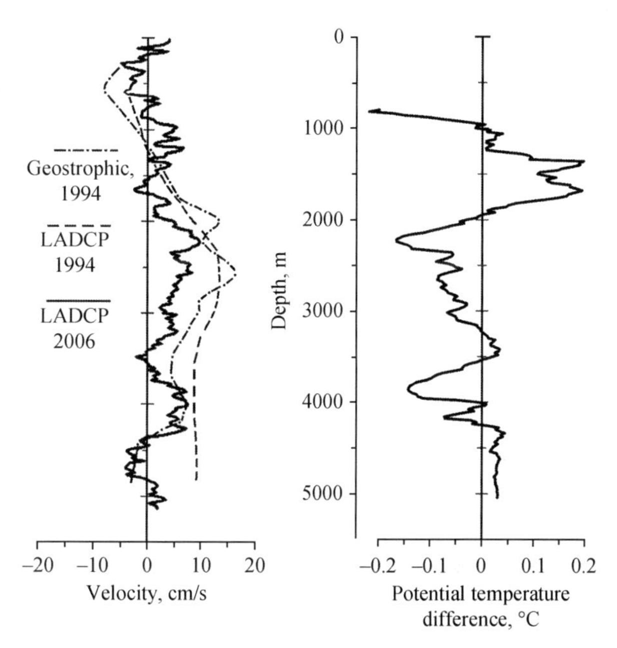

Fig. 5.25 Comparison of the vertical distribution of velocities (cm s^{-1}) and potential temperatures θ (°C) at a western distant station from the study region in 2006 and German station in 1994 (both at 10°47′ N, 41°56′ W)

3. Determination of the upper boundary of Antarctic Bottom Water is ambiguous and the total transport strongly depends on this fact. Therefore, we used in this work different versions of the location of this boundary coinciding with isotherms of potential temperature θ = 1.5, 1.8, 1.9, and 2.0°C.

Transports of bottom water in the Vema Fracture Zone are given in Table 5.11. In general, the estimates of transport appeared significantly smaller than reported in the literature with the upper boundary of Antarctic Bottom Water assumed as an isotherm of potential temperature θ = 2°C. The maximum estimate of transport reached only 0.64 Sv.

If we compare two major channels through which bottom water flows from the West Atlantic to the East Atlantic (Vema and Romanche fracture zones), we conclude that a smaller amount of cold waters (with θ < 1.5°C) is transported through the Vema Fracture Zone than through the Romanche Fracture Zone, because these waters almost do not reach the Vema Fracture Zone along the longer pathway in the western part of the Atlantic.

In order to have statistical representation of our estimates, we calculated bottom water transport across three sections in the main channel (Table 5.11). It is our opinion that the western section is most representative for characterizing the transport in the fracture zone, because it is located near the main sill of the fracture and its stations are located closer to the channel walls.

The estimates of transport of Lower North Atlantic Deep Water are characterized by strong differences over five meridional sections in the study region in 2006 (Table 5.11). To our opinion, the best estimates were obtained at the western section

Table 5.11 Transport of bottom waters (Sv) in the layers of Lower North Atlantic Deep Water (*upper line*) and at different selection of the Antarctic Bottom Water upper boundary in the Vema fracture zone

Water mass	θ (°C)	Western section, stations 21–25	Central section, stations 11–15	Eastern section, stations 1–5
LNADW	From θ = 2.4 to 2.0	0.38–0.66	0.16–0.21	0.33–0.5
AABW	Below θ = 2.0	0.15–0.64	0.13–0.21	0.23–0.46
	Below θ = 1.9	0.14–0.51	0.13–0.21	0.19–0.40
	Below θ = 1.8	0.12–0.37	0.11–0.18	0.13–0.32
	Below θ = 1.5	0.01	0.04	0.02

due to the smallest area of the fracture zone channel section for the extrapolation of velocities.

5.6 Eastern Basin Pathways and Further Propagation of Antarctic Bottom Water in the East Atlantic

5.6.1 General Description

In this section we analyze the inflow of bottom waters to the Northeast Atlantic from the Vema, Romanche, and Chain fracture zones and their further propagation in the basin. Many years had passed since the *Meteor* expedition in 1927, when Wüst (1936) found the coldest waters at the bottom (θ = 1.74°C) in the Northeast Atlantic near 12° N, 34° W. He thought that cold waters spread to the East Atlantic through the Romanche Fracture Zone, as well as through an unknown passage in the Mid-Atlantic Ridge at 8–11° N. The Vema Fracture Zone remained unknown and was found only in 1956. During the long history of analysis of bottom water properties all researchers agreed that these waters are of Antarctic origin as was initially suggested by Wüst. However, there were different concepts about the inflow of these waters through the fractures in the Mid-Atlantic Ridge. Alternative opinions about possible candidates among Vema and Romanche fracture zones were published in Worthington and Wright (1970a, b), Purdy et al. (1979), Mantyla and Reid (1983), and Heezen et al. (1964a).

The present-day concept was for the first time suggested in Mantyla and Reid (1983). They wrote that bottom waters propagating through the Romanche Fracture Zone influence only the equatorial and southeastern part of the Atlantic Ocean and do not spread to the north through the Kane Gap, whereas the Vema Fracture Zone is the main pathway for bottom waters into the northeastern Atlantic.

On the basis of a large amount of measurements before 1991, a scheme of Antarctic Bottom Water (with potential temperature θ < 2°C) spreading in the Northeast was suggested in McCartney et al. (1991). In particular, they based their analysis on the distribution of potential temperature at the bottom, which is a good indicator for Antarctic Bottom Water transport. According to this scheme, Antarctic Bottom Water that propagated to the Gambia Abyssal Plain in the East Atlantic through the Vema Fracture Zone splits into two branches spreading in different directions (Fig. 5.26). Here we suggest a new scheme of the distribution of potential temperature at the bottom based on recent measurements, which appeared after 1991, with inclusion of all available historical data based on WODB-2005 (Fig. 5.27).

According to McCartney et al. (1991), one branch of the bottom water flow from the Vema Fracture Zone in the Gambia Abyssal Plane is directed to the north and waters of the Antarctic origin fill the deepest parts of the Northeast Atlantic basins including the Canary Basin. This flow transports less than 2 Sv of bottom water to the north. This value of northward transport exceeds the transport through the Vema Fracture Zone; therefore we have to suppose that this flow includes anticyclonic circulation in the Canary Basin. Slopes of the Mid-Atlantic Ridge are constraints to this flow from the west. The flow reaches the Discovery Gap at 37° N. This region is the boundary for the further northward transport of bottom water with potential temperatures below 2°C.

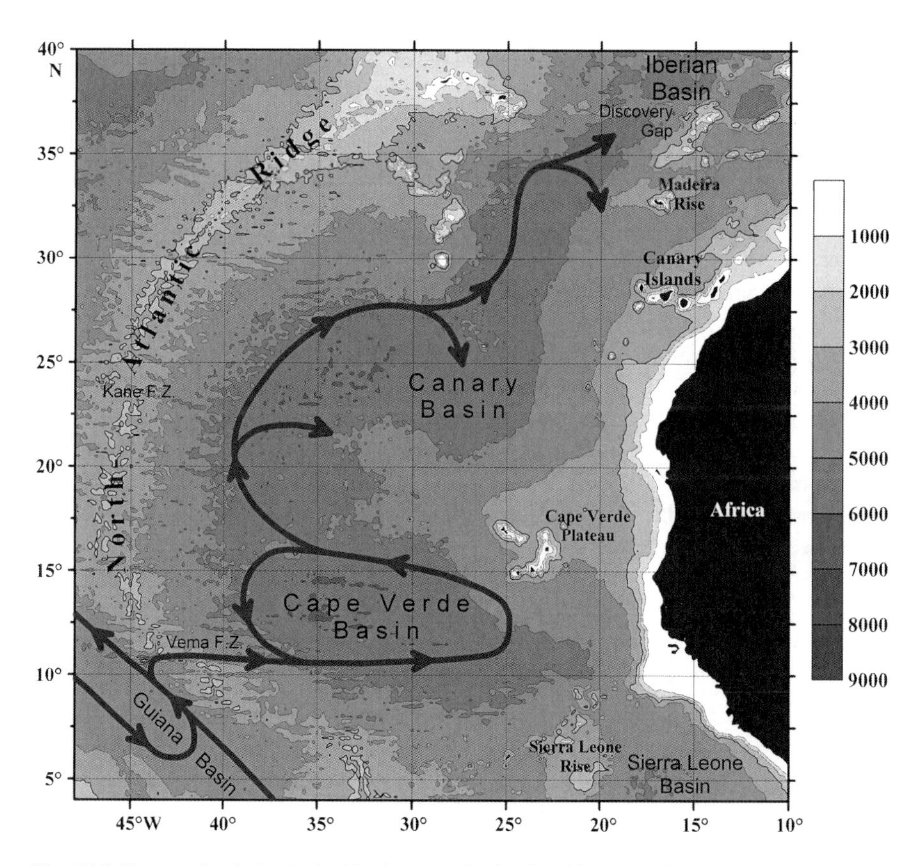

Fig. 5.26 Bottom circulation in the Northeastern Basin (Gambia Abyssal Plain, and Canary Basin) of the Atlantic Ocean. (Modified after McCartney et al. (1991))

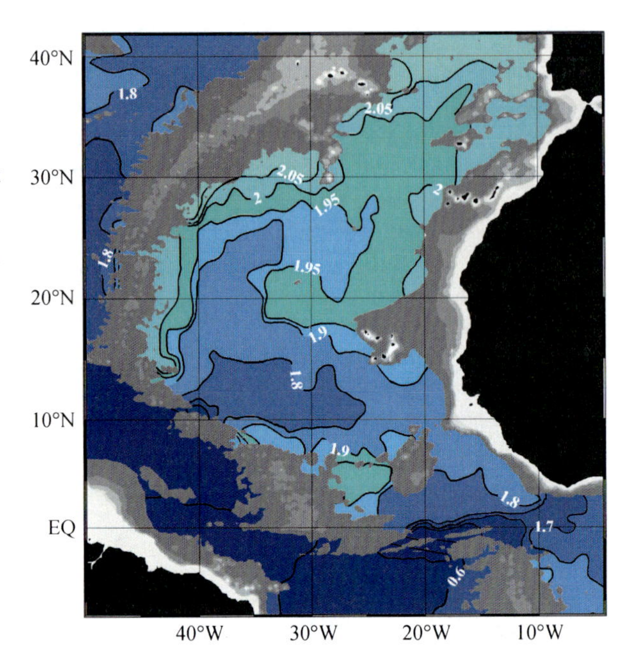

Fig. 5.27 Scheme of Antarctic waters spreading in the bottom layer of the Atlantic. Distribution of potential temperature (°C) at the bottom in the eastern part of the North Atlantic based on the WODB-2005 data. Only the stations deeper than 4,000 km were used. *Gray* shade shows the location of isobaths above 4 km

The second branch is directed from the Vema Fracture Zone to the southeast; the slopes of the Mid-Atlantic Ridge and Sierra Leone Rise are constraints for this flow from the south. This flow transports from 1.3 to 3 Sv of Antarctic Bottom Water (McCartney et al. 1991). Similarly to the northward flow, the value of southeastern transport exceeds the transport through the Vema Fracture Zone, therefore we have to suppose that here this flow includes cyclonic circulation in the Gambia Abyssal Plane (Cape Verde Basin). This branch reaches the Kane Gap near the coast of Guinea.

According to the CTD casts made in the last 20 years, waters with potential temperature less than $\theta = 1.85°C$ from the Romanche and Vema fracture zones merge in the region around the Kane Gap. Waters with $\theta = 1.80°C$ are located north and south of the Kane Gap. At the same time, isothermal surfaces $\theta = 2.00°C$ and $\theta = 1.90°C$ are not separated over the Kane Gap (Fig. 5.27), which indicates possible Antarctic Bottom Water exchange through this passage. However, according to McCartney et al. (1991) the southern part of the Gambia Abyssal Plain (the Kane Gap region) is occupied by a cyclonic gyre; therefore it is a cul-de-sac for the eastward flow of bottom waters from the Vema Fracture Zone. Thus, the large basin of the Northeast Atlantic including the Gambia Abyssal Plain and Canary Basin is filled with bottom water that propagated through the Vema Fracture Zone. The bottom water that propagated through the Romanche Fracture Zone is localized in relatively small basins: Sierra Leone and Guinea basins with a possible insignificant outflow to the Angola Basin.

Such localization seems surprising because Antarctic Bottom Water transports through the Romanche and Chain fracture zones are of the order of 1 Sv, which is almost the same as the water transport through the Vema Fracture Zone (see Sect. 4.3 and 4.4). We believe that it may be explained by stronger mixing in the Romanche and Chain Fracture Zones compared to the Vema Fracture Zone. This can be seen from enhanced transformation of bottom water properties in the equatorial channels. In particular before the entrance to the Romanche and Vema fracture zones, the minimum values of potential temperature at the bottom are 0.68 and 1.33°C, respectively (McCartney et al. 1991). The difference between the smallest values of potential temperature in these fracture zones becomes insignificant after they outflow to the East Atlantic from the fracture zones: $\theta = 1.66°$ (in Romanche) and $\theta = 1.69°$ (in Vema) (McCartney et al. 1991). In this case warming of bottom water during its propagation through the equatorial channels is not related to geothermal heat flux (see Fig. 5.12 and corresponding text). Mercier and Morin (1997) showed that benthal thermocline disappears east of the Romanche and Chain Fracture zones. In the region of the Vema Fracture Zone, benthal thermocline is found only at the stations near the sills. East of the fracture zone benthal thermocline disappears. We believe that this indicates ceasing of strong mixing near the bottom in the remote regions from the Mid-Atlantic Ridge.

The size of the Antarctic Bottom Water spreading region strongly depends on the method of determining its upper boundary. This effect is strongest in the basins east of the Mid-Atlantic Ridge. If the boundary of this water propagation is determined on the basis of isopycnals $\sigma_2 = 37.11$ ($\sigma_4 = 45.92$), Antarctic Bottom Water spreading region in the eastern basin is limited to a small space near the Romanche and Vema fracture zones (Fig. 3.4, brown boundary). If we take isotherm $\theta = 2°C$ as the boundary, Antarctic Bottom Water occupies the major part of the East Atlantic (Fig. 3.4 and 5.27) up to the Iberian Basin (van Aken 2000). Even more, if we use as a criterion $\sigma_4 = 45.87$ ($\gamma^n = 28.11$), which is equivalent to $\theta = 2°C$ in the region of the Brazil Basin (Fig. 3.4), the spreading region of Antarctic Bottom Water reaches 52° N (Fig. 3.4, yellow and orange boundaries).

The northward propagation of bottom waters from the Canary Basin to the northeastern Atlantic occurs through the Discovery Gap. This is a narrow passage in the East Azores Fracture Zone at 37° N between the Madeira and Iberian abyssal basins. Our knowledge of the flow through this passage is based on the research described in Saunders (1987) who named the passage 'Discovery Gap'. The passage is 150 km long. Its narrowest place is located at 37°20′ N, 15°40′ W. The width of the narrowest gap is 10 km and the depth of the sill is 4,800 m. The flow was measured from six moorings and supplemented by tracking floats at a depth of 4,700 m. The measured mean velocities were 5 cm s^{-1}. The flux of bottom water colder than potential temperature $\theta = 2.05°C$ was estimated at 0.2 Sv. Numerous CTD measurements around the Discovery Gap indicate that water with potential temperature below 2°C does not propagate through this passage.

According to Ambar (1983), Antarctic waters in the Eastern Basin join with Iceland Scotland Overflow Water and entrain Subpolar Modal Water. Together they flow through the Charlie Gibbs Fracture Zone (at 53° N), where they join the Antarctic waters propagating in the Northwestern basins of the Atlantic, and their mixture flows to the Labrador Basin region. Most numerical estimates of this

Table 5.12 Transport in the Eastern Basin

Authors	AABW boundary	Transport (Sv)	Comments
Saunders (1987)	$\theta = 2.05$	0.7	Current meter at 36° N
McCartney et al. (1991)	$\theta = 2$	0.83	Geostrophic transport at 36° N; reference surface 2–2.5°C
Lavin et al. (2003)	4,900 m	0.4	Geostrophic transport at 24° N
Lherminier et al. (2007)	$\sigma_4 = 45.85$	0.8	Combined LADCP and SADCP measurements; combined calculations of drift and geostrophic components and inverse model at 42° N in the Iberian Basin

transport (McCartney et al. 1991; Saunders 1994; Lherminier et al. 2007) are below 1 Sv (Table 5.12). Calculation of the amount of Antarctic waters in van Bennekom (1985) revealed that bottom waters in the Northeast Atlantic contain 29% of the Antarctic waters at 50° N, 9–11% at 57° N, and only a few percent in the Norwegian Sea. Table 5.10 summarizes results of different publications about the transport of bottom waters in the Northeast Atlantic (Canary Basin and Gambia Abyssal Plane).

In conclusion we note that the northeastern part of the Atlantic Ocean is a good example that the physical processes stand aside from terminology. Although the bottom water that propagated through the Vema Fracture Zone flows further to the north, gradually transforming and losing the extreme values of its properties, many authors give other names to the bottom water in the Northeast Atlantic. Antarctic Bottom Water east of the Mid-Atlantic Ridge is called Eastern Basin Bottom Water, Lower Deep Water (van Aken 2000), or Abyssal Water of the East Basin (Stephens and Marshall 2000), which is a product of strong transformation of Antarctic Bottom Water. Arhan et al. (1998) distinguished Antarctic waters that passed the Romanche and Chain fracture zones as a special water mass and identified it as Guinea Basin Bottom Water. These waters conserve all distinguishing indicators of Antarctic Bottom Water: low salinity, low concentration of dissolved oxygen, and high concentration of nutrients (van Aken 2000). However, some authors use the term Antarctic Bottom Water for the bottom waters in the Iberian Basin (Lherminier et al. 2007).

5.6.2 *Mixing Caused by the Barotropic Tide*

Here, we make an attempt to give a physical explanation why strong mixing occurs in the equatorial channel, which results in the isolation of different basins east of the Mid-Atlantic Ridge. We believe that this fact is explained by different intensities of mixing of Antarctic waters with the overlying North Atlantic Deep Water due to internal tides generated over the slopes of the Mid-Atlantic Ridge.

Morozov (1995) calculated the amplitudes of internal tidal waves in the World Ocean based on the integrated results of measurements and model calculations. According to these estimates, the amplitude of internal tidal waves reaches almost 50 m in the Romanche Fracture Zone region at the equator and only slightly exceeds 20 m in the Vema Fracture Zone region at 11° N.

Figure 5.28 shows a chart of amplitudes of internal tides in the Atlantic Ocean. Due to such strong difference in the amplitudes of waves in these regions, mixing of deep water masses will differ strongly. The main mixing occurs over the slopes of the submarine ridge, where amplitudes of internal waves are the greatest. Internal tides are intensely generated in the regions of strong barotropic tides if the currents are normal to the ridge. Generation is intensified if the inclination of the bottom coincides with the inclinations of characteristic curves of internal tides, which depend on stratification (Morozov 1995). The conditions that favor strong generation of internal tides are much better in the region of equatorial channels than in the Vema Fracture Zone.

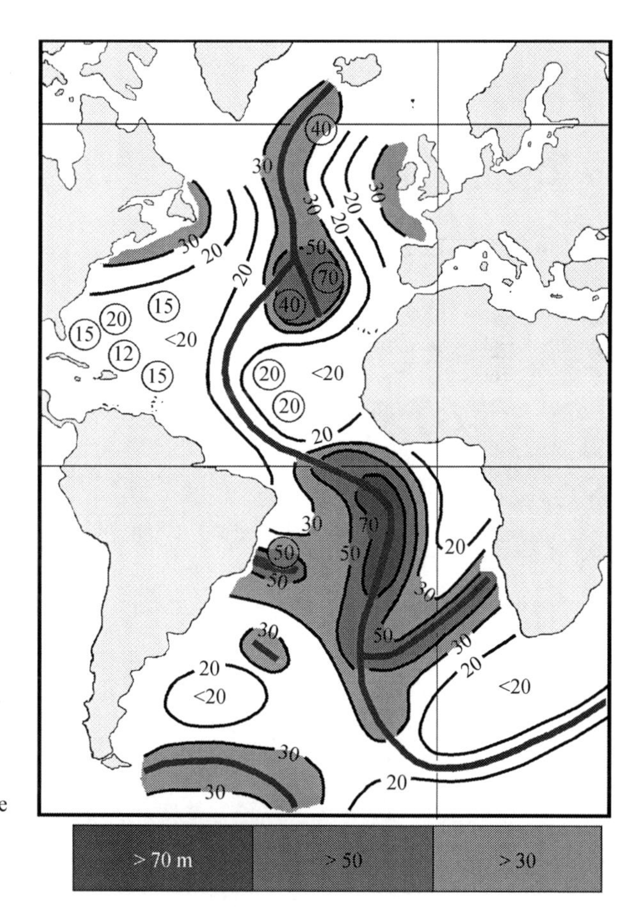

Fig. 5.28 Contour lines of amplitudes of internal tide in meters (Morozov 1995). *Thick lines* show submarine ridges. Numerals in *circles* denote actual measurements of amplitudes of internal tide. The internal tide amplitude exceeds 50 m in the Romanche Fracture Zone region and only slightly exceeds 20 m in the Vema Fracture Zone region

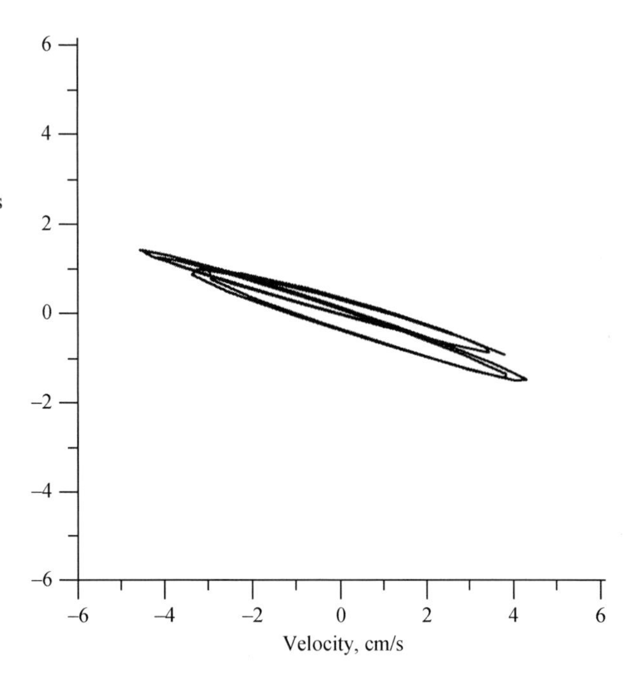

Fig. 5.29 Ellipses of M2 barotropic tide (cm s^{-1}) in the region east of the Chain Fracture Zone based on TOPEX/POSEIDON data. The ellipses are calculated for the time of measurements in the region (April 2009)

Velocity ellipses of the barotropic tide currents in the region of the bottom water outflow from the Romanche and Chain fracture zones were calculated from the TOPEX/POSEIDON satellite data by S. Erofeeva (Oregon State University) upon our request. Ellipses in Fig. 5.29 show two remarkable properties: (i) the velocities are quite high (5 cm s^{-1}) compared to usual 1–2 cm s^{-1} in the open ocean; (ii) the ellipse is strongly elongated in the E-W direction, which is close to the direction of the Romanche and Chain fracture zones. It is essential that the ellipses were calculated from the data of surface elevations and they represent the entire tidal structure in the water column, the largest part of which is located above the deep channel. Thus, such barotropic tides can induce strong generation of internal tides.

Characteristic curves for internal tides, where amplitudes of internal waves are maximal, are inclined near the slopes of the ridges. The results of numerical calculation of the field of horizontal velocities of particle motion in internal tidal waves over the slopes of the Mid-Atlantic Ridge in the Romanche Fracture Zone region are shown in Fig. 5.30. The results of numerical calculation are presented for the phase of barotropic tide when the velocities of the barotropic tide currents are zero. The calculation was made using the numerical model described in Vlasenko and Hutter (2002). Strong mixing in the Romanche Fracture Zone region transforms properties of the Antarctic water.

A similar calculation was performed for the Vema Fracture Zone region (Fig. 5.31). Velocities of internal tide are lower. Hence, internal waves provoke lesser mixing in this region.

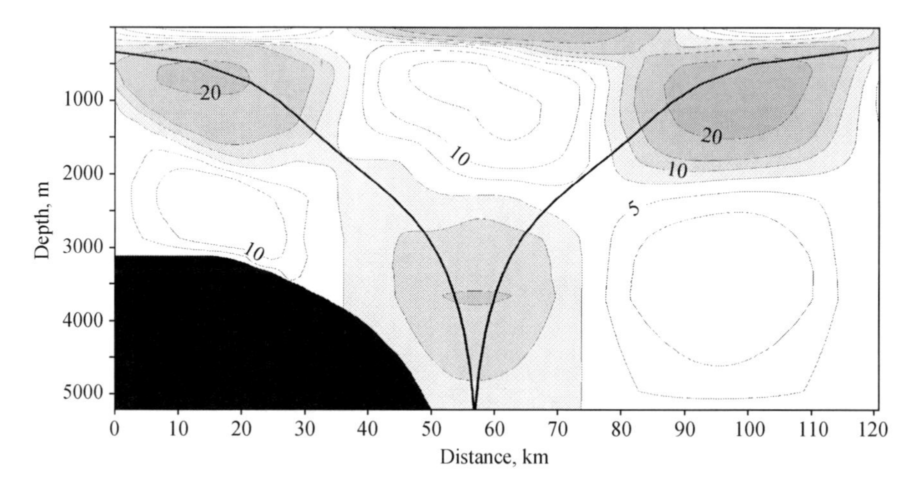

Fig. 5.30 Distribution of velocities during generation of internal tide over the slope of the Mid-Atlantic Ridge in the Romanche Fracture Zone region. Contour lines of velocities are given in cm s^{-1}. The *black color* denotes the ocean bottom. The *thick black line* shows the characteristic curve for internal tide M2

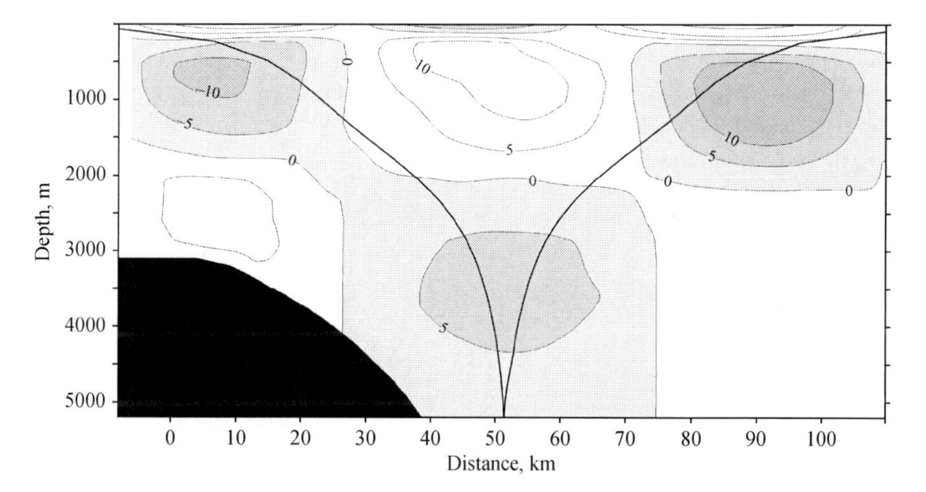

Fig. 5.31 Distribution of velocities during the generation of internal tide over the slope of the Mid-Atlantic Ridge in the Vema Fracture Zone region. Contour lines of velocities are given in cm s^{-1}. The *black color* denotes the ocean bottom. The *thick black line* shows the characteristic curve for internal tide M2

We summarize that mixing in the equatorial region of the East Atlantic is greater than in the region of the Vema Fracture Zone. This strong mixing is caused by a strong barotropic tide in the equatorial region of the East Atlantic compared to the region of the Mid-Atlantic Ridge near the Vema Fracture Zone.

Strong mixing explains the difference in the contribution of the equatorial channels and Vema Fracture Zone to the bottom water mass composition in the Northeast Atlantic. Without strong mixing in the equatorial region, the influence of the Romanche and Vema fracture zones on the Northeast Atlantic seems to be equal:

1. Antarctic Bottom Water transports through the Romanche and Chain fracture zones are of the order of 1 Sv, which is almost the same as the water transport through the Vema Fracture Zone (see Sects. 5.4 and 5.5).
2. The difference between the smallest values of potential temperature of bottom water is insignificant after the water outflows to the East Atlantic: $\theta = 1.66°$ (in Romanche) and $\theta = 1.69°$ (in Vema).
3. The depths of the main sills in the channels do not differ greatly: 4,359 m (Romanche FZ) and 4,571 m (Vema FZ).

Despite the fact that depths of the Kane Gap allow propagation of Antarctic Bottom Water and only 1.8°C isotherm is separated over the passage, the transport of bottom water through this channel is not strong.

Strong mixing with overlying waters results in strong transformation of properties of Antarctic Bottom Water after its outflow from the equatorial channels. In addition, the region of strong mixing includes also the Kane Gap, which makes this passage almost impossible for the strong northerly flow of bottom waters. Strong mixing closes this pathway for the flow of bottom water transported through the equatorial fracture zones. On the other hand, mixing conditions in the Vema Fracture Zone region are not as strong as in the equatorial East Atlantic and bottom water inflows are less transformed through the Vema Fracture Zone filling the entire Northeast Atlantic abyssal depths.

Different intensity of mixing results in different stratification at the outflow from the channels. Brunt-Väisälä frequency east of the Romanche Fracture Zone is $N = 0.14.10^{-3}$ s^{-1}, while east of the Vema Fracture Zone it is equal to $N = 0.80.10^{-3}$ s^{-1}. A simple relation in Gargett (1984) gives the dependence of vertical diffusivity k_z on stratification.

$$k_z = \frac{10^7 m^2 s^2}{N\ (z)}$$

According to this relation diffusivity in the deep layers of the Mid-Atlantic Ridge east of the Vema Fracture Zone is $k_z = 1.2 \times 10^{-4}$ m^2/s, and east of the Romanche Fracture Zone it is seven times greater ($k_z = 7.1 \times 10^{-4}$ m^2/s).

5.7 Kane Gap

The Kane Gap is located between the Grimaldi Mountains, which are a part of the Sierra Leone Rise and the Guinea Plateau near the African Continent (Fig. 5.32). The gap connects Gambia Abyssal Plain and Sierra Leone Basin. The sill depth in the gap is 4,502 m (Smith and Sandwell 1997). Another gap is located east of the main passage with depths exceeding 4,100 m (at longitude 18°40′ W).

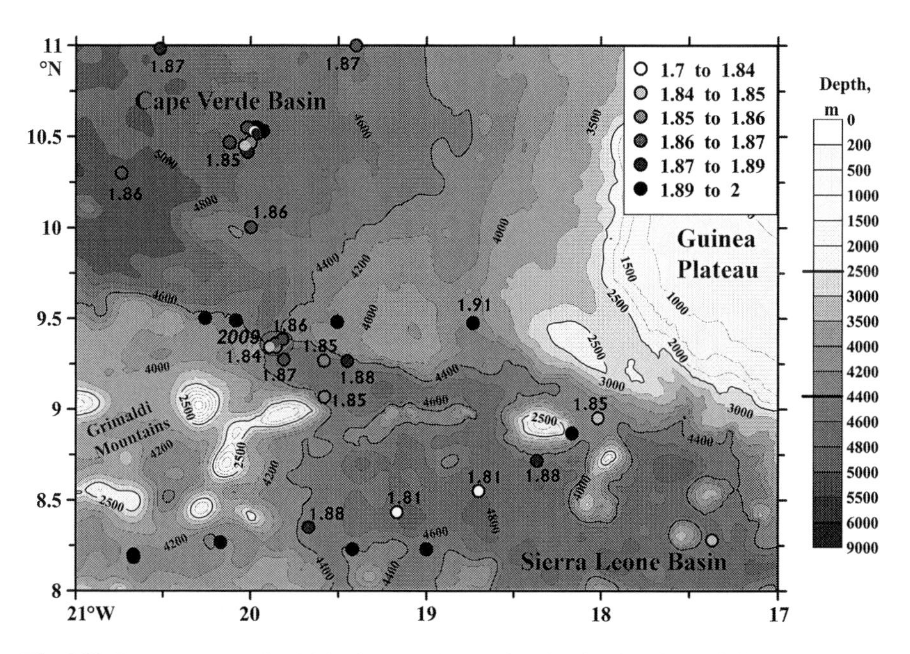

Fig. 5.32 Bottom topography (m) in the Kane Gap region showing locations of historical stations and stations in 2009 over the main sill of the gap. Bottom potential temperatures (°C) are indicated at deepest stations

According to World Ocean Database (WODB-2005), the first hydrographic stations were occupied in the Kane Gap region in 1963 during the cruise of R/V *Trident*. They were located mostly over the adjacent seamounts. Later, bottom sediments were studied during a cruise of R/V *Chain*. Several CTD casts were also made in the region (Hobart et al. 1975). In 1983, CTD stations were made from R/V *Knorr*. A few hydrographic sections were made close to the gap. In 1993, WOCE hydrographic section A06 was occupied along 7°30′ N south of the gap. In 2000, a hydrographic section was made from R/V *Akademik Ioffe* along 8° N close to the WOCE A06 section. During cruise 27 of R/V *Akademik Ioffe* in May 2009, a CTD station was occupied at the main sill of the gap at a depth of 4,552 m with the first LADCP measurements in the gap. Later, in October 2009, a section of three stations was made across the Kane Gap in the region of the sill. All available hydrographic stations in the regions of the Kane Gap and historical data from Hobart et al. (1975), which were not included in WODB-2005, are shown in Fig. 5.32. The bottom temperatures at the deepest stations are also indicated.

In the previous section we discussed a scheme of Antarctic Bottom Water spreading (with potential temperature $\theta < 2°C$) in the Northeast Atlantic. According to this scheme, the Kane Gap occupies the middle position between the Antarctic Bottom Water that propagated to the East Atlantic through the Vema Fracture Zone and the Antarctic Bottom Water that propagated through the Romanche and Chain fracture zones. Waters with potential temperature less than $\theta = 1.85°C$ from these two sources merge in the region around the Kane Gap. At the same time, isotherms θ = 1.90°C are not separated over the Kane Gap (Fig. 5.27).

The measurements in May 2009 recorded potential temperature θ = 1.857°C near the bottom in the Kane Gap, which agrees well with the previous measurements in the region. In October 2009 even lower temperatures were measured at the western slope of the gap (1.846°C). South and north of the Kane Gap potential temperatures at the bottom are cooler. We suppose that an abyssal front exists near the Kane Gap, where bottom waters from two different sources encounter.

In May 2009, an extremely low salinity (34.873 psu) was measured at the bottom in the Kane Gap. In October 2009, even lower salinities were measured in the Kane

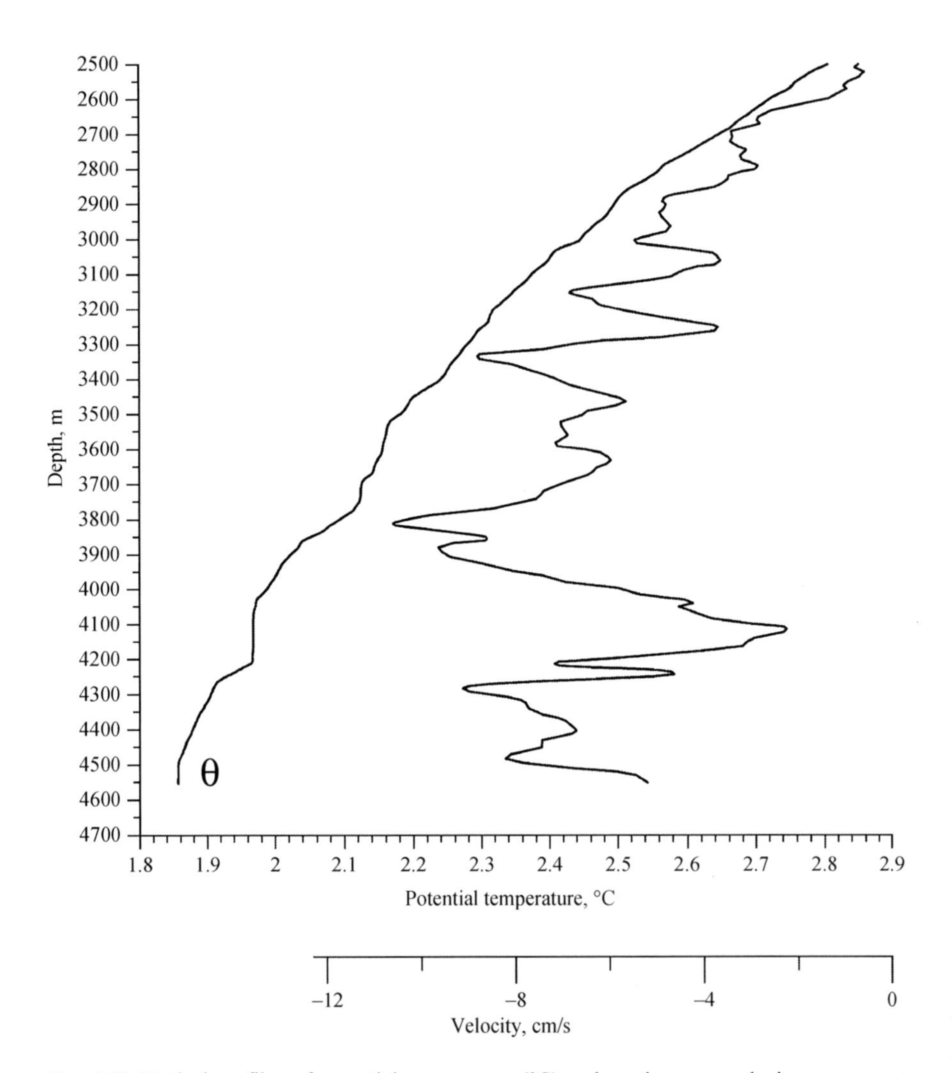

Fig. 5.33 Vertical profiles of potential temperature (°C) and southeastern velocity component (cm s^{-1}) measured in the Kane Gap in May 2009. Positive velocities in this figure are related to the northerly flow

Gap (34.869 psu). These values are close to the value measured in 2000 south of the gap at 8° N. It is worth noting that even lower salinities were found in the Gambia Abyssal Plain using hydrographic measurements.

Vertical distributions of potential temperature θ and velocity at the station in the Kane Gap (May 2009) are shown in Fig. 5.33. The bottom water structure is presented by a mixed layer between two gradient zones. Gradient zones are characterized by maximum velocities, while the velocity of the mixed layer is lower. Surprisingly, in May 2009, currents measured with LADCP were directed to the south at all depths below 2,500 m (Fig. 5.33). Thus, the bottom transport was directed from the Gambia Abyssal Plain to the Sierra Leone Basin. According to McCartney et al. (1991), bottom waters in the Kane Gap have more similar properties with the waters in the Gambia Abyssal Plain than with the waters in the Sierra Leone Basin, but the southeastern part of the Gambia Abyssal Plain is considered a cul-de-sac for further propagation of bottom waters to the south.

Three stations with CTD-profiling in October 2009 allowed us to plot the distributions of potential temperature and salinity across the Kane Gap. The graphs

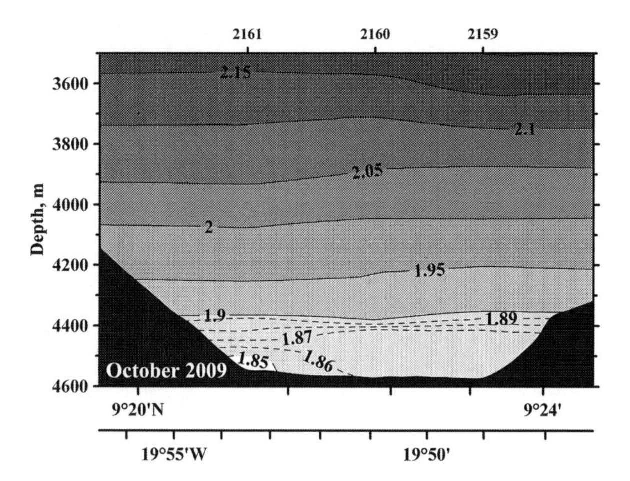

Fig. 5.34 Distribution of potential temperature (°C) across the Kane Gap in October 2009

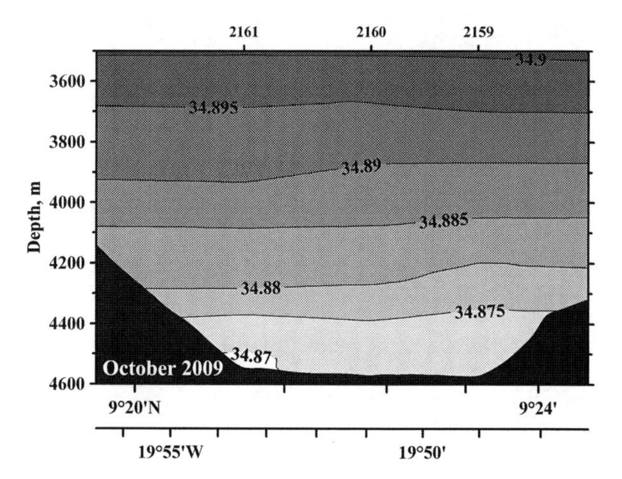

Fig. 5.35 Distribution of salinity (psu) across the Kane Gap in October 2009

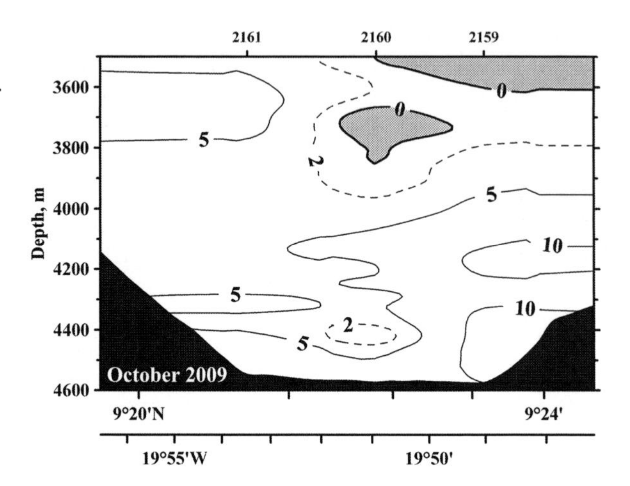

Fig. 5.36 Distribution of N-S velocities (cm s^{-1}) across the Kane Gap in October 2009. Positive velocities in this figure are related to the northerly flow

are shown in Figs. 5.34 and 5.35. In October 2009, measurements with an LADCP profiler demonstrated that the flow was directed to the northwest. Thus, the flow was opposite to the one recorded in May 2009. Distribution of velocities in the Kane Gap with the dominating northward transport is shown in Fig. 5.36. This result agrees with the idea discussed in Hobart et al. (1975) based on the photographs

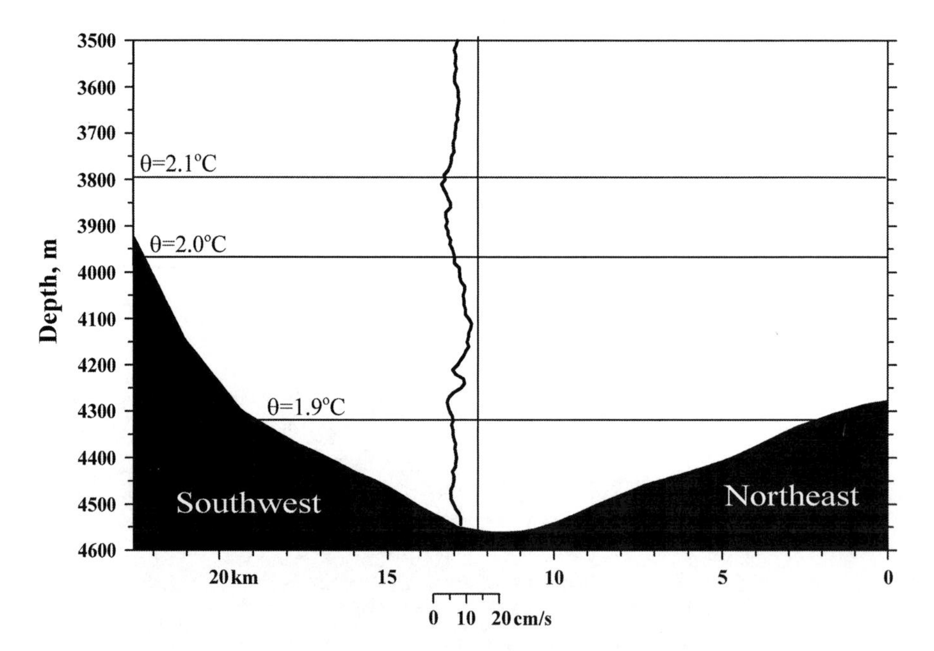

Fig. 5.37 Boundaries of bottom water transport calculations through the Kane Gap based on LADCP measurements. Bottom topography is based on Smith and Sandwell (1997). Velocity profile (cm s^{-1}) and depths (m) of potential temperature isotherms are shown. Negative velocities are related to the southerly flow. Width of the passage is laid at the abscissa axis (in km)

Table 5.13 Transport and extreme values of temperature in the Kane Gap

	Transport (Sv) (southeastern transport is positive)		Minimum potential temperature	Minimum salinity
Cruise	AABW upper boundary 1.9°C	AABW upper boundary 2.0°C	θ_{min}	S_{min}
May 2009	0.16	0.55	1.857	34.873
October 2009	−0.11	−0.29	1.846	34.869

of the bottom, from which they concluded that the bottom flow in the Kane Gap is directed to the north.

The temperature stratification of the flow is similar to the flow in the Vema Channel. The coolest and densest water of the flow is displaced to the western wall of the gap due to the Ekman friction. Lower salinities are also recorded here at the foot of the western slope (see Figs. 5.34 and 5.35). Since the Kane Gap is located in the Northern Hemisphere, Ekman friction displaces the densest water to the left wall of the channel (southwestern slope in our case).

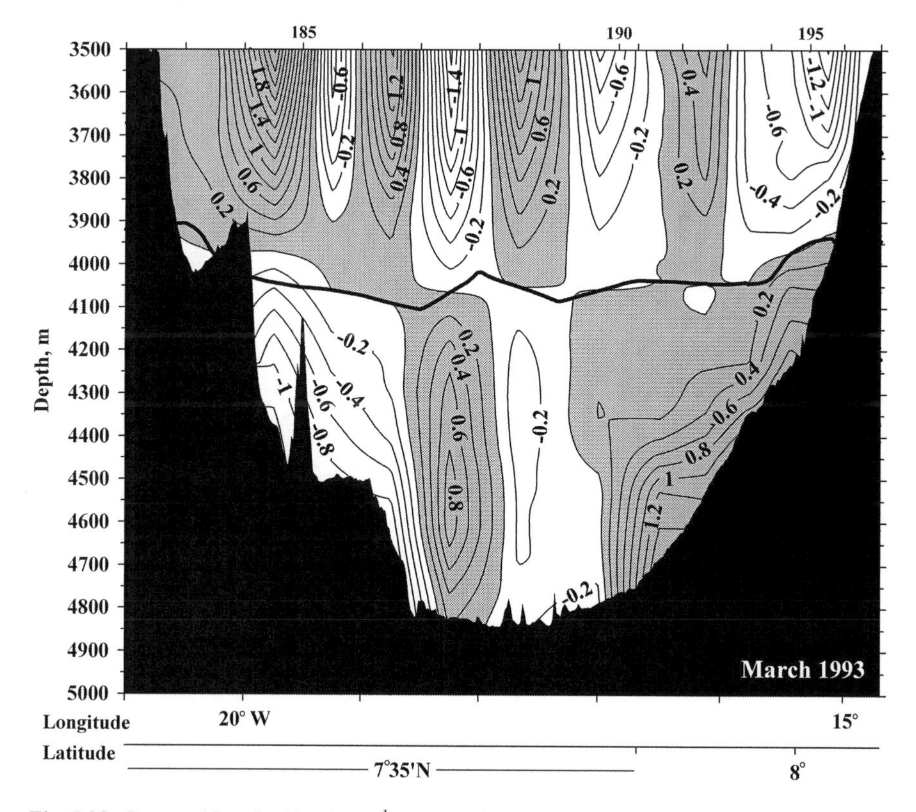

Fig. 5.38 Geostrophic velocities (cm s^{-1}) over section A6 in 1993

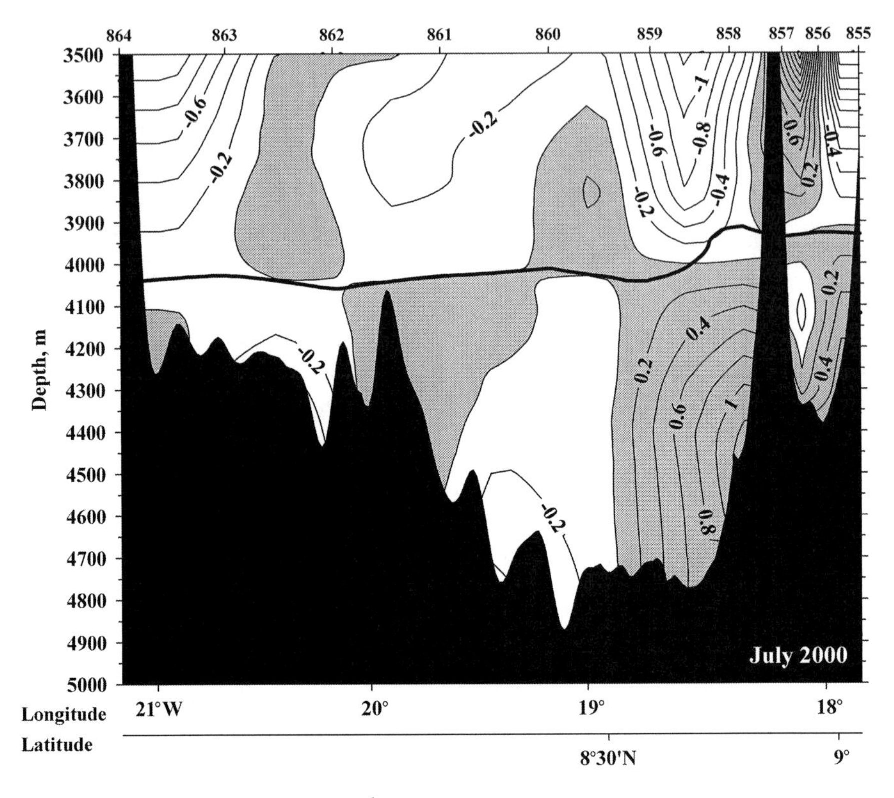

Fig. 5.39 Geostrophic velocities (cm s^{-1}) over the section in 2000

We want to estimate the transport of bottom waters through the Kane Gap based on measurements in May and October of 2009. We made only one profile in May 2009, thus knowing the velocity of the flow and boundaries formed by bottom topography and upper level of the bottom water we should assume constant velocity through the passage, which is actually not true. The upper boundary strongly influences calculation of the water transport. If we determine the upper boundary as the depth of potential temperature isotherm $\theta = 1.9$°C, the southeasterly flow is equal to 0.16 Sv (Fig. 5.37). If we accept the upper boundary as $\theta = 2.0$°C, bottom waters can propagate through the eastern passage (9°30′ N, 18°40′ W). No measurements were made in this passage in 2009, but bottom potential temperature was $\theta = 1.91$°C in 1983. In addition, one can see from Figs. 5.32 and 5.34 that isotherm $\theta = 2.0$°C at a depth of approximately 4,000 m encounters bottom only at a distance of 70 km from the Kane Gap, which makes calculation of water transport from the data of only one station absolutely incorrect. If we put a boundary in the east at a depth of 4,300 m (actually introducing a liquid vertical boundary at 23 km from the western slope), the southward transport would be as high as 0.55 Sv.

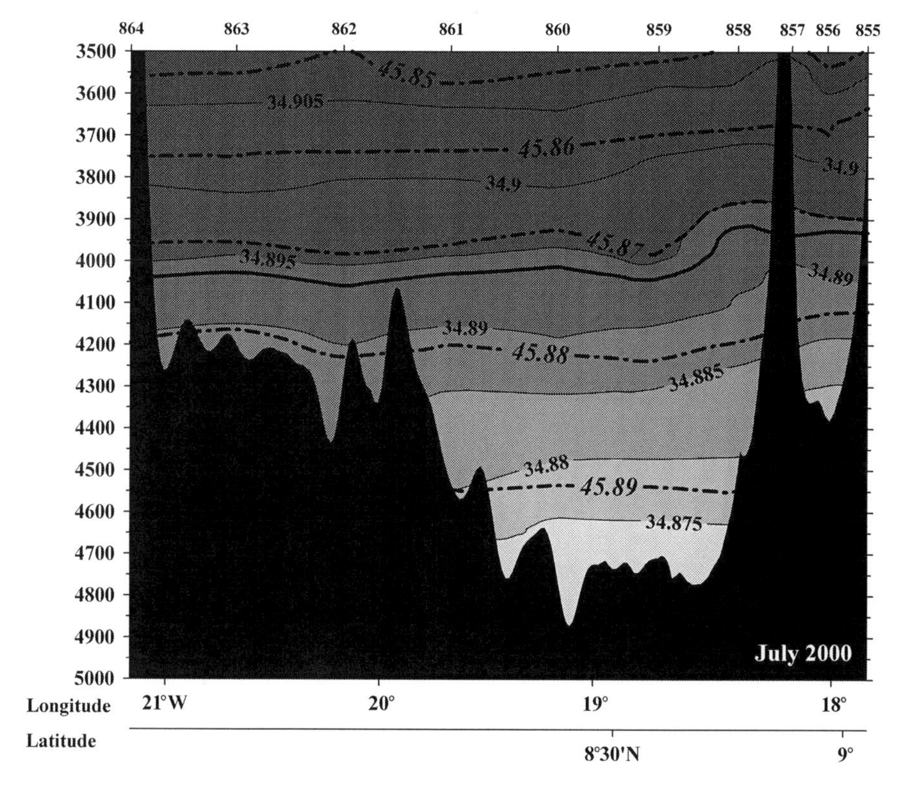

Fig. 5.40 Salinity (psu) distribution over the section in 2000. Isopycnals referenced to 4,000 m (σ_4 in kg/m^3) are shown

The total transport below 1.9 and 2.0°C potential temperature isotherm based on LADCP measurements and extreme values of temperature and salinity based on measurements in May and October 2009 are given in Table 5.13. These are the only measurements in the Kane Gap, which provide evidence of the existence of the alternating bottom water transport. Thus, the bottom water from the Vema Fracture Zone influences at least the northern part of the Sierra Leone Basin, while the bottom water from the Romanche Fracture Zone can spread to the north through the Kane Gap and influence the adjacent southern region of Cape Verde Basin. However, the bottom water transport is not very strong and can be influenced by tides.

We compare our direct measurements of velocity in the Kane Gap with geostrophic estimates based on results of other expeditions. We consider previous CTD-sections occupied close to the Kane Gap from the north and south of the passage. McCartney et al. (1991) demonstrated almost zero westerly geostrophic transport of bottom waters with $\theta < 2.0$°C north of the Kane Gap along 29° W. In 1993 and 2000, two hydrographic sections were occupied close to the Kane Gap. The WOCE A6 section was made by French oceanographers in 1993 and the section in 2000 was made by Russian scientists. These sections crossed the Sierra Leone Basin approximately at 8° N. Geostrophic velocities of bottom water across these sections were negligible. The sections are shown in Figs. 5.38 and 5.39.

If we take $\theta = 1.9°C$ as the upper boundary, the resulting transport will be almost zero. Calculation with boundary at isotherm $\theta = 2.0°C$ yields a northerly transport equal to 0.06 Sv in 1993 and 0.14 Sv in 2000, which are both close to zero. We can conclude that no significant transport exists here, because the estimates are within the limits of error for geostrophic calculations.

The geostrophic section in 2000 demonstrates a northerly flow along 18° W. The salinity section shown below in Fig. 5.40 indicates that salinity of the water transported with this flow is different from bottom salinities in the other part of the section. Salinity and density contour lines are elevated above the background level (passage at 18° W) by almost 300 m, indicating the existence of a southward bottom slope flow whose core is located in this passage.

5.8 Angola Basin

The Angola Basin located in the eastern South Atlantic is a deep basin. The major part of the basin is deeper than 5,000 m while the maximum depth exceeds 5,700 m (Fig. 5.41). It is bounded from the west by the Mid-Atlantic Ridge. Its

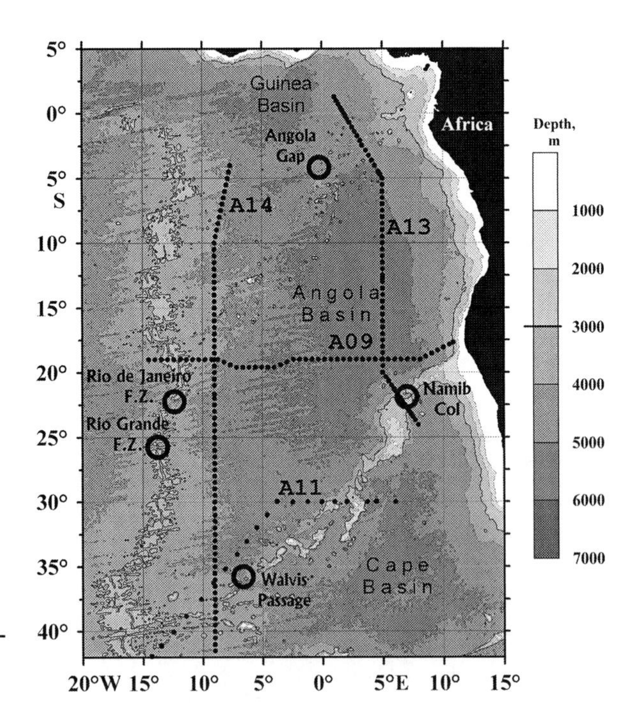

Fig. 5.41 Bathymetric chart of the Angola Basin and locations of WOCE stations. Possible passages for Antarctic Bottom Water inflow to the basin are indicated

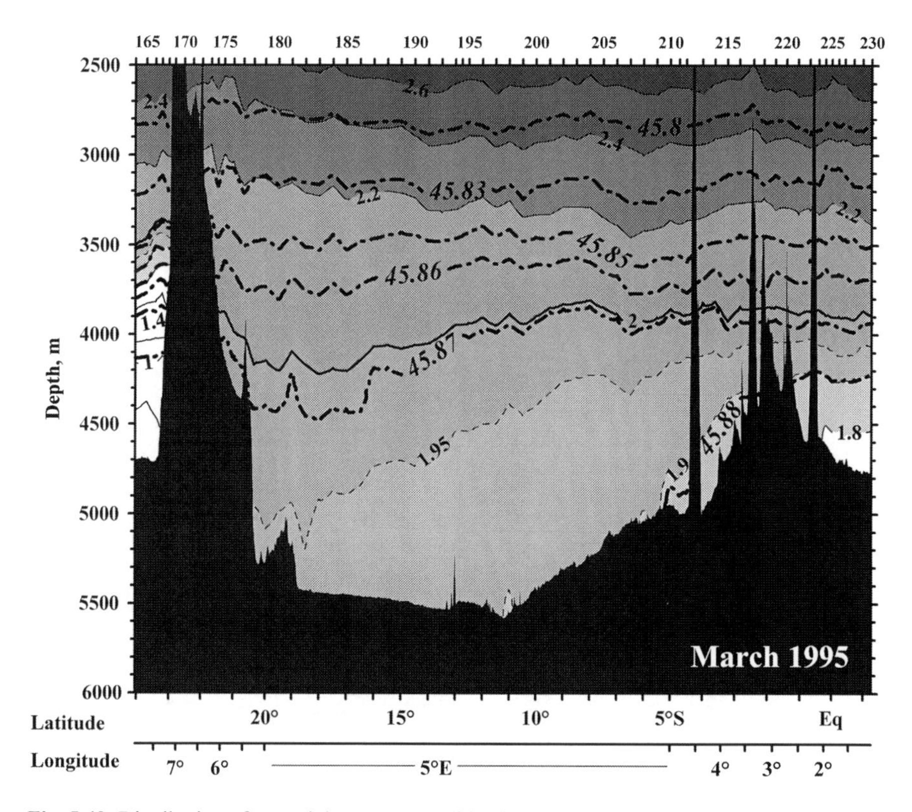

Fig. 5.42 Distribution of potential temperature (°C) along WOCE section A13

crest is crosscut by many narrow fractures not deeper than 3,500 m (Warren and Speer 1991). From the south the Angola Basin is bounded by the Walvis Ridge. The Guinea Rise bounding the basin from the north is the deepest barrier. Depths of the sills in the passages of the Guinea Rise exceed 4,000 m.

It is clearly seen from the distribution of potential temperature near the bottom in the Atlantic Ocean (Fig. 2.12) that bottom water with potential temperature less than 2°C penetrates through the passages in the Guinea Rise. However, ridges of the Guinea Rise almost block the water with potential temperature below 1.89°C from propagating south into the Angola Basin. The depth of this isotherm over WOCE section A13 (R/V *L'Atalante*, France, 1995) in the northern part of the Angola Basin suggests that the sill depth is about 4,300 m (Fig. 5.42).

Connary and Ewing (1974) showed the existence of a narrow passage (5–10 miles wide) in the Walvis Ridge in the southern part of the Angola Basin near the junction of the Walvis and Mid-Atlantic ridges at 36° S, 7° W. Depth of the sill is 4,120 m. This passage ('Walvis Passage' according to Connary and Ewing (1974)) is the main pathway for Antarctic Bottom Water (in terms of Wüst (1936)) transport from the Cape Basin to the Angola Basin. Figure 5.43 shows the poten-

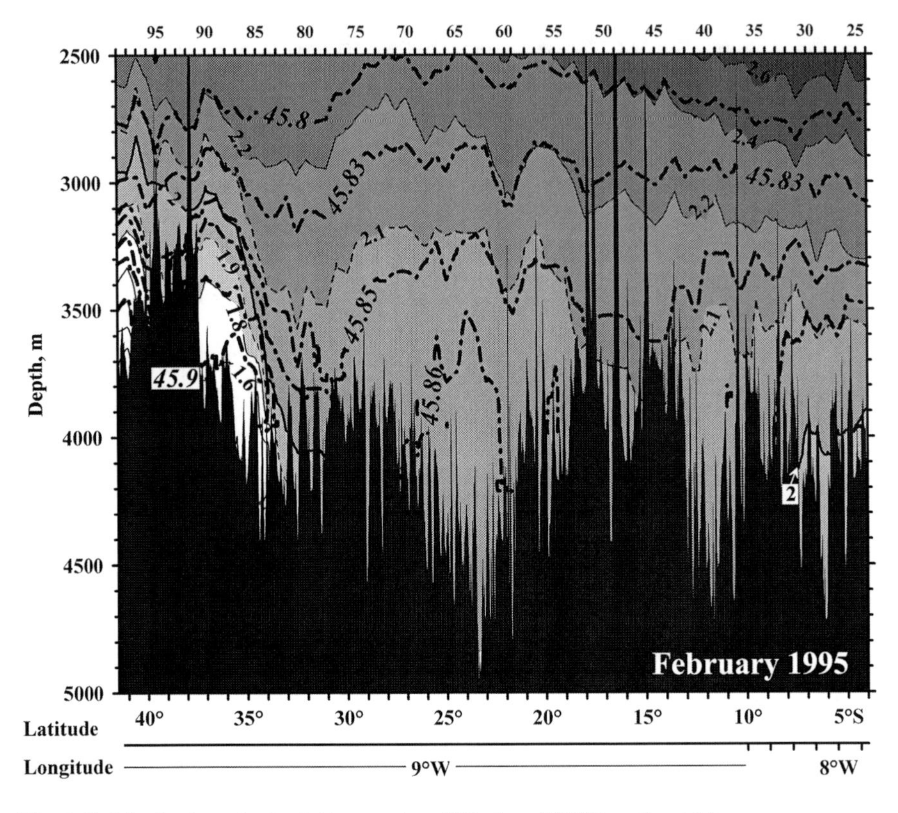

Fig. 5.43 Distribution of potential temperature (°C) along WOCE section A14

tial temperature distribution based on the WOCE A14 meridional CTD section approximately along 9° W (R/V *L'Atalante*, France, 1995). The coldest bottom water in the Angola Basin is found here south of 33° S. However, additional spreading of Antarctic Bottom Water to the Angola Basin through a small break in the Walvis Ridge at 31–32° S, 2° E was found in Shannon and Rijswijck (1969) and Shannon and Chapman (1991).

On the basis of the chart published in Needham et al. (1986), Warren and Speer (1991) reported the existence of a passage near 32°40′ S, 2°20′ W with a sill depth between 4,000 and 4,250 m, which can be the deepest corridor. However, the charts of the distribution of different parameters at the bottom show that this flow does not influence strongly the water properties in the Angola Basin. This is confirmed by the WOCE A11 section (R/V Discovery in 1993). Potential temperature distribution shows that the inflow of cold waters occurs near 7–8° W (Fig. 5.44).

Namib Col passage (22° S, 7° E) (Speer et al. 1995) is not a very deep passage that allows only the flow of North Atlantic Deep Water into or from the Angola Basin. The flow of North Atlantic Deep Water in the Angola Basin was also studied in Arhan et al. (2003).

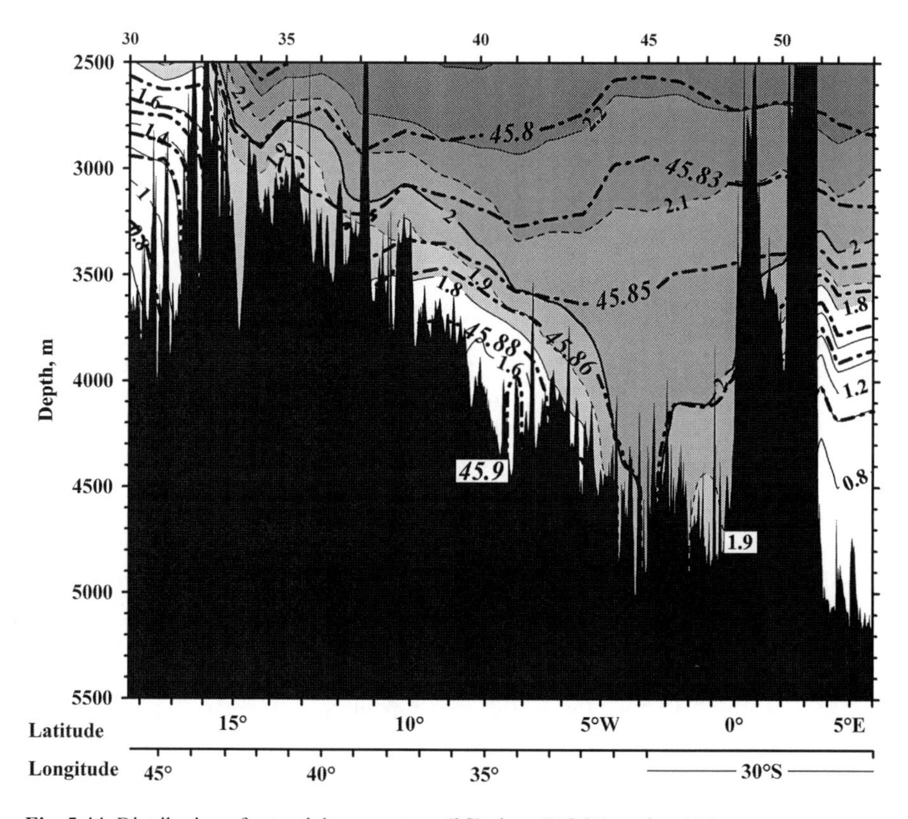

Fig. 5.44 Distribution of potential temperature (°C) along WOCE section A11

The depths of deep passages in the Mid-Atlantic Ridge south of the equatorial zone prevent the propagation of Antarctic waters. Analysis performed in Mercier et al. (2000) and Sandoval and Weatherly (2001) suggests that Antarctic waters propagate to the Southeast Atlantic through the Rio de Janeiro (22° S) fracture in the Mid-Atlantic Ridge. Mercier et al. (2000) analyze the signatures of through-flows at the Rio de Janeiro Fracture Zone (22° S) and the Rio Grande Fracture Zone (26° S) in the Mid-Atlantic Ridge in the tracer fields over a hydrographic section along 9° W. Bottom water flows are identified from anomalies of water properties in the Angola Basin. The through-flow water is supplied by a meridional band of cold and fresh water lying against the western flank of the Mid-Atlantic Ridge. The authors calculated a geostrophic transport, which appeared as high as 0.5 Sv. The Tristan de Cunha Fracture Zone remains unstudied as a possible pathway for bottom waters. Figure 5.45 shows the potential temperature distribution based on the WOCE A9 CTD section across the Angola Basin occupied along 18–19° S (R/V *Meteor*, Germany 1991). The figure demonstrates that there is no strong inflow of bottom waters from the west through the Mid-Atlantic Ridge at 18–19° S. Only a small deep region near the bottom is occupied with cold water with potential temperature 1.95°C.

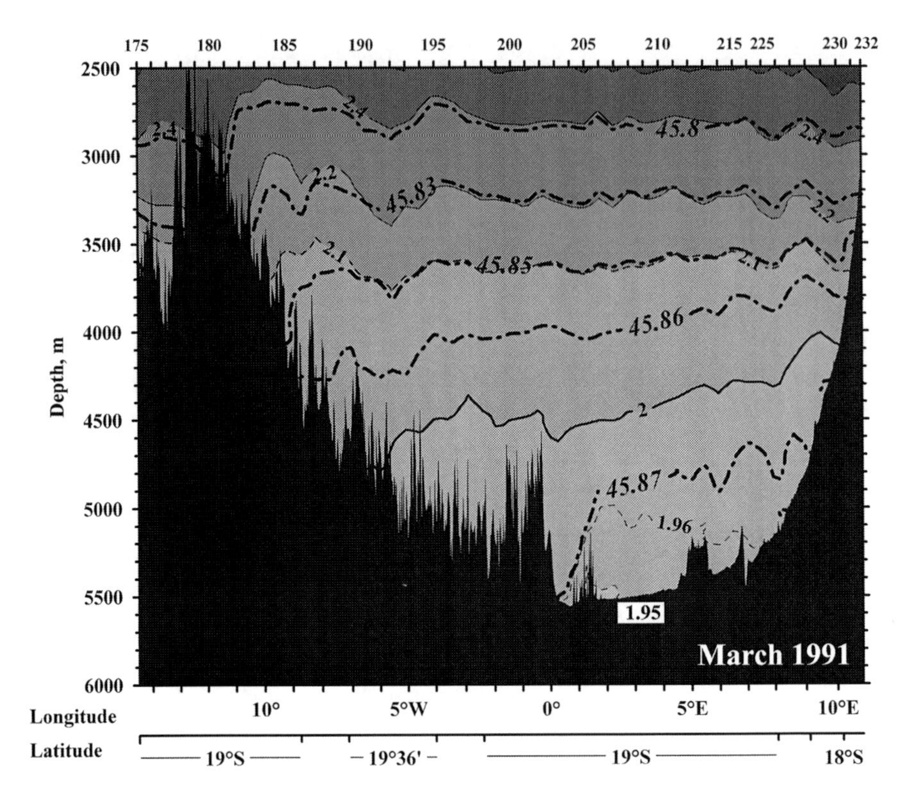

Fig. 5.45 Distribution of potential temperature (°C) along WOCE section A09

We can conclude that at depths exceeding 4,000 m, the bottom water spreading to the Angola Basin from the north across the Guinea Rise flows southward along the slopes of the Mid-Atlantic Ridge. In the southern part of the Angola Basin, a northward boundary current is supplied by the water propagating through the Walvis Passage.

Chapter 6
Flows through the Mid-Atlantic Ridge in the Northern Channels. Charlie Gibbs Fracture Zone and Other Fracture Zones[1]

In the winter period, cold dense water is formed in the Norwegian and Greenland seas due to severe cooling and intense heat release to the atmosphere. The water mass from the Greenland Sea overflows the shallow threshold between Greenland and Iceland and flows into the Irminger Basin. The water mass from the Norwegian Sea overflows the threshold between Iceland, Faeroe, and Shetland Islands and flows into the Iceland Basin. The water mass, which is formed during the overflow over the latter thresholds, is called Iceland Scotland Overflow Water (ISOW). In the Iceland Basin this water is mixed with the significantly warmer and more saline waters in the Northeastern Atlantic, which are subject to the influence of Mediterranean waters. As a result, Iceland Scotland Overflow Water appears more saline and warmer than the deep waters propagating into the Atlantic through the Denmark Strait between Greenland and Iceland (Denmark Strait Overflow Water, DSOW). Iceland Scotland Overflow Water flows first to the west with the Deep Northern Boundary Current and then to the south along the eastern slope of the Reykjanes Ridge. The circulation scheme of different layers of the North Atlantic is shown in Fig. 2.6.

Large portions of Iceland Scotland Overflow Water penetrate to the West Atlantic at 53° N through the deep Charlie Gibbs Fracture Zone and then flow predominantly to the north along the western slope of the Reykjanes Ridge. Following the 2,500-m isobath, Iceland Scotland Overflow Water reaches the northern boundary of the Labrador Basin, where it becomes part of the Deep Western Boundary Current (DWBC). Iceland Scotland Overflow Water in the North Atlantic is traced by a local salinity and silicate maximum between the cores of the Deep Western Boundary Current and Labrador Water. In the literature, Iceland Scotland Overflow Water is sometimes called Iceland Shetland water (in the eastern North Atlantic) or Charlie Gibbs Water (in the western basin). This issue was discussed in Saunders (1994) and Dickson et al. (1980). McCartney (1992) assumed that the flow through the Charlie Gibbs Fracture Zone should contain a component of ancient deep water from the eastern basin, which is of southern origin. The through-flow should consist of 40% of this water and 60% of overflow water of the northern origin.

---

[1] This chapter was written in cooperation with S. Dobrolyubov.

E. G. Morozov et al., *Abyssal Channels in the Atlantic Ocean*,
DOI 10.1007/978-90-481-9358-5_6, © Springer Science+Business Media B.V. 2010

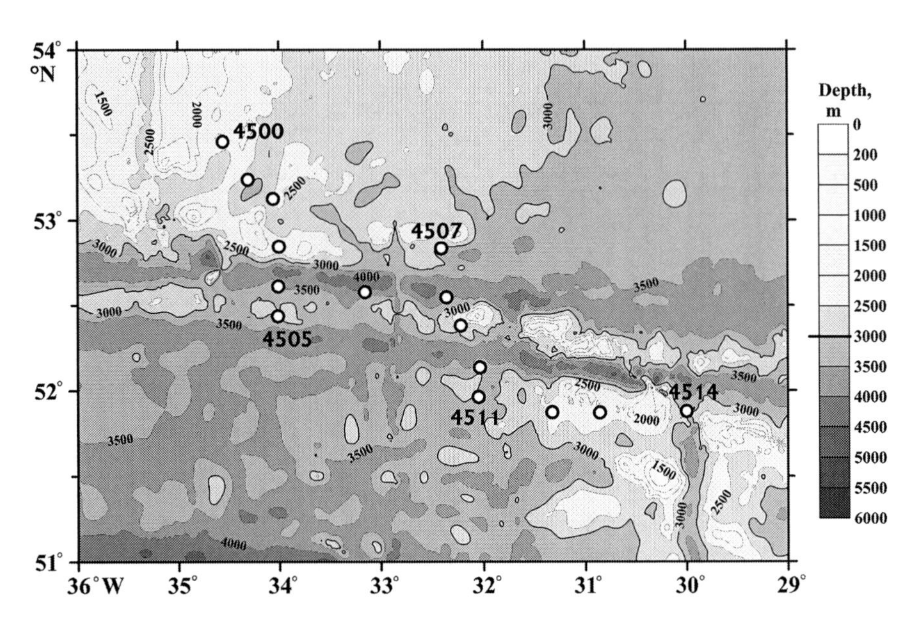

Fig. 6.1 A scheme of CTD-stations occupied in 2002 near the Charlie Gibbs Fracture Zone (4500–4511). A chart of bottom topography is shown in Fig. 8

The Charlie Gibbs Fracture Zone is located between 52–53° N and 30–35° W in the Mid-Atlantic Ridge. A bathymetric chart of this region is shown in Fig. 6.1. In the fracture zone region, the crest of the Mid-Atlantic Ridge and the rift valley are shifted by 350 km. The fracture zone comprises two channels separated by a median ridge, which is more than 400 km long in the west-east direction. Two transversal sills are located in each of the channels at depths of 3,600–3,700 m. The sill is located at 35° W in the northern channel and at 30° W in the southern channel.

In 2002, during cruise 48 of R/V *Akademik Mstislav Keldysh* a CTD section was occupied along 52–53° N, which crossed the Charlie Gibbs Fracture Zone. Water structure and transport were studied at eleven stations located in this region (Fig. 6.1). The stations were positioned in such a way as to cross the channels in the fracture zone twice. The first section was located along the rift valley and across the northern channel approximately along 35° W (stations 4500–4505); the second section crossed the northern and southern channels along 33–32° W (stations 4507–4511).

The western salinity section is shown in Fig. 6.2. Investigations in 2001 during cruise 9 of R/V *Akademik Ioffe* (Dobrolyubov et al. 2002) revealed that the main salinity maximum in the Iceland Scotland Overflow Water core is not located above the deepest channels (western section at 53°10′ N). Figure 6.3 shows θ/S-curves for the three deepest stations in the Charlie Gibbs Fracture Zone (4504, 4508, and 4510) and curves for the stations located west of the fracture zone (with the typical Labrador Sea Water and deep layers of Denmark Strait Overflow Water). East of the fracture zone, one can see typical curves for Iceland Scotland Overflow Water. The shape of θ/S-curves indicates that the water flowing west through the Charlie Gibbs

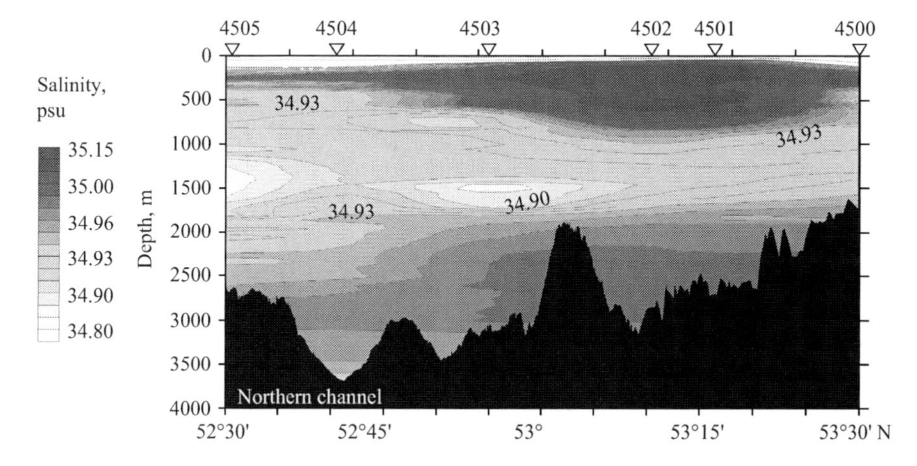

Fig. 6.2 Salinity section (psu) across the Charlie Gibbs Fracture Zone (western section along 35° W)

Fracture Zone is a product of mixing between Iceland Scotland Overflow Water and Denmark Strait Overflow Water. Thus, recirculation and intense mixing of deep waters of different origin are observed in this region.

Figure 6.4 shows a scheme of the geostrophic water transport through the northern and southern channels in the Charlie Gibbs Fracture Zone. Geostrophic calculations showed that the western water transport below isopycnal $\sigma_0 = 27.82$ (approximately between 1,800 m) and bottom is 1.2 Sv on the section along 35° W above the rift valley. On the same section, the westerly flow in the northern channel is

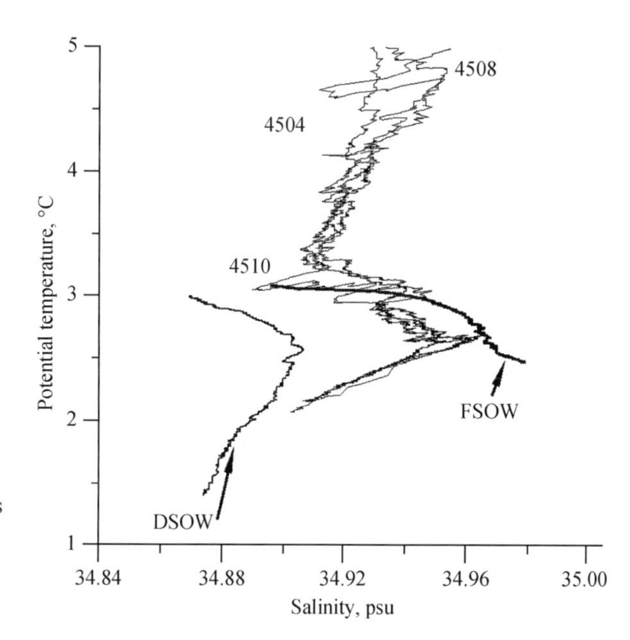

Fig. 6.3 The θ/S-curves for the Charlie Gibbs Fracture Zone (stations 4504, 4508, 4510) and curves for stations with a typical Iceland Scotland Overflow Water and Denmark Strait Overflow Water

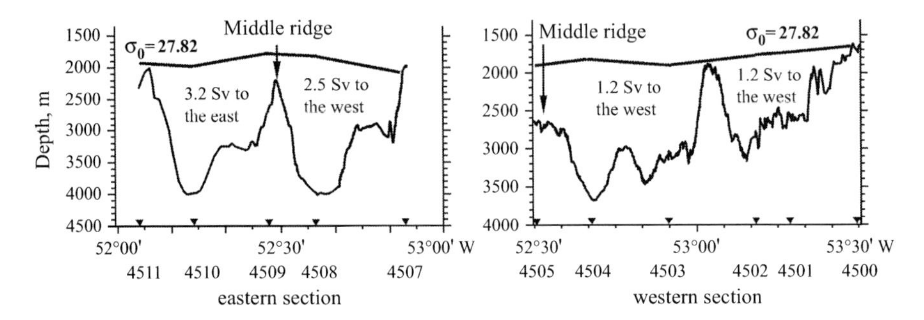

Fig. 6.4 Water transport through the Charlie Gibbs Fracture Zone. *Lines* in the upper part of the graph denote ISOW boundaries based on the position of potential density $\sigma_0 = 27.82$. Stations are indicated by *triangles*

1.2 Sv. On the section along 33° W, the westerly flow along the northern channel is 2.5 Sv, whereas the easterly flow along the southern channel (52°50′ N) is 3.2 Sv.

Let us compare the estimates of water transport in the Charlie Gibbs Fracture Zone in 2002 with instrumental measurements made in the previous years. Dickson et al. (1980) analyzed the results of instrumental measurements in the southern channel carried out from October 1977 to July 1978 and found a westerly transport along the northern slope of the southern channel and easterly transport along the southern slope. Both easterly and westerly mean transports are approximately 0.24 Sv. Saunders (1994) carried out measurements using CTD profiling and moor-

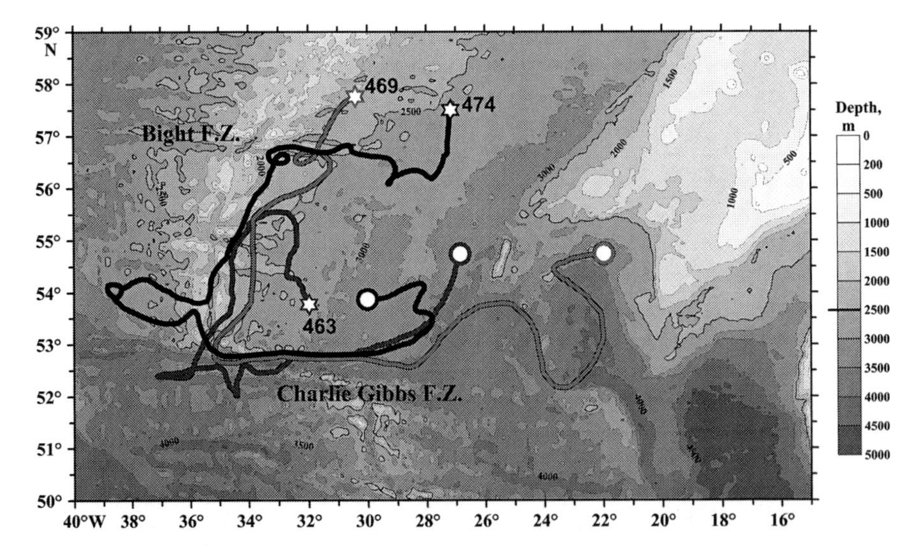

Fig. 6.5 Bathymetry of the Iceland Basin with the Mid-Atlantic Ridge on the western side and the Rockall and Porcupine Banks on the eastern side. Trajectories of three RAFOS floats (IfM tracking numbers 463, 469, and 474) are shown. The launching position and last fix before surfacing are marked with *stars* and *circles*, respectively. *Contour lines* of 1,000, 2,000, and 3,000 m are shown

ings deployed in June 1988 and recovered in September 1989. Only the westerly transport equal to 2.4 Sv was found in the northern channel. The measurements carried out during cruise 9 of R/V *Akademik Ioffe* in 2001 also found only the westerly transport in the northern channel. In this cruise, the number of stations located in the southern channel was not enough for the calculations.

Thus, the westerly water transport was always found. The easterly transport in the southern channel was inferred only in 1978 and 2002. These are the years of low North Atlantic Oscillation (NAO) and increased Arctic water transport through the Strait of Denmark, Faeroe-Shetland, and Faeroe-Iceland thresholds. In 1988–1989, the measurements were carried out during the high North Atlantic Oscillation index, decreased water transport from the Arctic Ocean (Denmark Strait Overflow Water and Iceland Scotland Overflow Water), and intense Labrador water formation. In 2002, the Iceland Scotland Overflow Water branch at 60° N was also weakened west of the Reykjanes Ridge and was not observed near the Greenland slope (Dobrolyubov et al. 2003).

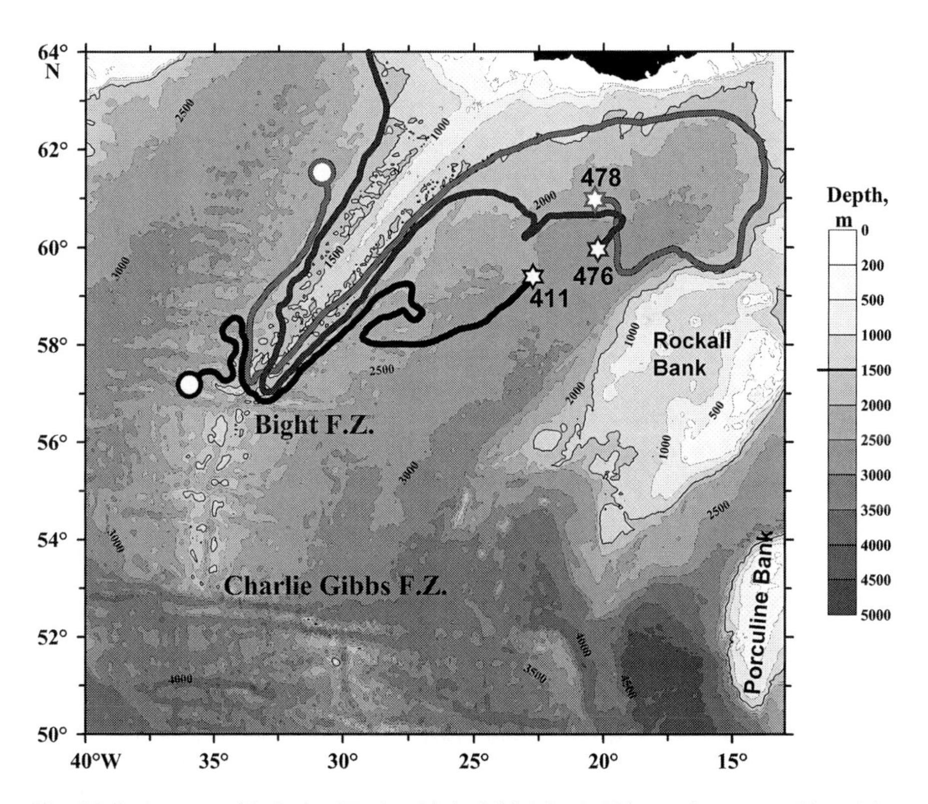

Fig. 6.6 Bathymetry of the Iceland Basin with the Mid-Atlantic Ridge on the western side and the Rockall and Porcupine Banks on the eastern side. Trajectories of three RAFOS floats are shown. The launching position and last fix before surfacing are marked with *asterisks* and *circles*, respectively. Contour lines of 1,000, 2,000, and 3,000 m are shown. *BFZ* Bight Fracture Zone, *CGFZ* Charlie Gibbs Fracture Zone

The flow through the Charlie Gibbs fracture Zone in the depth range of 1,500–1,600 m was also studied using RAFOS floats 463, 469, and 474 that were deployed at a depth of 1,530–1,560 m in August 1998. All of them passed the ridge in the western direction and then returned to the Iceland Basin through the northern channel of the Charlie Gibbs Fracture zone with the Labrador Sea Water flow. The floats popped up in February 2000, November 1999, and August 2000, respectively. Trajectories of the floats are shown in Fig. 6.5. A compilation of all relevant North Atlantic RAFOS float data from IFM-GEOMAR can be found on the internet (http://www.ifm-geomar.de/index.php?id=999&L=1).

Other fractures in the Reykjanes Ridge also provide water transport. The Bight Fracture Zone is likely the second most important channel after the Charlie Gibbs Fracture Zone. The Charlie Gibbs Fracture Zone undoubtedly plays the leading role for exchange of overflow waters between the eastern and western rims of the Mid-Atlantic Ridge. Nevertheless, Lagrangian observations by miscellaneous floats have demonstrated the importance of several other (secondary) channels. This category

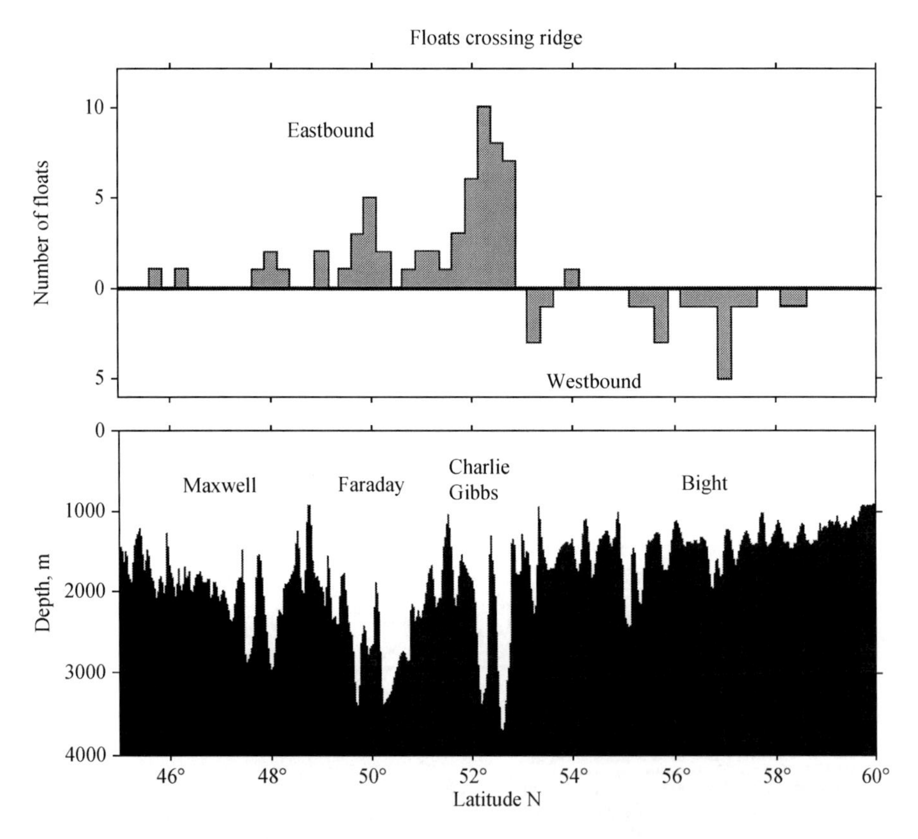

Fig. 6.7 *Upper panel*: number of floats crossing the Reykjanes Ridge in the western and eastern direction. *Lower panel*: Depth of the ridge crest as a function of latitude. (Reproduced from Bower et al. (2002) with the permission from the Nature Publishing Group)

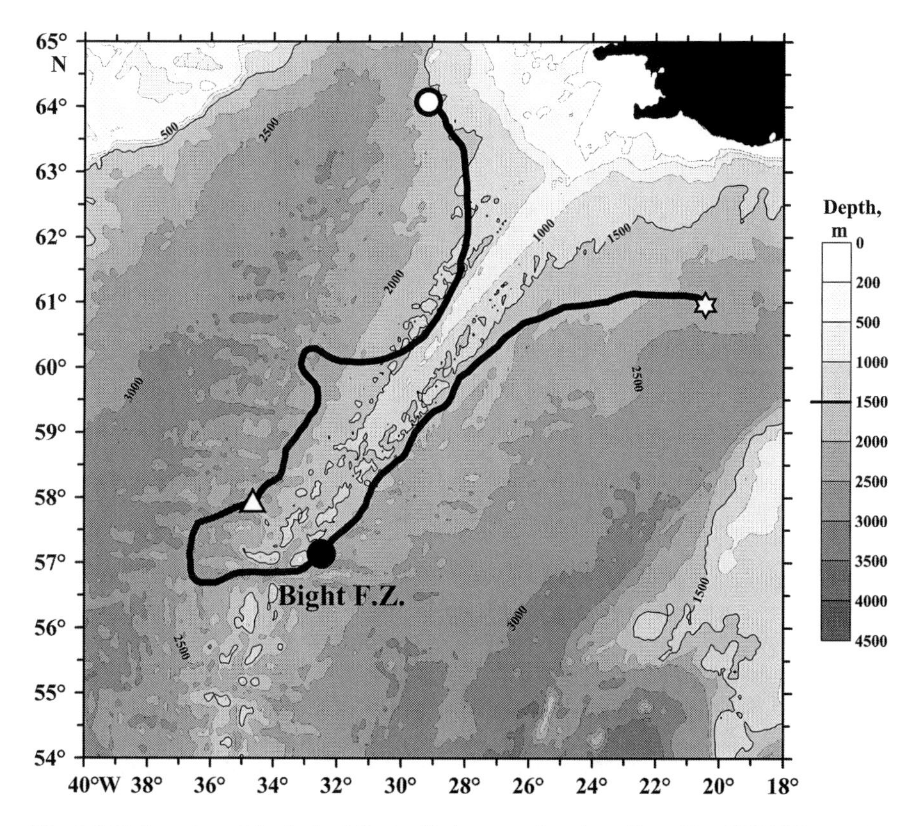

Fig. 6.8 Bathymetry of the Iceland Basin. Trajectory of an Argo float (No. 6900186) crossing the Reykjanes Ridge through the Bight Fracture Zone (57° N, 34° W) is shown. The launching position is marked with an *asterisk*. The last fix is marked with an *open circle*. The *white triangle* and *black circle* denote positions of selected profiles. (Modified and redrawn from Lankhorst and Zenk (2006))

includes the Faraday, Maxwell, and Bight fracture zones, as was shown in Bower et al. (2002) and Lankhorst and Zenk (2006).

The drifts of RAFOS floats through the Bight Fracture Zone are shown in Fig. 6.6. We compiled an ensemble of three eddy resolving RAFOS trajectories (with IfM tracking numbers 411, 476, 478) that all cross the Mid-Atlantic Ridge at ~57° N through the Bight Fracture Zone. The instruments were launched in the northern part of the Iceland Basin (star symbols). They were ballasted for a nominal pressure of 1,500 dbar, corresponding to a depth of ~1,500 m depth. Return positions are shown as circles. The eastern slope of the Mid-Atlantic Ridge is marked by the high speed path of Iceland Scotland Overflow Water (Lankhorst and Zenk 2006). On the western side, all floats reduce their drift speed on a northeasterly course. The image emphasizes the importance of this additional pathway through the ridge. In the older literature, this pathway was assumed as impermeable for Iceland Scotland Overflow Water north of the Charlie Gibbs Fracture Zone at ~52° N.

Fig. 6.9 Potential temperature vs. salinity sampled at two cycles between 2,000 m and the surface on each site of the Reykjanes Ridge shown with *triangles* (west of the ridge) and *circles* (east of the ridge). (Modified and redrawn from Lankhorst and Zenk (2006))

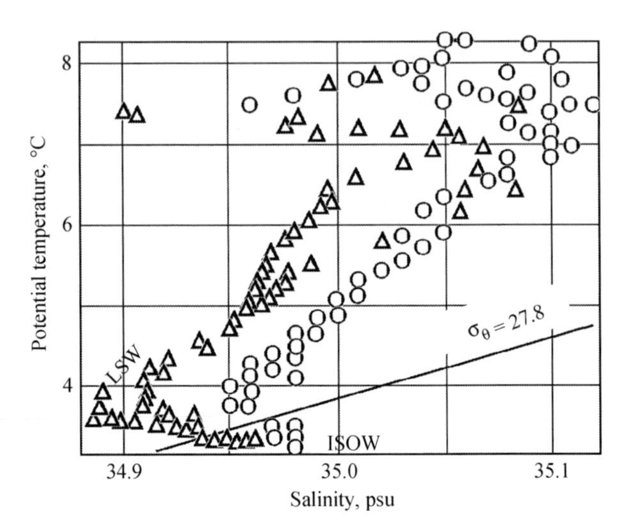

A diagram adopted from Bower et al. (2002) shows statistics of floats crossing the Reykjanes Ridge in both directions and depth of the ridge crest as a function of latitude (Fig. 6.7). Floats transported by the North Atlantic Current at the upper level crossed the ridge eastbound preferentially over the Charlie Gibbs fracture zone (31 of the total 61), the Faraday fracture zone (11/61), and, to a lesser extent, the Maxwell fracture zone (4/61) (float depths 200–800 m). The distribution of float deployment latitudes is significantly different from the distribution of crossing latitudes. Floats crossed the Reykjanes Ridge from east to west north of 53° N mainly through the Bight Fracture Zone and two unnamed gaps near 53.5° N and 55° N. North of 53° N, the gaps are not aligned east–west and the distribution of float crossing latitudes is therefore somewhat wider around these gaps.

Figure 6.8 displays the drift of an Argo float over almost 3 years from August 2002 to May 2005 at a density level $\sigma_\theta \sim 27.8$ corresponding to a nominal depth of 1,500 m. The instrument pathway is strongly guided by the bottom topography. The instrument was launched in the Iceland Basin. Then, it crossed the Mid-Atlantic Ridge at the latitude of Bight Fracture Zone (57° N) and reached the Irminger Sea, where it headed toward the Denmark Strait.

Additional potential temperature and salinity data (ARGO floats) were sampled at two cycles between 2,000 m and the surface on each site of the Mid-Atlantic Ridge. The potential temperature–salinity diagram of these four profiles is shown in Fig. 6.9. Note how the profiles converge at low temperatures (ISOW). The position of Labrador Sea Water (salinity minimum) and $\sigma_\theta = 27.80$ is also indicated in the figure. They demonstrate the mixing scheme between the incoming Labrador Sea Water from the west and the transgressing ISOW from the northeast resulting beneath the σ_θ surface approximately equal to 27.8. White triangles and circles denote positions of four selected profiles (two on either side of the ridge labeled by circles and triangles).

Integrated Conclusions

1. We compiled most of all available data on the flows in the main abyssal channels of the Atlantic Ocean and water masses transported by these flows. Transport of the bottom waters of the Antarctic origin is the main objective of our study. We emphasize that two principally different approaches to classification of these waters are used in literature. In the first approach, the term Antarctic Bottom Water unites all bottom waters in the Atlantic, which have Antarctic rather than North Atlantic origin (concept developed by Wüst (1936)). The second approach divides these waters into the waters of circumpolar and Antarctic origin (Reid et al. 1977). In this case, the term Antarctic Bottom Water unites all Antarctic waters in the Southern Ocean with density high enough not to overflow the threshold in the Drake Passage (approach developed in (Orsi et al. 1999)). Each of the approaches uses different characteristics or the upper boundaries of bottom water. Different approaches result in the fact that transports, and spreading regions for bottom water, differ significantly. Thus, ambiguity in classification, different (but sometimes the same) terminology, and different boundaries complicates comparison of quantitative characteristics of waters in publications by different authors.
2. As a test case we studied a strong flow of Antarctic Bottom Water (in the Wüst sense) through the Vema Channel from the Argentine to Brazil Basin on the basis of measurements during numerous visits to this study region. This channel is the only passage for Weddell Sea Deep Water which is the coldest and densest part of Antarctic Bottom Water. Part of Antarctic Bottom Water of lesser density propagates also through the Hunter Channel and over the Santos Plateau.
3. We studied the properties of bottom water along its pathway of northward propagation in the Atlantic Ocean. After spreading in the Brazil Basin, Antarctic Bottom Water propagates into the North Atlantic. Antarctic Bottom Water flows to the Eastern Basin of the North Atlantic through the Vema Fracture Zone (at 11° N), Romanche and Chain fracture zones (at the equator).
4. We studied variations in the properties of Antarctic Bottom Water in the course of its propagation on the basis of historical and recent measurements.

E. G. Morozov et al., *Abyssal Channels in the Atlantic Ocean*,
DOI 10.1007/978-90-481-9358-5, © Springer Science+Business Media B.V. 2010

5. Analysis of historical data and recent studies in the Scotia Sea and Drake Passage shows that the ridges of the Hero and Shackleton fracture zones strongly restrict spreading of Weddell Sea Deep Water and Circumpolar Bottom Water. These ridges are responsible for several semi-enclosed circulations in the adjacent basins. A new pathway for Weddell Sea Deep Water transit transport was found in the western part of the Scotia Sea. The ridge in the Shackleton Fracture Zone declines a part of the western flow of Weddell Sea Deep Water to the northeast and then the water merges with the easterly flow in the southern Antarctic Circumpolar Current branch. The total transport of Circumpolar Bottom Water between the Antarctica and South America is close to zero.
6. According to the moored measurements (two moorings), the mean transport of Antarctic Bottom Water (layer below 2 °C isotherm) through the Vema Channel is estimated at 3.5 Sv. The greatest velocities reach 60 cm s^{-1}. However, the instantaneous transport measured by LADCP instruments (five profiles) appears lower and fluctuates between 2.5 and 3.5 Sv. Usually, the jet core is vertically mixed in a layer approximately 150 m thick. Owing to the Ekman friction and veering, the coldest core of the flow in the Vema Channel is usually displaced to the eastern slope of the channel.
7. We studied the time variation of the bottom flow in the Vema Channel across the standard section at 31°14′ S. During the period from 1979 to 2003, a temperature increase was observed in the coldest jet in the Vema Channel. The temperature increased from −0.18 to −0.12°C. In the end of 2004, this warming stopped and temperature fluctuations with an amplitude of 0.02°C were observed. Thus, we observed a general trend of warming of Weddell Sea Deep Water with slight fluctuations over a period greater than 30 years.
8. We suggest a concept explaining warming of the jet in the Vema Channel and associate it with the processes in the Weddell Sea. Variable wind stress over the Weddell Gyre leads to trapping of the coldest water south of the South Scotia Rise and periodical release of this trapping. This process is responsible for temperature fluctuations of Weddell Sea Deep Water volumes in the Argentine Basin, which later flow through the Vema Channel and others.
9. We studied variability of the flow along the Vema Channel. The temperature in the jet increases, while the flow propagates from south to north in the Vema Channel due to mixing with the overlying North Atlantic Deep Water.
10. A cold jet of Weddell Sea Deep Water with a potential temperature of −0.1°C was observed above the western slope of the channel at a depth of 4,200 m. Actually this flow was observed at the topography break and can be characterized as "coastal" intensification. A southerly countercurrent was found over the eastern slope of the Vema Channel. This countercurrent transports part of Antarctic Bottom Water back in the southern direction.
11. The Antarctic Bottom Water ($\theta < 2$°C) flow through the Vema Fracture Zone (11° N) is estimated at 0.5 Sv. The mean velocity is 10 cm s^{-1}, while the greatest velocity reaches 30 cm s^{-1}. The Vema Fracture Zone is the main pathway for Antarctic Bottom Water to the Northeast Atlantic.

12. The Antarctic Bottom Water ($\theta < 2°C$) flow through the Romanche and Chain fracture zones is estimated at 0.5–0.7 Sv in each channel. The mean velocities are 10–20 cm s^{-1}. Velocities measured by current meters on moorings in 1991–1992 and using LADCP are very close. The bottom water passing through the Romanche and Chain fracture zones spreads only to the southeastern and equatorial parts of the Atlantic. Its further propagation to the north is almost limited by the Kane Gap. Strong mixing at the Mid-Atlantic Ridge in the equatorial zone induced by internal tides causes a rapid temperature increase of bottom waters.
13. A cataract was found in the eastern part of the Chain Fracture Zone. Bottom water descends from a sill at 4,000 m down to 4,500 m. The flow splits into a dense water downstream along the bottom and a mixed layer flow continuation at a depth of 4,000 m. This phenomenon is similar to laboratory experiments.
14. We studied the flow in the Kane Gap. A southerly flow with velocity of 10 cm s^{-1} and transport of 0.15 Sv was observed in the Kane Gap in May of 2009; a similar northerly flow was observed in October 2009 indicating that the direction of the flow in the Kane Gap is alternating.
15. We explained why the Vema Fracture Zone is the main pathway for the further propagation of bottom waters to the Northeast Atlantic. This fact is determined by a stronger mixing in the region of the Romanche and Chain fracture zones due to a stronger generation of internal tides.
16. Data analysis of the flows across the Mid-Atlantic Ridge in the North Atlantic demonstrates that the Charlie Gibbs Fracture Zone and also the Bight Fracture Zone are important channels for bottom water flow.

It was the intention of our contribution to provide an overview of the widely spread information on the structure of deep and bottom currents in major abyssal channels of the Atlantic Ocean. We have included new ship-borne and mooring observations collected in the recent past. The presented data sets shed new light on the representativeness of fixed-point stations and repeat sections. We showed numerous examples for natural changes in the long-term behavior of physical parameters in the abyssal. Our compilation may help to facilitate an advanced merging of observational and modeled information. Such a blend can improve the ability of climate prediction models by taking measurement errors and an updated statistics into account. During and after the field phase of the World Ocean Circulation Experiment (WOCE), the concept of a mean oceanographic state was ruled out step by step by a scenario of rich spatial and temporal variability (Smith 2001). We therefore plead not to interrupt existing long-term time series in the abyssal Atlantic Ocean, but to transfer them into a global network of sustained observations. Such observations provide a unique view of inter-annual variability of water mass properties and transports, enabling a better understanding of the deep circulation and improved oceanic prediction skills.

References

Ambar I (1983) A shallow core of Mediterranean Water off western Portugal. Deep Sea Res 30:677–680

Andrie C, Ternon JF, Messias MJ, Memery L, Bourlès B (1998) Chlorofluormethane distribution in the deep Equatorial Atlantic during January–March 1993. Deep Sea Res I 45:903–930

Arhan M, Mercier H, Bourlès B, Gouriou Y (1998) Hydrographic section across the Atlantic at 7°30′ N and 4°30′ S. Deep Sea Res I 45:829–872

Arhan M, Heywood KJ, King BA (1999) The deep waters from the Southern Ocean at the entry to the Argentine Basin. Deep Sea Res II 46:475–499

Arhan M, Carton X, Piola A, Zenk W (2002a) Deep lenses of circumpolar water in the Argentine Basin. J Geophys Res 107(C1):101029–101040

Arhan M, Naveira Garabato AC, Heywood KJ, Stevens DP (2002b) The Antarctic Circumpolar Current between the Falkland Islands and South Georgia. J Phys Oceanogr 32(6):1914–1931

Arhan M, Mercier H, Park Y-H (2003) On the deep water circulation of the eastern South Atlantic Ocean. Deep Sea Res I 50:889–916

Armi L (1978) Some evidence for boundary mixing in the deep ocean. J Geophys Res 83:1971–1979

Armi L, D'Asaro E (1980) Flow structures of the benthic ocean. J Geophys Res 85:469–484

Bacon S, Gould WJ, Jia Y (2003) Open-ocean convection in the Irminger Sea. Geophys Res Lett 30(5):1246. doi:10.1029/2002GL016271

Bader R, Gerard RD et al (1970) Initial reports of the deep sea drilling project, vol 4. U.S. Government Printing Office, Washington, DC, pp 77–84

Baines PG, Condie S (1998) Observation and modelling of Antarctic downslope flows: a review, ocean, ice, and atmosphere: interactions at the Antarctic continental margin. Antarct Res Ser 75:29–49

Barnier B, Marchesiello P, De Miranda AP, Molines JM, Coulibaly M (1998) A sigma-coordinate primitive equation model for studying the circulation in the South Atlantic. Part I: Model configuration with error estimates. Deep Sea Res I 45:543–572

E. G. Morozov et al., *Abyssal Channels in the Atlantic Ocean,*
DOI 10.1007/978-90-481-9358-5,

Barre N, Provost C, Sennechael N, Lee JH (2008) Circulation in the Ona Basin, southern Drake Passage. J Geophys Res 113:C04033. doi:10.1029/2007JC004549

Baum SK (2004) Glossary of physical oceanography and related disciplines. http://stommel.tamu.edu/~baum/paleo/ocean/ocean.html

Belderson RH, Jones FJW, Gorini MA, Kenyon NH (1984) A long-range sidescan sonar (Gloria) survey of the Romanche active transform in the Equatorial Atlantic. Mar Geol 56:65–78

Biscaye PE, Eittreim SL (1977) Suspended particulate loads and transports in the nepheloid layer of the abyssal Atlantic Ocean. Mar Geol 23:155–172

Blazhchishin AI, Lukashina NP (1977) Stratification of bottom sediments in the Charlie Gibbs Fracture Zone. In: Study of the open part of the Atlantic Ocean, GO USSR, Leningrad, pp 126–131 [in Russian]

Bonatti E (1973) Origin of offsets of the Mid-Atlantic Ridge in structure zones. J Geol 81:144–156

Bonatti E (1976) Serpentinite protrusions in the oceanic crust, Earth Planet. Sci Lett 32:107–113

Bonatti E (1996) Origin of the large fracture zones offsetting the Mid-Atlantic Ridge. Geotectonics 6:430–440

Bonatti E, Chermak A (1981) Formerly emerging crustal blocks in the Equatorial Atlantic. Tectonophysics 72:165–180

Bonatti E, Fisher D (1971) Oceanic basalts: chemistry versus distance from oceanic ridge, Earth, Planet. Sci Lett 11:307–311

Bonatti E, Honnorez J (1976) Sections of the Earth's crust in the Equatorial Atlantic. J Geophys Res 81(23):4104–4116

Bonatti E, Sarnthein M, Boersma A, Gorini M, Honnorez J (1977) Neogene crustal emersion and subsidence at the Romanche Fracture Zone, Equatorial Atlantic, Earth, Planet. Sci Lett 35:369–383

Bonatti E, Daniele D, Peyve A (1992) Peridotites from the Chain fracture zone in the Equatorial Atlantic: a preliminary report. Acta Vulcanologica 2:65–71

Bonatti E, Ligi M, Gasperini L, Peyve A, Raznitsin YU, Chen YJ (1994) Transform migration and vertical tectonics at the Romanche Fracture Zone, Equatorial Atlantic. J Geophys Res 99(11):21779–21802

Bonatti E, Brunelli D, Fabretti P, Ligi M, Portaro RA, Seyler M (2001) Steady state creation of crust free lithosphere at cold spots in mid ocean ridges. Geology 29(11):979–982

Bonatti E, Brunelli D, Buck WR, Cipriani A, Fabretti P, Ferrante V (2005) Flexural uplift of a lithospheric slab near the Vema transform (central Atlantic): timing and mechanisms, Earth and Planet. Sci Lett 240:642–655

Bower AS, le Cann B, Rossby T, Zenk W, Gould J, Speer K, Richardson PL, Prater MD, Zhang H-M (2002) Directly measured mid-depth circulation in the northeastern north Atlantic Ocean. Nature 419:603–607

Brennecke WS (1921) Die ozeanographischen Arbeiten der Deutchen Antarktischen Expedition, 1911–1912. Arch Dtsch Seewarte 39(1):1–216

Broecker WS (1991) The great ocean conveyor. Oceanography 4(2):79–89

Broecker WS, Takahashi T, Li YH (1976) Hydrography of the central Atlantic (I). The two-degree discontinuity. Deep Sea Res 23:1083–1104
Broecker WS, Peacock SL, Walker S, Weiss R, Fahrbach E, Schroeder M, Mikolajewicz U, Heinze C, Key R, Peng T-H, Rubin S (1998) How much deep water is formed in the Southern Ocean? J Geophys Res 103(C8):15833–15843
Bulychev AA, Gilod DA, Kulikov EY, Schereider AA (1997) A method for determining magnetization in sediment layers, Vestnik of the Moscow State University, Series 4. Geology 5:59–67
Bulychev AA, Gilod DA, Kulikov EY, Schreider AA (2004) First data on the geochronology of the oceanic lithosphere in the region of the Romanche Fracture Zone (Equatorial Atlantic). Oceanology 44(4):574–578
Cai W, Greatbatch R (1995) Compensation for the NADW outflow in a global ocean general circulation model. J Phys Oceanogr 25(2):226–241
Cande S, Labreque J, Haxby W (1988) Plate kinematics of the South Atlantic: chron C34 to present. J Geophys Res 93:13479–13492
Canuto VM, Howard A, Cheng Y, Dubovikov MS (2001) Ocean turbulence. Part I: one-point closure model – momentum and heat vertical diffusivities. J Phys Oceanogr 31:1413–1426
Carmack EC (1977) Water characteristics of the Southern Ocean south of the polar front. In: Deacon G, Angel M (eds) A voyage of discovery, 70th Anniversary Volume. Pergamon Press, Oxford, pp 15–41
Carmack EC, Foster TD (1975) On the flow of water out of the Weddell Sea. Deep Sea Res 22:711–724
Cherkis N, Fleming N, Massingill J (1973) Is the Gibbs Fracture Zone a westward projection of the Hercynian front into North America? Nat Phys Sci 245:113–115
Clarke RA, Gascard JC (1983) The formation of Labrador Sea Water, Part I: Large-scale processes. J Phys Oceanogr 13(10):1764–1778
Cochran JR (1973) Gravity and magnetic investigations in the Guiana Basin, western Equatorial Atlantic. Geol Soc Am Bull 84:3249–3268
Coles VJ, McCartney MS, Olson BD, Smethie WJ Jr (1996) Changes in Antarctic Bottom Water properties in the western South Atlantic in the late 1980s. J Geophys Res 101(C4):8957–8970
Comiso JC (2000) Variability and trends in Antarctic surface temperatures from in situ and satellite inferred measurements. J Clim 13:1674–1696
Comiso JC, Gordon AL (1998) Interannual variability in summer sea ice minimum, coastal polynyas and bottom water formation in the Weddell Sea. In: Jeffries MO (ed) Antarctic sea ice: physical processes, interactions and variability, Antarctic res. Ser., vol 74. AGU Publishers, Washington, DC, pp 293–315
Connary SD, Ewing M (1974) Penetration of Antarctic Bottom Water from the Cape Basin into the Angola Basin. J Geophys Res 79:463–469
Davis RE (1998) The mid-depth circulation in the tropical and South Pacific from ALACE floats, U.S. WOCE Implementation Report No 10, pp 18–21
Deacon GER (1937) The hydrology of the Southern Ocean. Discov Rep 15:1–124

De Madron XD, Weatherly G (1994) Circulation, transport and bottom boundary layers of the deep currents in the Brazil Basin. J Mar Res 52:583–638
Demidov AN (2003) Distinguishing the intermediate and deep water masses in the South Atlantic. Oceanology 43:153–163
Demidov AN, Morozov EG, Neiman VG (2006) Structure and variability of deep waters in the Romanche Fracture Zone. Dokl Earth Sci 410:1136–1140
Demidov AN, Dobrolyubov SA, Morozov EG, Tarakanov RY (2007a) Transport of bottom waters through the Vema Fracture Zone in the Mid-Atlantic Ridge. Dokl Earth Sci 416:1120–1124
Demidov AN, Tarakanov RY, Morozov EG, Gangnus IA (2007b) Water masses of the oceans and seas. Maks-Press, Moscow, pp 55–91
Denker C (2007) Schwankungen von Wassermasseneigenschaften an der Schwelle des Vema-Kanals. Diploma Thesis, Christian-Albrechts-Universitaet, Kiel, p 63
Dickson RR, Brown J (1994) The production of North Atlantic Deep Water: sources, rates, and pathways. J Geophys Res 99:12319
Dickson RR, Gurbutt PA, Medler KJ (1980) Long-term water movements in the southern trough of the Charlie-Gibbs Fracture Zone. J Mar Res 18(3):571–583
Dickson RR, Gmitrowicz EM, Watson AJ (1990) Deep-water renewal in the northern North Atlantic. Nature 344:848–850
Dobrolyubov SA, Lappo SS, Morozov EG, Sokov AV, Tereschenkov VP, Shapovalov SM (2002) Water structure in the North Atlantic in 2001 based on a transatlantic section along 53° N. Dokl Earth Sci 382(4):120–123
Dobrolyubov SA, Lappo SS, Morozov EG, Pisarev SV, Sokov AV (2003) Variability of water masses in the North Atlantic based on hydrographic sections along 60° N. Dokl Earth Sci 390(2):566–570
Dobrovolsky AD, Zalogin BS (1992) Regional oceanography. MGU Publishers, Moscow
Dubinin EP (1987) Transform fractures of the oceanic lithosphere. MGU Publishers, Moscow [in Russian]
Dubinin EP, Ushakov SA (2001) Oceanic riftogenesis. GEOS, Moscow [in Russian]
Eittreim S, Ewing J (1978) Vema Fracture Zone transform fault. Geology 3:555–558
Eittreim SL, Biscaye PE, Jacobs SS (1983) Bottom-water observations in the Vema Fracture Zone. J Geophys Res 88:2609–2614
Emelyanov E (2008) Sedimentation and near-bottom currents in the south-western Atlantic. Geologija 50(4):275–289
Emery K, Uchupi E, Phillips J, Bowin C, Mascle J (1975) Continental margin off Western Africa: Angola to Sierra Leone. Am Assoc Pet Geol Bull 59(12):2209–2265
Fahrbach E, Rohardt G, Schröder J, Strass V (1994) Transport and structure of the Weddell Gyre. Ann Geophysicae 12:840–855
Fahrbach E, Rohardt G, Scheele N, Schroeder M, Strass V, Wisotzki A (1995) Formation and discharge of deep and bottom water in the northwestern Weddell Sea. J Mar Res 53:515–538

Fahrbach E, Meyer R, Rohardt G, Schroder M, Woodgate RA (1998) Gradual warming of the Weddell Sea deep and bottom water. In: Oerter H (ed) Filchner – Ronne Ice Shelf Programme (FRISP). Alfred-Wegener-Institut, Bremerhaven, Report No. 12, pp 24–34

Fahrbach E, Harms S, Rohardt G, Schroder M, Woodgate R (2001) Flow of bottom water in the northwestern Weddell Sea. J Geophys Res 106:2761–2778

Fahrbach E, Hoppema M, Rohardt G, Schroder M, Wisotzki A (2004) Decadal-scale variations of water mass properties in the deep Weddell Sea. Ocean Dyn 54:77–91. doi:10.1007/s10236-003-0082-3

Ferron B, Mercier H, Speer K, Gargett A, Polzin K (1998) Mixing in the Romanche Fracture Zone. J Phys Oceanogr 28:1929–1945

Fischer J, Rhein M, Schott F, Stramma L (1996) Deep water masses and transports in the Vema Fracture Zone. Deep Sea Res I 43(7):1067–1074

Fofonoff NP (1956) Some properties of sea water influencing the formation of Antarctic Bottom Water. Deep Sea Res 4:32–35

Fogelqvist E, Blindheim J, Tanhua T, Osterhus S, Buch E, Rey F (2003) Greenland-Scotland overflow studied by hydro-chemical multivariate analysis. Deep Sea Res I 50:73–102

Foldvik A, Gammelsrød T, Tørresen T (1985) Circulation and water masses on the southern Weddell Sea shelf. In: Jacobs S (ed) Oceanology of the Antarctic Continental shelf, vol 43. Antarctic Res Ser. AGU Publishers, Washington, DC, pp 5–20

Foldvik A, Gammelsrød T, Østerhus S, Fahrbach E, Rohardt G, Schröder M, Nicholls KW, Padman L, Woodgate RA (2004) Ice shelf water overflow and bottom water formation in the southern Weddell Sea. J Geophys Res 109:C02015. doi:10.1029/2003JC002008

Foster TD (1995) Abyssal water mass formation off the eastern Wilkes Island coast of Antarctica. Deep Sea Res 42:501–522

Foster TD, Carmack EC (1976) Frontal zone mixing and Antarctic Bottom Water formation in the southern Weddell Sea. Deep Sea Res 23:301–317

Friedrichs MA, Hall MM (1993) Deep circulation in the tropical North Atlantic. J Mar Res 51(4):697–736

Friedrichs MA, McCartney MS, Hall MM (1994) Hemispheric asymmetry of deep water transport modes in the western Atlantic. J Geophys Res 99(C12):25165–25179

Fu LL (1981) The general circulation and meridional heat transport of the Subtropical South Atlantic determined by inverse methods. J Phys Oceanogr 11(9):1171–1193

Fuglister FC (1960) Atlantic Ocean atlas of temperature and salinity profiles and data from the international geophysical year of 1957–1958. Woods Hole Oceanographic Institution, Atlas Series 1, Woods Hole, MA, p 209

Fukasawa M, Freeland H, Perkin R, Watanabe T, Uchida H, Nishina A (2004) Bottom water warming in the North Pacific Ocean. Nature 427:825–827

Ganachaud A (2003) Large-scale mass transports, water mass formation, and diffusivities estimated from World Ocean Circulation Experiment (WOCE) hydrographic data. J Geophys Res 108(C7):3213. doi:10.1029/2002JC001565
Gargett AE (1984) Vertical eddy diffusivity in the ocean interior. J Mar Res 42:359–393
Gasperini L, Bonatti E, Ligi M, Sartori R, Borsetti A, Negri A, Ferrari A, Sokolov S (1997) Stratigraphic numerical modelling of a carbonate platform on the Romanche Transverse Ridge, Equatorial Atlantic. Mar Geol 136:245–257
Georgi DT (1981) On the relationship between the large-scale property variations and fine structure in the circumpolar deep water. J Geophys Res 86(C7):6556–6566
Gill A (1973) Circulation and bottom water production in the Weddell Sea. Deep Sea Res 20:111–140
Girton JB, Sanford TB, Käse R (2000) Synoptic sections of the Denmark Strait overflow. Geophys Res Lett 28(8):1619–1622
Golivets SV, Koshkyakov MN (2004) Eddy formation at the Subantarctic front based on the data of satellite observations and formation of Antarctic Intermediate Water. Oceanology 44(4):485–494
Gordon AL (1966) Potential temperature, oxygen and circulation of bottom water in the Southern Ocean. Deep Sea Res 13:1125–1138
Gordon AL (1967) Structure of Antarctic waters between 20° W and 170° W. In: Bushell VC (ed) Antarctic Map Folio Series (folio 6). American Geographic Society, New York
Gordon AL (1971) Oceanography of Antarctic waters. In: Reid JL (ed) Antarctic oceanography, vol 15. Antarctic Research Series. American Geophysical Union, Washington, DC, pp 169–203
Gordon AL (1972) Spreading of Antarctic Bottom Waters II. Studies in physical oceanography – a tribute of George Wüst on his 80th birthday, vol 2. Gordon and Breach, New York, pp 1–17
Gordon AL (1978) Deep Antarctic convection west of Maud Rise. J Phys Oceanogr 8:600–612
Gordon AL (1998) Western Weddell Sea thermohaline stratification. In: Jacobs SS, Weiss RF (eds) Ocean, ice, and atmosphere: interactions at the Antarctic continental margin. Antarctic Research Series, vol 75. American Geophysical Union, Washington, DC, pp 215–240
Gordon AL, Huber BA, Hellmer H, Ffield A (1993) Deep and bottom water of the Weddell Sea's western rim. Science 262:95–97
Gordon AL, Mensch M, Dong Z, Smethie WM, de Bettencourt J (2000) Deep and bottom water of the Bransfield Strait eastern and central basins. J Geophys Res 105(C5):11337–11346
Gordon AL, Visbeck M, Huber B (2001) Export of Weddell Sea deep and bottom water. J Geophys Res 106(C5):9005–9017
Gouretski V, Koltermann KP (2004) WOCE global hydrographic climatology. Ber Bundesamtes Seeschiffahrt Hydrogr 35:1–52

Gradstein F, Ogg J, Smith A (2006) A geologic time scale 2004. Cambridge University Press, Cambridge

Haine TWN, Watson AJ, Liddicoat MI, Dickson RR (1998) The flow of Antarctic Bottom Water to the southwest Indian Ocean estimated using CFCs. J Geophys Res 103(C12):27637–27653

Hall MM, McCartney MS, Whitehead JA (1997) Antarctic Bottom Water flux in the Equatorial Western Atlantic. J Phys Oceanogr 27(9):1903–1926

Hall MM, Joyce TM, Pickart RS, Smethie WM, Torres DJ (2004) Zonal circulation across 52° W in the North Atlantic. J Geophys Res 109:C11008. doi:10.1029/2003JC002103

Harms S, Fahrbach E, Strass VH (2001) Sea ice transports in the Weddell Sea. J Geophys Res 106(C5):9057–9073

Harvey JG (1980) Deep and bottom water in the Charlie-Gibbs Fracture Zone. J Mar Res 38:172–173

Harvey J, Arhan M (1988) The water masses of the central North Atlantic in 1983–1984. J Phys Oceanogr 18(12):1855–1874

Heezen BC, Gerard RD, Tharp M (1964a) The Vema Fracture Zone in the Equatorial Atlantic. J Geophys Res 69:733–739

Heezen B, Bunce E, Hersey J, Tharp M (1964b) Chain and Romanche Fracture Zones. Deep Sea Res 11:30–33

Hellmer HH, Beckmann A (2001) The Southern Ocean: A ventilation contributor with multiple sources. Geophys Res Lett 28(15):2927–2930

Hiller W, Käse RH (1983) Objective analysis of hydrographic data sets from mesoscale surveys. Berichte Aus Dem Institut fur Meereskunde 116:1–78

Hobart MA, Bunce ET, Sclater JG (1975) Bottom water flow through the Kane Gap, Sierra Leone Rise, Atlantic Ocean. J Geophys Res 80:5083–5088

Hogg NG (1983) Hydraulic control and flow separation in a multi-layered fluid with application to the Vema Channel. J Phys Oceanogr 13:695–708

Hogg NG (2001) Quantification of the deep circulation. In: Siedler G, Church J, Gould J (eds) Ocean circulation and climate. Academic Press, San Diego, CA, pp 259–270

Hogg NG, Owens WB (1999) Direct measurement of the deep circulation within the Brazil Basin. Deep Sea Res II 46:335–353

Hogg NG, Thurnherr M (2005) A zonal pathway for NADW in the South Atlantic. J Oceanogr 61:493–507

Hogg NG, Zenk W (1997) Long-period changes in the bottom water flowing through Vema Channel. J Geophys Res 102(C7):15639–15646

Hogg N, Biscaye P, Gardener W, Schmitz WJ (1982) On the transport and modification of Antarctic Bottom Water in the Vema Channel. J Mar Res 40(Suppl):231–263

Hogg NG, Owens WB, Siedler G, Zenk W (1996) Circulation in the deep Brazil Basin. In: Wefer G, Berger WH, Siedler G, Webb DJ (eds) The South Atlantic: present and past circulation. Springer, Berlin Heidelberg, pp 249–260

Hogg N, Siedler G, Zenk W (1999) Circulation and variability at the southern boundary of the Brazil Basin. J Phys Oceanogr 29:145–157

Holfort J, Siedler G (2001) The meridional oceanic transport of heat and nutrients in the South Atlantic. J Phys Oceanogr 31(1):5–28

Hollister CD and Heezen BC (1967) The floor of the Bellingshausen Sea. In: Hersey JB (ed) Deep-Sea photography, The John Hopkins studies, No. 3, Chap. 17. The John Hopkins Press, Baltimore, MD, pp 177–189

Hoppema M, Klatt O, Roether W, Fahrbach E, Bulsiewicz K, Rodehacke C, Rohardt G (2001) Prominent renewal of Weddell Sea deep water from a remote source. J Mar Res 59:257–279

Iorga M, Lozier MS (1999) Signatures of the Mediterranean outflow from a North Atlantic climatology 2. Diagnostic velocity fields. J Geophys Res 104(26):26011–26029

Jackett DR, McDougall TJ (1997) A neutral density variable for the World's Ocean. J Phys Oceanogr 27(2):237–263

Jacobs SS, Giulivi CF (1998) In: Jacobs SS, Weiss RF (eds) Interannual ocean and sea ice variability in the Ross Sea, in ocean, ice, and atmosphere: interactions at the Antarctic continental margin. Antarctic Research Series, vol 75. AGU Publishers, Washington, DC, pp 135–150

Johnson DA, McDowell SE, Sullivan LG, Biscaye PE (1976) Abyssal topography, nephelometry, currents, and benthic boundary layer structure in the Vema Channel. J Geophys Res 81:5771–5786

Johnson GC (2008) Quantifying Antarctic Bottom Water and North Atlantic deep water volumes. J Geophys Res 113:C05027. doi:10.1029/2007JC004477

Johnson GC, Doney SC (2006) Recent western South Atlantic bottom water warming. Geophys Res Lett 33:L14614. doi:10.1029/2006GL026769

Johnson JC (1998) Deep water properties, velocities, and dynamics over ocean trenches. J Mar Res 56:329–347

Jungclaus J, Vanicek M (1999) Frictionally modified flow in a deep ocean channel: application to the Vema Channel. J Geophys Res 104(C9):21123–21136

Jungclaus JH, Hauser J, Käse RH (2001) Cyclogenesis in the Denmark Strait overflow plume. J Phys Oceanogr 31(11):3214–3229

Kanamori H, Stewart G (1976) Mode of the strain release along the Gibbs Fracture Zone: Mid-Atlantic Ridge, Physics of the Earth Planet. Inter. No 11, pp 312–332

Käse RH, Oschlies A (2000) Flow through Denmark Strait. J Geophys Res 105:28 527–28 546

Kashintsev GL, Schreider AA, Maksimochkin VI, Bulychev AA, Gilod DA (2008) Transtension and alkaline magmatism of the Romanche Fracture Zone. Geotectonics 4:318–323

Kastens K, Bonatti E, Caress D, Carrara G, Dauteuil O, Frueh-Green G, Ligi M (1998) The Vema transverse ridge (central Atlantic). Mar Geophys Res 20:533–556

Keeling RF, Peng T (1995) Transport of heat, CO2, and O2 by the Atlantic's thermohaline circulation. Philos Trans R Soc Lond B 348:133–142

Khain VE (2001) Tectonics of continents and oceans. Nauchny Mir, Moscow

Klatt O, Roether W, Hoppema M, Bulsiewicz K, Fleischmann U, Rodehacke C, Fahrbach E, Weiss RF, Bullister JL (2002) Repeated CFC sections at the Greenwich Meridian in the Weddell Sea. J Geophys Res 107(C4):3030. doi:10.1029/2000JC000731

Klatt O, Fahrbach E, Hoppema M, Rohardt G (2005) The transport of the Weddell Gyre across the Prime Meridian. Deep Sea Res II 52:513–528

Klein B, Molinari RL, Muller TJ, Seidler G (1995) A transatlantic section at 14.5°N: meridional volume and heat fluxes. J Mar Res 53:929–957

Koltermann KP, Sokov AV, Tereschenkov VP, Dobroliubov SA, Lorbacher K, Sy A (1999) Decadal changes in the thermohaline circulation of the North Atlantic. Deep Sea Res II 46:109–138

Koshlyakov MN, Sazhina TG (1995) Meridional water and heat transport by large-scale geostrophic currents in the Pacific sector of Antarctica. Oceanology 35(6):842–853

Koshlyakov MN, Tarakanov RY (1999) Water masses of the Pacific Antarctic. Oceanology 39(1):5–15

Koshlyakov MN, Tarakanov RY (2003a) Antarctic Bottom Water in the Pacific sector of the Southern Ocean. Oceanology 43(1):5–20

Koshlyakov MN, Tarakanov RY (2003b) Antarctic Circumpolar Water in the southern part of the Pacific Ocean. Oceanology 43(5):607–621

Koshlyakov MN, Tarakanov RY (2004) Pacific Deep Water in the Southern Ocean. Oceanology 44(3):325–340

Koshlyakov MN, Lisina II, Morozov EG, Tarakanov RY (2007) Absolute geostrophic currents in the Drake Passage based on observations in 2003 and 2005. Oceanology 47(4):451–463

Kuksa VI (1983) Intermediate waters of the World Ocean. Gidrometeoizdat, Leningrad [in Russian]

Langseth M, Hobart M (1976) Interpretation of heat flow measurements in the Vema Fracture Zone. Geophys Res Lett 3(5):241–249

Lankhorst M, Zenk W (2006) Lagrangian observations of the middepth and deep velocity fields of the northeastern Atlantic Ocean. J Phys Oceanogr 36:43–63

Lappo SS (1984) On the origin of the northward cross-equatorial heat advection in the Atlantic Ocean. In: Ocean-atmosphere interaction studies. Gidrometeoizdat, Moscow, pp 125–129 [in Russian]; translated in Lappo SS (2006) Clivar Exchanges 11(2):28–10

Larque L, Maamaatuaiahutapu K, Garcon VC (1997) On the intermediate and deep water flow in the South Atlantic Ocean. J Geophys Res 102(C6):12425–12440

Lavin AM, Bryden HL, Parrilla G (2003) Mechanisms of heat, freshwater, oxygen and nutrient transports and budgets at 24.5° N in the subtropical North Atlantic. Deep Sea Res I 50:1099–1128

Lenz B (1997) Bodenwassertransporte über die Rio-Grande-Schwelle. Diplom thesis, Kiel University, p 68. [Available from Library, Christian-Albrechts-Universität, D-24098 Kiel, Germany]

Leontieva VV (1985) Hydrology of trenches in the World Ocean. Nauka, Moscow [in Russian]

Le Pichon X, Ewing M, Truchan M (1971) Sediment transport and distribution in the Argentine Basin, 2, Antarctic bottom current passage into the Brazil Basin. In: Ahrens LH, Press F, Runcorn SK, Urey HC (eds) Physics and chemistry of the Earth, vol 8. Pergamon Press, Oxford, pp 29–48

Lherminier P, Mercier H, Gourcuff C, Alvarez M, Bacon S, Kermabon C (2007) Transports across the 2002 Greenland-Portugal Ovide section and comparison with 1997. J Geophys Res 112:C07003. doi:10.1029/2006JC003716

Lilwall R, Kirk R (1985) Ocean bottom seismograph observations on the Charlie-Gibbs Fracture Zone. Geophys J R Astron Soc 80:195–208

Limeburner R, Whitehead JA, Cenedese C (2005) Variability of Antarctic Bottom Water flow into the North Atlantic. Deep Sea Res II 52:495–512

Locarnini RA, Whitworth T III, Nowlin WD Jr (1993) The importance of the Scotia Sea on the outflow of Weddell Sea Deep Water. J Mar Res 51:135–153

Lonsdale P (1994) Structural geomorphology of the Eltanin fault system and adjacent transform faults of the Pacific-Antarctic plate boundary. Mar Geophys Res 16:105–143

Ludwig WJ, Rabinowitz PD (1980) Structure of the Vema Fracture Zone. Mar Geol 35:99–110

Lukas R, Santiago-Mandujano F, Bingham F, Mantyla A (2001) Cold bottom water events observed in the Hawaii Ocean time-series: implications for vertical mixing. Deep Sea Res I 48:995–1021

Luyten J, McCartney MS, Stommel H, Dickson R, Gmitrowicz E (1993) On the sources of North Atlantic Deep Water. J Phys Oceanogr 23:1885–1892

Macdonald A (1993) Property fluxes at 30° S and their implications for the Pacific-Indian throughflow and the global heat budget. J Geophys Res 98(C4):6851–6868

Macdonald A (1998) The global ocean circulation a hydrographic estimate and regional analysis. Prog Oceanogr 41:281–382

Macrander A, Käse RH, Send U, Valdimarsson H, Jónsson S (2007) Spatial and temporal structure of the Denmark Strait overflow revealed by acoustic observations. Ocean Dyn 57:75–89. doi:10.1007/s10236-007-0101-x

Mamayev OI (1992) Abyssal waters of the World Ocean. IRO Publishers, Moscow [in Russian]

Mantyla AW, Reid JL (1983) Abyssal characteristics of the World Ocean waters. Deep Sea Res 30(8):805–833

Marsh R, de Cuevas BA, Coward AC, Nurser AJG, Josey SA (2005) Water mass transformation in the North Atlantic over 1985–2002 simulated in an eddy-permitting model. Ocean Sci 1:127–144

Martineau DP (1953) The influence of the current systems and lateral mixing upon Antarctic Intermediate Water in the South Atlantic, Ref. No 53–72. Woods Hole Oceanographic Institution, Woods Hole, MA, p 12

Maurer H, Stocks TH (1933) Die Echolotungen des "METEOR". In: Defant A (ed) Wissenschaftliche Ergebnisse, Deutsche Atlantische Expedition auf dem Forschungs und Vermessungsschiff "METEOR" 1925–1927, vol 2. Walter de Gruyter & Co, Berlin

McCartney MS (1977) Mode water. In: Angel M (ed) A voyage of discovery, George Deacon 70th Anniversary volume. Pergamon Press, Oxford, pp 103–119
McCartney MS (1992) Recirculating components to the deep boundary current of the northern North Atlantic. Prog Oceanogr 29:283–382
McCartney MS, Curry RA (1993) Transequatorial flow of Antarctic Bottom Water in the western Atlantic Ocean: abyssal geostrophy at the equator. J Phys Oceanogr 23:1264–1276
McCartney MS, Bennet SL, Woodgate-Jones ME (1991) Eastward flow through the Mid-Atlantic Ridge at 11° N and its influence on the abyss of the eastern basin. J Phys Oceanogr 21(8):1089–1121
McCave IN (1986) Local and global aspects of the bottom nepheloid layers in the world ocean. Neth J Sea Res 20(2/3):167–181
McDonagh EL, Arhan M, Heywood KJ (2002) On the circulation of bottom water in the region of the Vema Channel. Deep Sea Res I 49:1119–1139
Memery L, Arhan M, Alvarez-Salgado XA, Messias M-J, Mercier H, Castro CG, Rios AF (2000) The water masses along the western boundary of the south and Equatorial Atlantic. Prog Oceanogr 47:69–98
Menard H (1966) Fracture zones and offsets of the east Pacific rise. J Geophys Res 71:682–685
Menard H, Chase T (1970) Fracture zones. Sea 4:421–453
Mensch M, Bayer R, Bullister JL, Schlosser P, Weiss RF (1997) The distribution of tritiumand CFCs in the Weddell Sea during the mid-1980s. Prog Oceanogr 38:377–414
Mensch M, Simon A, Bayer R (1998) Tritium and CFC functions for the Weddell Sea. J Geophys Res 103(C8):15923–15937
Mercier H, Morin P (1997) Hydrography of the Romanche and Chain fracture zones. J Geophys Res 102(C5):10373–10389
Mercier H, Speer KG (1998) Transport of bottom water in the Romanche Fracture Zone and the Chain fracture zone. J Phys Oceanogr 28(5):779–790
Mercier H, Weatherly G, Arhan M (2000) Bottom water throughflows at the Rio de Janeiro and Rio Grande fracture zones. Geophys Res Lett 27:1503–1506
Meredith MP, Locarnini RA, van Scoy KA, Watson AJ, Heywood KJ, King BA (2000) On the sources of Weddell Gyre Antarctic Bottom Water. J Geophys Res 105(C1):1093–1104
Meredith MP, Naveira Garabato AC, Stevens DP, Heywood KJ, Sanders RJ (2001) Deep and bottom waters in the eastern Scotia Sea: rapid changes in properties and circulation. J Phys Oceanogr 31(8):2157–2168
Meredith MP, Hughes CW, Foden PR (2003) Downslope convection north of Elephant Island, Antarctica: influence on deep waters and dependence on ENSO. Geophys Res Lett 30(9):1462. doi:10.1029/2003GL017074
Meredith MP, Naveira Garabato AC, Gordon AL, Johnson GC (2008) Evolution of the deep and bottom waters of the Scotia Sea, Southern Ocean, during 1995–2005. J Clim 21:3327–3343

Messias M-J, Andrie C, Memery L, Mercier H (1999) Tracing the North Atlantic Deep Water through the Romanche and Chain fracture zones with chlorofluoromethanes. Deep Sea Res 46:1247–1278

Metcalf WG, Heezen BC, Stalcup MC (1964) The sill depth of the Mid-Atlantic Ridge in the equatorial region. Deep Sea Res 11:1–10

Molinari RL, Johns E, Festa JF (1990) The annual cycle of meridional heat flux in the Atlantic Ocean at 26° N. J Phys Oceanogr 20(3):476–482

Molinari RL, Fine RA, Johns E (1992) The Deep Western Boundary Current in the tropical North Atlantic. Deep Sea Res 39:1967–1984

Morozov EG (1995) Semidiurnal internal wave global field. Deep Sea Res 42(1):135–148

Morozov EG, Demidova TA, Lappo SS, Pisarev SV, Sokov AV (2003) Spreading of Antarctic Bottom Water in the Vema Channel. Dokl Earth Sci 390:593–596

Morozov EG, Demidov AN, Tarakanov RY (2008) Transport of Antarctic waters in the deep channels of the Atlantic Ocean. Dokl Earth Sci 423(8):1286–1289

Morris MY, Hall MM, St. Laurent LC, Hogg NG (2001) Abyssal mixing in the Brazil Basin. J Phys Oceanogr 31:3331–3348

Mosby H (1934) The waters of the Atlantic Antarctic Ocean. In: Scientific Results of the Norwegian Antarctic Expedition 1927–1928, pp 1–131

Muench RD, Gordon AL (1995) Circulation and transport of water along the western Weddell Sea margin. J Geophys Res 100:18503–18515

Munk WH (1966) Abyssal recipes. Deep Sea Res 13:707–730

Munk WH, Wunsch C (1998) Abyssal recipes II: energetics of tidal and wind mixing. Deep Sea Res 45:1977–2010

Naveira Garabato AC, Heywood KJ, Stevens DP (2002a) Modification and pathways of Southern Ocean deep waters in the Scotia Sea. Deep Sea Res 49:681–705

Naveira Garabato AC, McDonagh EL, Stevens DP, Heywood KJ, Sanders RJ (2002b) On the export of Antarctic Bottom Water from the Weddell Sea. Deep Sea Res II 49:4715–4742

Naveira Garabato AC, Stevens DP, Heywood KJ (2003) Water mass conversion, fluxes, and mixing in the Scotia Sea diagnosed by an inverse model. J Phys Oceanogr 33(12):2565–2587

Needham HD, Carré D, Sibuet J-C (1986) Carte Bathymétrique de la Ride de Walvis: Océan Atlantique Sud. Echelle 1/4. 382, 832. IFREMER, Direction de l'Environnement et des Recherches Océaniques, Département Géosciences Marines

Neiman VG, Burkov VA, Shcherbinin AD (1997) Dynamics of the Indian Ocean waters. Nauchnyi Mir, Moscow

New M, Hulme M, Jones P (2000) Representing twentieth-century space-time variability, part II: Development of 1901–1996 monthly grids of terrestrial surface climate. J Clim 13:2217–2238

Nowlin WD, Zenk W (1988) Westward bottom currents along the margin of the south Shetland Island arc. Deep Sea Res I 35:269–301

Oliver JL, LePichon X, Monti S, Sichler B (1974) Charlie-Gibbs Fracture Zone. J Geophys Res 79:2059–2065

Olsen SM, Hansen B, Quadfasel D, Østerhus S (2008) Observed and modelled stability of overflow across the Greenland–Scotland Ridge. Nature 455:519–522. doi:10.1038/nature07302

Onken R (1995) The spreading of Lower Circumpolar Water in the Atlantic Ocean. J Phys Oceanogr 25(12):3051–3063

Orsi AH, Nowlin WD Jr, Whitworth T III (1993) On the circulation and stratification of the Weddell Gyre. Deep Sea Res I 40:169–203

Orsi AH, Whitworth T, Nowlin WD (1995) On the meridional extent and fronts of the Antarctic Circumpolar Current. Deep Sea Res 42(5):641–673

Orsi AH, Johnsson GC, Bullister JL (1999) Circulation, mixing, and production of Antarctic Bottom Water. Prog Oceanogr 43:55–109

Orsi AH, Smethie WM, Bullister JL (2002) On the total input of Antarctic waters to the deep ocean: A preliminary estimate from chlorofluorocarbon measurements. J Geophys Res 107(C8). doi:10.1029/2001JC000976

Ostlund HG, Broecker WS, Spencer D (1987) GEOSECS Atlantic, Pacific and Indian Ocean expeditions. Shore based data and graphics, vol 17. U.S. Govt. Printing Office, Washington, DC

Outdot C, Morin P, Baurand F, Wafar M, Le Corre P (1998) Northern and southern water masses in the Equatorial Atlantic: distribution of nutrients on the WOCE A6 and A7 lines. Deep Sea Res 45:873–902

Park Y-H, Charriaud E, Craneguy P, Kartavtseff A (2001) Fronts, transport, and Weddell Gyre at 30° E between Africa and Antarctica. J Geophys Res 106(C2):2857–2879

Patterson SL, Whitworth T (1990) Physical oceanography. In: Glasby GP (ed) Antarctic sector of the Pacific. Elsevier Oceanographic Series, vol 51, pp 55–93

Pätzold J, Heidland K, Zenk W, Siedler G (1996) On bathymetry of the Hunter Channel. In: Wefer G, Berger WH, Siedler G, Webb DJ (eds) The South Atlantic: present and past circulation. Springer, Berlin, pp 355–361

Pätzold J et al (1999) Report and preliminary results of Meteor-Cruise 41/3, Vityria – Salvador, 18.4.–15.5.1998. Berichte, Fachbereich Geowissenschaften, Universität Bremen, pp 129, 160

Perez FF, Mintrop L, Llinas O, Gonzalez-Davila M, Castro C, Alvarez M, Kortzinger A, Santana-Casiano M, Rueda M, Ros A (2001) Mixing analysis of nutrients, oxygen and inorganic carbon in the Canary Islands region. J Mar Syst 28:183–201

Peterson RG, Whitworth T III (1989) The Subantarctic and Polar fronts in relation to deep water masses through the southwestern Atlantic. J Geophys Res 94(C8):10817–10838

Pickart RS (1992) Water mass components of the North Atlantic Deep Western Boundary Current. Deep Sea Res 39:1553–1572

Pickart RS, Straneo F, Moore GWK (2003) Is Labrador Sea Water formed in the Irminger Basin? Deep Sea Res I 50:23–52

Polzin KL, Speer KG, Toole JM, Schmitt RW (1996) Intense mixing of Antarctic Bottom Water in the Equatorial Atlantic Ocean. Nature 380:54–57

Popova AK, Smirnov YB, Khutorskoi MD (1984) Geothermal field of transform fractures. Deep fractures of the Oceanic bottom. Nauka, Moscow [in Russian]

Purdy GM, Rabinowitz PD, Schouten H (1979) The Mid-Atlantic Ridge at 23° N: bathymetry and magnetics, Init. Repts. DSDP, vol 45. U.S. Govt. Printing Office, Washington, DC, pp 119–128

Pushcharovsky YM (2005) Tectonics of the Earth. Vol. 2. Tectonics of the oceans. Nauka, Moscow [in Russian]

Reid JL (1965) Intermediate waters of the Pacific Ocean. The Johns Hopkins Press, Baltimore, MD

Reid JL (1989) On the total geostrophic circulation of the South Atlantic Ocean: flow pattern, tracers and transports. Prog Oceanogr 23:149–244

Reid JL, Nowlin WD, Patzert WC (1977) On the characteristics and circulation of the southwestern Atlantic Ocean. J Phys Oceanogr 7(1):62–91

Rhein M, Stramma L, Send U (1995) The Atlantic Deep Western Boundary Current: water masses and transports near the equator. J Geophys Res 100:2441–2457

Rhein M, Schott F, Stramma L, Fischer J, Plahn O, Send U (1996) The Deep Western Boundary Current in the tropical Atlantic: deep water distribution and circulation off Brazil. Int. WOCE Newsl. 23:11–14

Rhein M, Stramma L, Krahmann G (1998) The spreading of Antarctic Bottom Water in the tropical Atlantic. Deep Sea Res I 45:507–527

Richards KJ (1982) Modeling the benthic boundary layer. J Phys Oceanogr 12:428–439

Richardson MJ, Gardner WD (1985) Analysis of suspended particle-size distribution over the Nova Scotian continental rise. Mar Geol 66:189–203

Rintoul SR (1998) On the origin and influence of Adelie land bottom water, ocean, ice, and atmosphere: interactions at the Antarctic continental margin. Antarctic Research Series, vol 75. American Geophysical Union, Washington, DC, pp 151–171

Rintoul SR (2007) Rapid freshening of Antarctic Bottom Water formed in the Indian and Pacific oceans. Geophys Res Lett 34:L06606. doi:10.1029/2006GL028550

Rintoul SR, Bullister JL (1999) A late winter hydrographic section from Tasmania to Antarctica. Deep Sea Res I 46:1417–1454

Robb JM, Kane MF (1975) Structure of the Vema Fracture Zone from gravity and magnetic intensity profiles. J Geophys Res 80(32):4441–4449

Robertson R, Visbeck M, Gordon A, Fahrbach E (2002) Long-term temperature trends in the deep waters of the Weddell Sea. Deep Sea Res II 49:4791–4806

Roemmich D (1983) The balance of geostrophic and Ekman transports in the tropical Atlantic Ocean. J Phys Oceanogr 13(8):1534–1539

Rudnick DL (1997) Direct velocity measurements in the Samoan Passage. J Geophys Res 102(C2):3293–3302

Ruth C, Well R, Roether W (2000) Primordial 3He in South Atlantic deep waters from sources on the Mid-Atlantic Ridge. Deep Sea Res I 47:1059–1075

Sandoval FJ, Weatherly GL (2001) Evolution of the Deep Western Boundary Current of Antarctic Bottom Water in the Brazil Basin. J Phys Oceanogr 31(6):1440–1460

Saunders PM (1987) Flow through Discovery Gap. J Phys Oceanogr 17(5):631–643

Saunders PM (1994) The flux of overflow water through the Charlie-Gibbs Fracture Zone. J Geophys Res 99(C6):12343–12355

Saunders PM (2001) The dense northern overflows. In: Siedler G, Church J, Gould J (eds) Ocean circulation and climate. Academic Press, San Diego, CA, pp 401–417

Saunders PM, King B (1995) Oceanic fluxes on the WOCE A11 section. J Phys Oceanogr 25(9):1942–1958

Saunders PM, Thompson SR (1993) Transport, heat, and freshwater fluxes within a diagnostic numerical model (FRAM). J Phys Oceanogr 23:452–464

Schlitzer R (1987) Renewal rates of east Atlantic Deep Water estimated by inversion of 14C data. J Geophys Res 92:2953–2980

Schlitzer R (1996) Mass and heat transports in the South Atlantic derived from historical hydrographic data. In: Wefer G, Berger WH, Siedler G, Webb DJ (eds) The South Atlantic: present and past circulation. Springer, Berlin, pp 83–104

Schmid C, Siedler G, Zenk W (2000) Dynamics of intermediate water circulation in the subtropical South Atlantic. J Phys Oceanogr 30(12):3191–3211

Schmitz WJ Jr (1996a) On the World Ocean circulation: volume I, some global features. North Atlantic circulation. Technical Report WHOI-96-03. Woods Hole Oceanographic Institution, Woods Hole, MA

Schmitz WJ Jr (1996b) On the World Ocean circulation: volume II. The Pacific and Indian Oceans, a global update. Technical Report WHOI-96-08. Woods Hole Oceanographic Institution, Woods Hole, MA

Schmitz WJ Jr, McCartney MS (1995) On the interbasin-scale thermohaline circulation. Rev Geophys 33(5):151–173

Schodlock MP, Hellmer HH, Beckmann A (2002) On the transport, variability and origin of dense water masses crossing the South Scotia Ridge. Deep Sea Res II 49:4807–4825

Schott FA, Dengler M, Brandt P, Affler K, Fischer J, Bourle's B, Gouriou Y, Molinari R., Rhein M, (2003) Geophys Res Lett 30(7):1349–1352

Schreider AA (2001) Geomagnetic investigations of the Indian Ocean. Nauka, Moscow [in Russian]

Schreider AA, Schreider AlA, Bulychev AA, Galindo-Zaldivar J, Maldonado A, Kashintsev GL (2006a) Geochronology of the American–Antarctic Ridge. Oceanology 46(1):114–122

Schreider AA, Schreider AlA, Bulychev AA, Lodolo E, Kashintsev GL (2006b) Transtension of the Romanche transform fault. Oceanology 46(5):683–691

Sclater L, Bowin C, Hey R, Tapscott C (1976) The Bouvet triple junction. J Geophys Res 81(11):1857–1869

Searle R (1980) The active part of the Charlie-Gibbs Fracture Zone: a study using sonar and other geophysical techniques. J Geophys Res 86(B1):243–262

Shannon LV, Chapman P (1991) Evidence of Antarctic Bottom Water in the Angola Basin at 32°S. Deep Sea Res 38(10):1299–1304

Shannon LV, van Rijswijck M (1969) Physical oceanography of the Walvis Ridge region. Investigational Report No 70, Division of Se Fisheries of RSA [Available from the library of the University of Cape Town]

Sheldon RW, Parsons TR (1967) A practical manual on the use of the Coulter counter in marine science. Coulter Electronics, Toronto

Siedler G, Church J, Gould J (eds) (2001) Ocean circulation and climate. Academic Press, San Diego, CA

Sievers HA, Nowlin WD Jr (1984) The stratification and water masses at Drake Passage. J Geophys Res 89(C6):10489–10514

Sloyan BM, Rintoul SR (2001a) Circulation, renewal, and modification of Antarctic mode and intermediate water. J Phys Oceanogr 31(4):1005–1030

Sloyan BM, Rintoul SR (2001b) The Southern Ocean limb of the global deep overturning circulation. J Phys Oceanogr 31(1):143–173

Smith IJ, Stevens DP, Heywood KJ, Meredith MP (2010) The flow of the Antarctic Circumpolar Current over the North Scotia Ridge. Deep Sea Res 57(1):14–28

Smith N (2001) Ocean and climate prediction – the WOCE legacy. In: Siedler G, Church J, Gould J (eds) Ocean circulation and climate. Academic Press, San Diego, CA, pp 585–602

Smith WHF, Sandwell DT (1997) Global sea floor topography from satellite altimetry and ship depth soundings. Science 277:1956–1962. http://topex.ucsd.edu/cgi-bin/get_data.cgi

Smoot N, Sharman G (1985) Charlie-Gibbs: a fracture zone ridge. Tectonophysics 116:137–142

Smythe-Wright D, Boswell S (1998) Abyssal circulation in the Argentine Basin. J Geophys Res 103(C8):15845–15851

Speer KG, McCartney MS (1991) Tracing lower North Atlantic Deep Water across the equator. J Geophys Res 96(C11):20443–20448

Speer KG, McCartney MS (1992) Bottom water circulation in the western North Atlantic. J Phys Oceanogr 22(1):83–92

Speer KG, Zenk W (1993) The flow of Antarctic Bottom Water into the Brazil Basin. J Phys Oceanogr 23:2667–2682

Speer KG, Zenk W, Siedler G, Pätzold J, Heidland C (1992) First resolution of flow through the Hunter Channel in the South Atlantic, Earth Planet. Sci Lett 113:287–292

Speer KG, Siedler G, Talley L (1995) The Namib Col Current. Deep Sea Res 42:1933–1950

Speer KG, Rintoul SR, Sloyan BM (2000) The diabatic Deacon cell. J Phys Oceanogr 30:3212–3222

Spiess F (1928) The Meteor expedition. Scientific results of the German Atlantic expedition, 1925–1927. D. Reimer, Berlin. Translated by W. J. Emery, Amerind Publishing Co., New Delhi, p 429

Spiess F (1932) Das Forschungsschiff und seine Reise. Dt. Atl. Exp. "Meteor" 1925–1927, vol 1, p 442

Stephens JC, Marshall DP (2000) Dynamical pathways of Antarctic Bottom Water in the Atlantic. J Phys Oceanogr 30(3):622–640
Stommel H, Arons AB (1960) On the abyssal circulation of the World Ocean. II. An idealized model of the circulation pattern and amplitude in oceanic basins. Deep Sea Res 6(3):217–233
Stramma L (1991) Geostrophic transports of the South Equatorial Current in the Atlantic. J Mar Res 49:281–294
Stramma L, England M (1999) On the water mass and mean circulation of the South Atlantic Ocean. J Geophys Res 104(C9):20863–20883
Sverdrup H, Johnson M, Fleming R (1942) The Oceans. Their physics, chemistry and general biology. Prentice Hall Inc, NY
Swift JH, Aagaard K, Malmberg S-A (1980) The contribution of the Denmark Strait overflow to the deep North Atlantic. Deep Sea Res 27:29–42
Sykes L (1970) Focal mechanism solutions for earthquakes along the world rift system. Bull Seismol Soc Am 60(5):1749–1752
Syrsky VI, Greku RKH (1975) Investigations of the shear zones in the Equatorial Atlantic. Multidisciplinary geophysical studies of the Mid-Atlantic Ridge. MGI AN USSR, Sebastopol, pp 42–51 [in Russian]
Tarakanov RY (2009) Antarctic Bottom Water in the Scotia Sea and the Drake Passage. Oceanology 49(5):607–621
Tarakanov RY (2010) Circumpolar Bottom Water in the Scotia Sea and Drake Passage. Oceanology 50(1):5–22
Thorpe SA (1978) On the shape and breaking of finite amplitude internal gravity waves in a shear flow. J Fluid Dyn 85:7–31
Tomczak M (1981) A multi-parameter extension of temperature/salinity diagram techniques for the analysis of nonisopycnal mixing. Prog Oceanogr 10:147–171
Tomczak M, Godfrey SR (1994) Regional oceanography: an introduction. Pergamon Press, London, p 422
Trukhin VI, Schreider AA, Zhilyaeva VA, Bulychev AA, Maksimochkin AA (2005) Seafloor magnetism in the Romanche transform zone (the Equatorial Atlantic), Izv. RAS. Phys Solid Earth 41(3):179–192
Tsuchiya M (1989) Circulation of the Antarctic Intermediate Water in the North Atlantic Ocean. J Mar Res 47(4):747–755
Tsuchiya M, Talley LD, McCartney MS (1992) An eastern Atlantic section from iceland southward across the equator. Deep Sea Res 39:1885–1918
Tsuchiya M, Talley LD, McCartney MS (1994) Water mass distribution in the western South Atlantic: A section from South Georgia Island (54°S) northward across the equator. J Mar Res 52:55–81
Tucholke BE, Embley RW (1984) Cenozoic regional erosion of the abyssal sea floor off South Africa. In: Schlee JS (ed) Interregional unconformities and hydrocarbon accumulation. AAPG Memoir 36, Tulsa, pp 145–164
Tucholke BE, Wright WR, Hollister CD (1973) Abyssal circulation over the Greater Antilles Outer Ridge. Deep Sea Res 20:973–995

van Andel T, von Herzen R, Phillips I (1971) The Vema Fracture Zone and the tectonics of transverse shear zones in oceanic crustal plates. Mar Geophys Res 1(3):78–97

van Aken HM (2000) The hydrography of the mid-latitude northeast Atlantic Ocean I: The deep water masses. Deep Sea Res 47:757–788

van Aken H.M (2007) The oceanic thermohaline circulation: an introduction. Springer, New York

van Aken HM, de Boer CJ (1995) On the synoptic hydrography of intermediate and deep water masses in the Iceland Basin. Deep Sea Res I 42(2):165–189

van Bennekom AJ (1985) Dissolved silica as an indicator of Antarctic Bottom Water penetration, and the variability in the bottom layers of the Norwegian and Iceland Basins. Rit Fiskideildar 9:101–109

Vangriesheim A (1980) Antarctic Bottom Water flow through the Vema Fracture Zone. Oceanol Acta 3:199–207

Vanicek M, Siedler G (2002) Zonal fluxes in the deep water layers of the western South Atlantic Ocean. J Phys Oceanogr 32(8):2205–2235

Visbeck M (2002) Deep velocity profiling using lowered acoustic Doppler current profiler: bottom track and inverse solution. J Atmos Oceanic Technol 19(5):794–807

Vlasenko V, Hutter K (2002) Numerical experiments on the breaking of solitary internal waves over a slope-shelf topography. J Phys Oceanogr 32:1779–1793

von Gyldenfeldt AB, Fahrbach E, García MA, Schröder M (2002) Flow variability at the tip of the Antarctic Peninsula. Deep Sea Res II 49:4743–4766

Walkden GJ, Heywood KJ, Stevens DP (2008) Eddy heat fluxes from direct current measurements of the Antarctic, polar front in Shag Rocks Passage. Geophys Res Lett 35:L06602

Warren BA, Speer KG (1991) Deep circulation in the eastern South Atlantic Ocean. Deep Sea Res 38(Suppl 1):281–322

Weatherly GL, Kelley EA (1982) 'Too cold' bottom layers at the base of the Scotian Rise. J Mar Res 40(4):985–1012

Weatherly GL, Martin PJ (1978) On the structure and dynamics of the oceanic bottom boundary layer. J Phys Oceanogr 8:557–570

Weatherly GL, Kim YY, Kontar EA (2000) Eulerian measurements of the North Atlantic Deep Water Deep Western Boundary Current at 18°S. J Phys Oceanogr 30:971–986

Weiss RF, Östlund HG, Craig H (1979) Geochemical studies of the Weddell Sea. Deep Sea Res 26:1093–1120

Well R, Roether W, Stevens DP (2003) An additional deep-water mass in Drake Passage as revealed by 3He data. Deep Sea Res I 50:1079–1098

Weppernig R, Schlosser P, Khatiwala S, Fairbanks RG (1996) Isotope data from Ice Station Weddell: implications for deep water formation in the Weddell Sea. J Geophys Res 101(C9):25723–25739

Whitehead JA (1989) Surges of Antarctic Bottom Water into the North Atlantic. J Phys Oceanogr 19(6):853–861

Whitehead JA, Wang W (2008) A laboratory model of vertical ocean circulation driven by mixing. J Phys Oceanogr 38(5):1091–1106

Whitehead JA, Worthington LV (1982) The flux and mixing rates of Antarctic Bottom Water within the North Atlantic. J Geophys Res 87:7903–7924

Whitworth T, Nowlin WD, Worley SJ (1982) The net transport of the Antarctic Circumpolar Current through Drake Passage. J Phys Oceanogr 12(9):960–971

Whitworth T, Nowlin WD, Pillsbury RD, Moore MI, Weiss RF (1991) Observations of the Antarctic Circumpolar Current and Deep Boundary Current in the Southwest Atlantic. J Geophys Res 96(15):105–118

Whitworth T, Nowlin WD, Orsi AH, Locarnini RA, Smith SG (1994) Weddell Sea Shelf Water in the Bransfield Strait and Weddell-Scotia confluence. Deep Sea Res I 41:629–641

Whitworth T, Orsi AH, Kim S-J, Nowlin WD, Locarnini RA (1998) Water masses and mixing near the Antarctic slope front. In: Jacobs S, Weiss R (eds) Ocean, ice, and atmosphere: interactions at the Antarctic continental margin. Antarctic Research Series, vol 75. American Geophysical Union, Washington, DC, pp 1–27

Whitworth T, Warren BA, Nowlin WD Jr, Rutz SB, Pillsbury RD, Moore MI (1999) On the Deep Western Boundary Current in the southwest Pacific Basin. Progr Oceanogr 43:1–54

Wienders N, Arhan M, Mercier H (2000) Circulation at the western boundary of the south and Equatorial Atlantic: exchanges with the ocean interior. J Mar Res 58:1007–1039

Wilson T (1965) A new class of faults and their bearing of continental drift. Nature 207(4995):343–347

Wittstock R-R, Zenk W (1983) Some current observations and surface T/S distribution from the Scotia Sea and the Bransfield Strait during early austral summer 1980/81, Meteor Forschung-Ergebnisse, Reihe A/B. No 24, pp 77–86

WODB (2005) World Ocean Data Base (CD-version), 2001. http://www.nodc.noaa.gov/OC5/WOD05

Worthington LV (1969) An attempt to measure the volume transport of Norwegian Sea Overflow Water through the Denmark Strait. Deep Sea Res 16(Suppl):421–432

Worthington LV (1970) The Norwegian Sea as a Mediterranean basin. Deep Sea Res 17:77–84

Worthington LV, Volkmann GH (1965) The volume transport of the Norwegian Sea Overflow Water in the North Atlantic. Deep Sea Res 12:667–676

Worthington LV, Wright WR (1970a) North Atlantic Ocean atlas of potential temperature, salinity, and oxygen profiles from the Erika Dan Cruise of 1962. Woods Hole Oceanographic Institution, Atlas Series 2, Woods Hole, MA, pp 1–82

Worthington LV, Wright WR (1970b) North Atlantic Ocean atlas of potential temperature and salinity in the deep water. Woods Hole Oceanographic Institution, Atlas Series 2, Woods Hole, MA, p 24

Wright WR (1970) Northward transport of Antarctic Bottom Water in the western Atlantic Ocean. Deep Sea Res 17:367–371

Wunsch C (1984) An eclectic Atlantic Ocean circulation model. Part I: The meridional flux of heat. J Phys Oceanogr 14(11):1712–1733

Wüst G (1936) Schichtung und Zirkulation des Atlantischen Ozeans, Das Bodenwasser und die Stratosphäre. In: Defant A (ed) Wissenschaftliche Ergebnisse, Deutsche Atlantische Expedition auf dem Forschungs – und Vermessungsschiff "Meteor" 1925–1927, 6(1). Walter de Gruyter & Co, Berlin, p 411

Yaremchuk M, Nechaev D, Schroter J, Fahrbach E (1998) A dynamically consistent analysis of circulation and transports in the southwestern Weddell Sea. Ann Geophys 16:1024–1038

You Y (2005) Proceedings of IAPSO International Assembly "Dynamic Planet." Cairns, Australia

Zemba JC (1991) The structure and transport of the Brazil Current between 27° and 36°S. Ph.D. dissertation, WHOI-91-37, p 160 [Available from Institut für Meereskunde, Düstenbrooker Weg 20, 24105 Kiel, Germany]

Zenk W (1981) Detection of overflow events in the Shag Rocks Passage, Scotia Ridge. Science 213(4512):1113–1114

Zenk W (2008) Temperature fluctuations and current shear in Antarctic Bottom Water at the Vema Sill. Prog Oceanogr 77:276–284

Zenk W, Hogg NG (1996) Warming trend in Antarctic Bottom Water flowing into the Brazil Basin. Deep Sea Res I 43(9):1461–1473

Zenk W, Morozov EG (2007) Decadal warming of the coldest Antarctic Bottom Water flow through the Vema Channel. Geophys Res Lett 34:L14607. doi:10.1029/2007GL030340

Zenk W, Speer KG, Hogg NG (1993) Bathymetry at the Vema Sill. Deep Sea Res 40:1925–1933

Zenk W, Becker S, Jungclaus J, Link R (1998) Physical oceanography. In: Schulz HD, Devey CW, Pätzold J, Fischer G (eds) GEO Bremen/GPI Kiel South Atlantic, Cruise No 41, METEOR-Berichte, Universität Hamburg, 99-3, pp 341

Zenk W, Siedler G, Lenz B, Hogg NG (1999) Antarctic Bottom Water flow through the Hunter channel. J Phys Oceanogr 29(11):2785–2801

Zenk W, Morozov E, Sokov A, Müller TJ (2003) Vema Channel: Antarctic Bottom Water temperatures continue to rise. CLIVAR Exchanges 8(1):24–26

Zubov NN (1947) Dynamic oceanology. Gidrometeoizdat, Moscow